The UHMWPE Handbook

The UHMWPE Handbook

Ultra-High Molecular Weight Polyethylene in Total Joint Replacement

Steven M. Kurtz, Ph.D.
Principal Engineer, Exponent, Inc.
Research Associate Professor, Drexel University
3401 Market Street, Suite 300
Philadelphia, PA 19104

ELSEVIER
ACADEMIC
PRESS

AMSTERDAM • BOSTON • HEIDELBERG • LONDON
NEW YORK • OXFORD • PARIS • SAN DIEGO
SAN FRANCISCO • SINGAPORE • SYDNEY • TOKYO

Elsevier Academic Press
525 B Street, Suite 1900, San Diego, California 92101-4495, USA
84 Theobald's Road, London WC1X 8RR, UK

This book is printed on acid-free paper. ♾

Library of Congress Cataloging-in-Publication Data

Kurtz Steven M., 1968-
  The UHMWPE handbook: ultra-high molecular weight polyethylene in total joint replacement/Steven M. Kurtz.
    p. cm.
  Includes index.
  ISBN 0-12-429851-6
  1. Artificial joints. 2. Polyethylene—Therapeutic use. 3. Plastics in medicine. I. Title: Ultra-high molecular weight polyethylene in total joint replacement. II. Title.

RD686.K87 2004
617.5'80592--dc22   2003070856

British Library Cataloging-in-Publication Data
A catalogue record for this book is available from the British Library

ISBN: 0-12-429851-6

For all information on all Academic Press publications
visit our Web site at www.academicpress.com

Printed in China

04 05 06 07 08 9 8 7 6 5 4 3 2 1

To Karen, for listening to all my stories.

Contents

Contributors xi
Preface xiii

1. A Primer on UHMWPE **1**
    Introduction 1
    What Is a Polymer? 2
    What Is Polyethylene? 4
    Crystallinity 6
    Thermal Transitions 7
    Overview of the *Handbook* 9

**2. From Ethylene Gas to UHMWPE Component: The Process
of Producing Orthopedic Implants** **13**
    Introduction 13
    Polymerization: From Ethylene Gas to UHMWPE Powder 14
    Conversion: From UHMWPE Powder to Consolidated Form 22
    Machining: From Consolidated Form to Implant 31
    Conclusion 32

3. Packaging and Sterilization of UHMWPE **37**
    Introduction 37
    Gamma Sterilization in Air 38
    Gamma Sterilization in Barrier Packaging 41
    Ethylene Oxide Gas Sterilization 44
    Gas Plasma Sterilization 45
    Shelf Life of UHMWPE Components for Total Joint Replacement 47
    Overview of Current Trends 48

4. The Origins of UHMWPE in Total Hip Arthroplasty **53**
    Introduction and Timeline 53
    The Origins of a Gold Standard (1958–1982) 55
    Charnley's First Hip Arthroplasty Design with PTFE (1958) 56
    Implant Fixation with Pink Dental Acrylic Cement (1958–1966) 56
    Interim Hip Arthroplasty Designs with PTFE (1958–1960) 58

Final Hip Arthroplasty Design with PTFE (1960–1962)   58
Implant Fabrication at Wrightington   61
The First Wear Tester   62
Searching to Replace PTFE   64
UHMWPE Arrives at Wrightington   66
Implant Sterilization Procedures at Wrightington   66
Overview   68

5. The Clinical Performance of UHMWPE in Hip Replacements   71
Introduction   71
Joint Replacements Do Not Last Forever   73
Range of Clinical Wear Performance in Cemented
    Acetabular Components   75
Wear Versus Wear Rate of Hip Replacements   77
Comparing Wear Rates Between Different Clinical Studies   79
Comparison of Wear Rates in Clinical and Retrieval Studies   82
Current Methods for Measuring Clinical Wear in
    Total Hip Arthroplasty   83
Range of Clinical Wear Performance in Modular Acetabular
    Components   85
Conclusion   86

6. Alternatives to Conventional UHMWPE for Hip Arthroplasty   93
Introduction   93
Metal-on-Metal Alternative Hip Bearings   94
Ceramics in Hip Arthroplasty   101
Highly Crosslinked and Thermally Stabilized UHMWPE   109
Summary   114

**7. The Origins and Adaptations of UHMWPE for Knee
    Replacements   123**
Introduction   123
Frank Gunston and the Wrightington Connection to
    Total Knee Arthroplasty   126
Polycentric Knee Arthroplasty   129
Unicondylar Polycentric Knee Arthroplasty   132
Bicondylar Total Knee Arthroplasty   134
Patello–Femoral Arthroplasty   141
UHMWPE with Metal Backing   142
Conclusion   146

8. The Clinical Performance of UHMWPE in Knee Replacements   151
Introduction   151
Biomechanics of Total Knee Arthroplasty   153
Clinical Performance of UHMWPE in Knee Arthroplasty   160
Osteolysis and Wear in Total Knee Arthroplasty   172
UHMWPE Is the Only Alternative for Knee Arthroplasty   182

9. **The Clinical Performance of UHMWPE in Shoulder Replacements** 189

Stefan Gabriel

Introduction 189
The Shoulder Joint 190
Shoulder Replacement 191
Biomechanics of Total Shoulder Replacement 195
Contemporary Total Shoulder Replacements 197
Clinical Performance of Total Shoulder Arthroplasty 203
Controversies in Shoulder Replacement 207
Future Directions in Total Shoulder Arthroplasty 211
Conclusion 213

10. **The Clinical Performance of UHMWPE in the Spine** 219

Marta L. Villarraga and Peter A. Cripton

Introduction 219
Biomechanical Considerations for UHMWPE in the Spine 222
Total Disc Replacement Designs Using UHMWPE 226
Clinical Performance of UHMWPE in the Spine 237
Alternatives to UHMWPE for Total Disc Arthroplasty in the Spine 239
Conclusion 240

11. **Mechanisms of Crosslinking and Oxidative Degradation of UHMWPE** 245

Luigi Costa and Pierangiola Bracco

Introduction 245
Mechanisms of Crosslinking 245
UHMWPE Oxidation 250
Oxidative Degradation after Implant Manufacture 256
In Vivo Absorption of Lipids 257

12. **Characterization of Physical, Chemical, and Mechanical Properties of UHMWPE** 263

Stephen Spiegelberg

Introduction 263
What Does the Food and Drug Administration Require? 264
Physical Property Characterization 265
Intrinsic Viscosity 269
Chemical Property Characterization 274
Mechanical Property Characterization 280
Other Testing 284
Conclusion 284

13. **Development and Application of the Small Punch Test to UHMWPE** 287

Avram Allan Edidin

Introduction 287

Overview and Metrics of the Small Punch Test   288
Accelerated and Natural Aging of UHMWPE   291
In Vivo Changes in Mechanical Behavior of UHMWPE   294
Effect of Crosslinking on Mechanical Behavior and Wear   295
Shear Punch Testing of UHMWPE   298
Fatigue Punch Testing of UHMWPE   301
Conclusion   305

14. Computer Modeling and Simulation of UHMWPE   **309**
Jörgen Bergström
Introduction   309
Overview of Available Modeling and Simulation Techniques   310
Characteristic Material Behavior of UHMWPE   311
Material Models for UHMWPE   317
Discussion   334

**15. Compendium of Highly Crosslinked and Thermally Treated
UHMWPEs**   **337**
Introduction   337
Honorable Mention   338
Crossfire   339
DURASUL   342
Longevity   345
Marathon   348
Prolong   351
XLPE   352
Current Trends and Prevalence in Total Hip and
    Total Knee Arthroplasty   353
The Future for Highly Crosslinked UHMWPE   357

Appendix   **365**
Index   **369**

Contributors

Jörgen Bergström Ph.D.
Senior Engineer
Exponent, Inc.
21 Strathmore Rd.
Natick, MA 01760

Pierangiola Bracco
Researcher
Dipartimento di Chimica IFM
Università di Torino
Via Giuria 7
10125 Torino (Italy)

Luigi Costa Ph.D.
Professor
Dipartimento di Chimica IFM
Università di Torino
Via Giuria 7
10125 Torino (Italy)

Peter Cripton Ph.D.
Assistant Professor
Department of Mechanical
   Engineering
University of British Columbia
2054-2324 Main Mall
Vancouver, B.C.
V6T1Z4 Canada

Avram Allan Edidin Ph.D.
Research Associate Professor
School of Biomedical Engineering,
   Science, and Health Systems
Drexel University
3141 Chestnut St.
Philadelphia, PA 19104

Stefan Gabriel Ph.D., P.E.
Principal Engineer
Smith & Nephew Endoscopy
130 Forbes Blvd.
Mansfield, MA 02048

Stephen Spiegelberg Ph.D.
President
Cambridge Polymer Group, Inc.
52-R Roland St.
Boston, MA 02129

Marta Villarraga Ph.D.
Managing Engineer, Exponent, Inc.
Research Associate Professor
School of Biomedical Engineering,
   Science, and Health Systems
Drexel University
3401 Market St., Suite 300
Philadelphia, PA 19104

Preface

This book has its origins in a review article, "Advances in the Sterilization, Processing, and Crosslinking of Ultra-High Molecular Weight Polyethylene for Total Joint Arthroplasty" (*Biomaterials* 1999; 20: 1659–1688), which I coauthored with Orhun Muratoglu, Mark Evans, and Av Edidin. Our review was written between 1997 and 1998, at a time when highly crosslinked and thermally treated UHMWPE materials were about to be clinically introduced for total hip replacement. Several other important milestones in the clinical use of UHMWPE had occurred, including the abandonment of gamma sterilization in air (at least in the United States) and the trend to reduce calcium stearate (a processing additive) in the UHMWPE powder.

Since 1998, a major shift in the orthopedic application of UHMWPE has occurred. Currently six different highly crosslinked UHMWPE formulations are in clinical use for both hip and knee replacements. Although intermediate-term clinical follow-up data are not yet available, the adoption by surgeons of highly crosslinked UHMWPE for total hip replacements has been pervasive. A similar, but somewhat slower, trend has been initiated with the use of these new materials in total knee replacement starting in 2001.

If we include the alternative bearing solutions (e.g., metal on metal and ceramic on ceramic) for total joint arthroplasty, the number of choices available to the orthopedic surgeon is far greater today than in 1998. Yet, with all these advanced technologies available, we should still be mindful that for very elderly patients, artificial joints incorporating conventional UHMWPE will continue to afford long-term clinical benefits that could potentially last the rest of these patients' natural lives. On the other hand, the more recently introduced alternative bearing technologies, including crosslinked UHMWPE, should provide the greatest benefit to young patients (younger than 60 years) who lead an active lifestyle and who need a total hip replacement. For patients in need of knee arthroplasty, shoulder arthroplasty, or total disc replacement, conventional UHMWPE continues to prevail as the polymeric bearing material of choice.

The latest generation of new processing technology has contributed to some confusion, among surgeons and researchers alike, regarding the technical details associated with highly crosslinked UHMWPE. My goal in writing the *UHMWPE Handbook* is to explain the common concepts underlying all UHMWPE materials, as well as the technical differences in specific formulations for an audience of surgeons, researchers, and students who may be starting work in this field. Some of the early chapters in this book may also be of interest to current or prospective patients who are motivated to learn more about current treatment options. Because all of the alternative bearing solutions (both highly crosslinked UHMWPE as well as hard-on-hard bearings) have their origins in the 1950s, 1960s, and 1970s, it is important to understand the historical context in which the first generation of highly crosslinked UHMWPEs were originally developed. With this in mind, I have also taken some care to review the historical development of UHMWPE bearings for joint replacement.

The story of UHMWPE in orthopedics, seemingly immutable and static, still continues to evolve. Early in 2002, I undertook the task of expanding my website, the UHMWPE Lexicon (www.uhmwpe.org), with an online monograph of six introductory chapters covering the basic scientific principles and clinical performance of UHMWPE in hip replacement. The response to these online chapters was overwhelmingly positive and encouraged me to revise the first six chapters for hardcopy publication and to develop the additional nine chapters for the *UHMWPE Handbook*.

Despite my diligent efforts to summarize the state of the art in this field, I appreciate that current trends related to UHMWPE in orthopedics will give way to new ideas as further clinical data becomes available in the future. For this reason, I hope that the UHMWPE Lexicon website will continue to disseminate new research findings to the orthopedic and polymer research communities. The Lexicon website and this current *Handbook* play complementary roles. When the accumulation of new ideas and findings has diffused sufficiently into the orthopedic clinical and research practice, it will be time to update this written work. I look forward to your comments and suggestions for future expansion of the UHMWPE Lexicon website and the *UHMWPE Handbook*.

This monograph would not have been possible without the suggestions and advice of many supportive colleagues. I am especially grateful to my coauthors, experts in the fields of joint arthroplasty, spine, as well as in the testing and evaluation of UHMWPE, who have cheerfully contributed chapters to this book on relatively short notice. I have also included acknowledgements at the end of chapters to thank my many friends and associates for their contributions.

—*Steven Kurtz, Ph.D.*
Philadelphia, PA
October 3, 2003

UHMWPE
Lexicon

Visit

www.UHMWPE.org

www.UHMWPE.org

to learn about:

**THE LATEST SCIENTIFIC & CLINICAL RESEARCH
NEW STANDARDS FOR MEDICAL GRADE UHMWPE**

The UHMWPE Lexicon website provides:

- Online Reference for Ultra-High Molecular Weight Polyethylene used in Total Joint Replacements

- Overview of State-of-the-Art Research in Several Key Polyethylene Related Problems of Clinical Significance

- Ideas for Hypothesis Driven Polyethylene Research

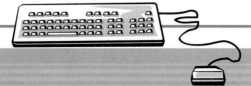

Chapter *1*

A Primer on UHMWPE

Introduction

Ultra-high molecular weight polyethylene (UHMWPE) is a unique polymer with outstanding physical and mechanical properties. Most notable are its chemical inertness, lubricity, impact resistance, and abrasion resistance. These characteristics of UHMWPE have been exploited since the 1950s in a wide range of industrial applications (Figure 1.1), including pickers for textile machinery, lining for coal chutes and dump trucks, runners for bottling production lines, as well as bumpers and siding for ships and harbors. More than 90% of the UHMWPE produced in the world is used by industry.

Since 1962, UHMWPE also has been used in orthopedics as a bearing material in artificial joints. Each year, about 1.4 million joint replacement procedures are performed around the world. Despite the success of these restorative procedures, UHMWPE implants have only a finite lifetime. Wear and damage of the UHMWPE components is one of the factors limiting implant longevity.

UHMWPE comes from a family of polymers with a deceptively simple chemical composition, consisting of only hydrogen and carbon. However, the simplicity inherent in its chemical composition belies a more complex hierarchy of organizational structures at the molecular and supermolecular length scales. At a molecular level, the carbon backbone of polyethylene can twist, rotate, and fold into ordered crystalline regions. At a supermolecular level, the UHMWPE consists of powder (also known as resin or flake) that must be consolidated at elevated temperatures and pressures to form a bulk material. Further layers of complexity are introduced by chemical changes that arise in UHMWPE due to radiation sterilization and processing.

The purpose of this *Handbook* is to explore the complexities inherent in UHMWPE and to provide the reader with a background in the terminology, history, and recent advances related to its use in orthopedics. A monograph such as this is helpful in several respects. First, it is important that members of the surgical community have access to up-to-date knowledge about the properties of UHMWPE so that this information can be more accurately communicated to

Figure 1.1
Dump truck liner of UHMWPE, an example of an industrial application for the polymer.

their patients. Second, members of the orthopedic research community need access to timely synthesis of the existing literature so that future studies are more effectively planned to fill in existing gaps in our current understanding. Finally, this handbook may also serve as a resource for university students at both the undergraduate and graduate levels.

This introductory chapter starts with the basics, assuming the reader is not familiar with polymers, let alone polyethylene. This chapter provides basic information about polymers in general, describes the structure and composition of polyethylene, and explains how UHMWPE differs from other polymers (including high-density polyethylene [HDPE]) and from other materials (e.g., metals and ceramics). The concepts of crystallinity and thermal transitions are introduced at a basic level. Readers familiar with these basic polymer concepts may want to consider skipping ahead to the next chapter.

What Is a Polymer?

The ultra-high molecular weight polyethylene used in orthopedic applications is a type of polymer generally classified as a linear homopolymer. Our first task is to explain what is meant by all of these terms. Before proceeding to a definition of UHMWPE, one needs to first understand what constitutes a linear homopolymer.

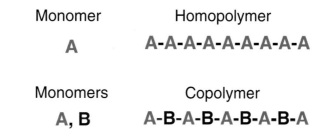

Figure 1.2
Schematics of homopolymer and copolymer structure.

A *polymer* is a molecule consisting of many (*poly-*) parts (*-mer*) linked together by chemical covalent bonds. The individual parts, or monomer segments, of a polymer can all be the same. In such a case, we have a homopolymer as illustrated in Figure 1.2. If the parts of a polymer are different, it is termed a *copolymer*. These differences in chemical structure are also illustrated in Figure 1.2, with generic symbols (A, B) for the monomers.

Polymers can be either linear or branched as illustrated in Figure 1.3. The tendency for a polymer to exhibit branching is governed by its synthesis conditions. Keep in mind that the conceptual models of polymer structure illustrated in Figures 1.2 and 1.3 have been highly simplified. For example, it is possible for a copolymer to have a wide range of substructural elements giving rise to an impressive range of possibilities. In industrial practice, polyethylenes, including UHMWPE, are often copolymerized with other monomers (e.g., polypropylene) to achieve improved processing characteristics or to alter the physical and mechanical properties of the polymer. For example, according to ISO 11542, which is the industrial standard for UHMWPE, the polymer can contain a large concentration of copolymer (up to 50%) and still be referred to as UHMWPE. However, most of the UHMWPEs used to fabricate orthopedic implants are homopolymers, and so we will restrict our further discussion to polymers with only a single type of monomer.

The principal feature of a polymer that distinguishes it from other materials, such as metals and ceramics, is its molecular size. In a metallic alloy or ceramic,

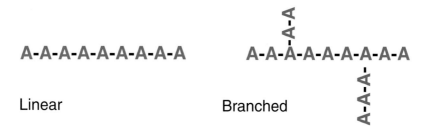

Figure 1.3
Schematics of linear and branched polymer structures.

the elemental building blocks are individual metal atoms (e.g., Co, Cr, Mo) or relatively small molecules (e.g., metal carbides or oxides). However, in a polymer, the molecular size can comprise more than a 100,000 monomer units, with molecular weights of up to millions of g/mol.

The molecular chain architecture of a polymer also imparts many unique attributes, including temperature dependence and rate dependence. Some of these unique properties are further illustrated in the specific case of UHMWPE in subsequent sections of this chapter. For further background on general polymer concepts, the reader is referred to textbooks by Rodriguez (1989) and Young (1983).

What Is Polyethylene?

Polyethylene is a polymer formed from ethylene (C_2H_4), which is a gas having a molecular weight of 28. The generic chemical formula for polyethylene is $-(C_2H_4)_n-$, where n is the degree of polymerization. A schematic of the chemical structures for ethylene and polyethylene is shown in Figure 1.4.

For UHMWPE, the molecular chain can consist of as many as 200,000 ethylene repeat units. Put another way, the molecular chain of UHMWPE contains up to 400,000 carbon atoms.

There are several kinds of polyethylene (LDPE, LLDPE, HDPE, UHMWPE), which are synthesized with different molecular weights and chain architectures. LDPE and LLDPE refer to low-density polyethylene and linear low-density polyethylene, respectively. These polyethylenes generally have branched and linear chain architectures, respectively, each with a molecular weight of typically less than 50,000 g/mol.

HDPE is a linear polymer with a molecular weight of up to 200,000 g/mol. UHMWPE, in comparison, has a viscosity average molecular weight of up to 6 million g/mol. In fact, the molecular weight is so ultra-high that it cannot be measured directly by conventional means and must instead be inferred by its intrinsic viscosity (IV).

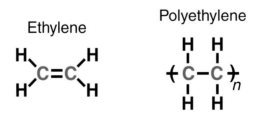

Figure 1.4

Schematic of the chemical structures of ethylene and polyethylene.

Table 1.1 summarizes the physical and mechanical properties of HDPE and UHMWPE. As shown in Table 1.1, UHMWPE has a higher ultimate strength and impact strength than HDPE.

Perhaps more relevant from a clinical perspective, UHMWPE is significantly more abrasion resistant and wear resistant than HDPE. The following wear data for UHMWPE and HDPE was collected using a contemporary, multidirectional hip simulator (Edidin and Kurtz 2000). Based on hip simulator data, shown in Figure 1.5, the volumetric wear rate for HDPE is 4.3 times greater than that of UHMWPE.

In the early 1960s, UHMWPE was classified as a form of HDPE among members of the polymer industry (Chubberley 1965). Thus, Charnley's earlier references to UHMWPE as HDPE are technically accurate for his time (Charnley 1963), but they have contributed to some confusion over the years as to exactly what kinds of polyethylenes have been used clinically. From a close reading of Charnley's works, it is clear that HDPE is used synonymously with RCH-1000, the trade name for UHMWPE produced by Hoechst in Germany (Charnley 1979). With the exception of a small series of 22 patients who were implanted with silane-crosslinked HDPE at Wrighington (Wroblewski et al. 1996), there is no evidence in the literature that lower molecular weight polyethylenes have been used clinically.

Table 1.1

Typical Average Physical Properties of HDPE, UHMWPE

Property	HDPE	UHMWPE
Molecular weight (10^6 g/mole)	0.05–0.25	2–6
Melting temperature (°C)	130–137	125–138
Poisson's ratio	0.40	0.46
Specific gravity	0.952–0.965	0.932–0.945
Tensile modulus of elasticity* (GPa)	0.4–4.0	0.8–1.6
Tensile yield strength* (MPa)	26–33	21–28
Tensile ultimate strength* (MPa)	22–31	39–48
Tensile ultimate elongation* (%)	10–1200	350–525
Impact strength, Izod* (J/m of notch; 3.175 mm thick specimen)	21–214	>1070 (No Break)
Degree of crystallinity (%)	60–80	39–75

*Testing conducted at 23°C.

From Edidin A.A., and S.M. Kurtz. 2000. The influence of mechanical behavior on the wear of four clinically relevant polymeric biomaterials in a hip simulator. *J Arthroplasty* 15:321–331.

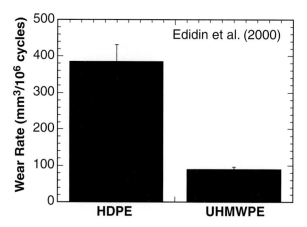

Figure 1.5

Comparison of wear rates of HDPE and UHMWPE in a multidirectional hip simulator. (From Edidin A.A., and S.M. Kurtz. 2000. The influence of mechanical behavior on the wear of four clinically relevant polymeric biomaterials in a hip simulator. *J Arthroplasty* 15:321–331.)

Crystallinity

One can visualize the molecular chain of UHMWPE as a tangled string of spaghetti, more than a kilometer long. Because the chain is not static, but imbued with internal (thermal) energy, the molecular chain can become mobile at elevated temperatures. When cooled below the melt temperature, the molecular chain of polyethylene has the tendency to rotate about the C–C bonds and create chain folds. This chain folding, in turn, enables the molecule to form local ordered, sheetlike regions known as crystalline lamellae. These lamellae are embedded within amorphous (disordered) regions and may communicate with surrounding lamellae by tie molecules. All of these morphological features of UHMWPE are shown schematically in Figure 1.6.

The degree and orientation of crystalline regions within a polyethylene depends on a variety of factors, including its molecular weight, processing conditions, and environmental conditions (such as loading), and will be discussed in later chapters.

The crystalline lamellae are microscopic and invisible to the naked eye. The lamellae diffract visible light, giving UHMWPE a white, opaque appearance at room temperature. At temperatures above the melt temperature of the lamellae, approximately 137°C, UHMWPE becomes translucent. The lamellae are on the order of 10–50 nm in thickness and 10–50 μm in length (Kurtz et al. 1999). The average spacing between lamellae is on the order of 50 nm (Bellare, Schnablegger, and Cohen 1995).

The crystalline morphology of UHMWPE can be visualized using transmission electron microscopy (TEM), which can magnify the polymer by up to

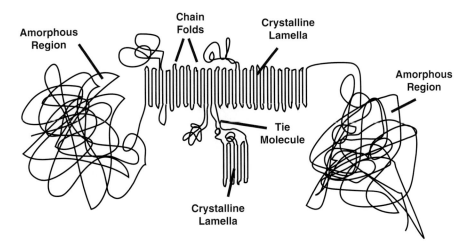

Figure 1.6
Morphological features of UHMWPE.

16,000 times. An ultramicrotomed slice of the polymer is typically stained with uranyl acetate to improve contrast in the TEM. The staining procedure makes the amorphous regions turn gray in the micrograph. The lamellae, which are impervious to the contrast agent, appear as white lines with a dark outline. From the TEM micrograph in Figure 1.7, one can appreciate the composite nature of UHMWPE as an interconnected network of amorphous and crystalline regions.

Thermal Transitions

As already indicated, one of the distinguishing characteristics of polymeric materials is the temperature dependence of their properties. Returning to our conceptual model of an UHMWPE molecule as a mass of incredibly long spaghetti, one must also imagine it to be jiggling and writhing with thermal energy. Generally speaking, many polymers undergo three major thermal transitions: the glass transition temperature (T_g), the melt temperature (T_m), and the flow temperature (T_f).

The glass transition (T_g) is the temperature below which the polymer chains behave like a brittle glass. Below T_g, the polymer chains have insufficient thermal energy to slide past one another, and the only way for the material to respond to mechanical stress is by stretching (or rupture) of the bonds constituting the molecular chain. In UHMWPE, the glass transition occurs at around $-160°C$.

As we raise the temperature above T_g, the amorphous regions within the polymer gain increased mobility. When the temperature of UHMWPE rises above 60–90°C, the smaller crystallites in the polymer begin to melt. The melting behavior of semicrystalline polymers, including UHMWPE, is typically

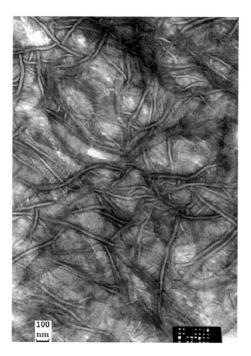

Figure 1.7
TEM micrograph of UHMWPE showing amorphous and crystalline regions (lamellae).

measured using differential scanning calorimetry (DSC). DSC measures the amount of heat needed to increase the temperature of a polymer sample. Some representative DSC data for UHMWPE is shown in Figure 1.8.

The DSC trace for UHMWPE shows two key features. The first feature of the DSC curve is the peak melting temperature (T_m), which occurs at around 137°C and corresponds to the point at which the majority of the crystalline regions have melted. The melt temperature reflects the thickness of the crystals, as well as their perfection. Thicker and more perfect polyethylene crystals will tend to melt at a higher temperature than smaller crystals.

In addition, the area underneath the melting peak is proportional to the crystallinity of the UHMWPE. DSC provides a measure of the total heat energy per unit mass (also referred to as the change in enthalpy, ΔH) required to melt the crystalline regions within the sample. By comparing the change in enthalpy of an UHMWPE sample to that of a perfect 100% crystal, one can calculate the degree of crystallinity of the UHMWPE. Most bulk UHMWPEs are approximately 50–55% crystalline.

As the temperature of a semicrystalline polymer is raised above the melt temperature, it may undergo a flow transition and become liquid. Polyethylenes with a molecular weight of less than 500,000 g/mol can be observed to undergo such a flow transition (T_f). However, when the molecular weight of polyethylene increases above 500,000 g/mol, the entanglement of the

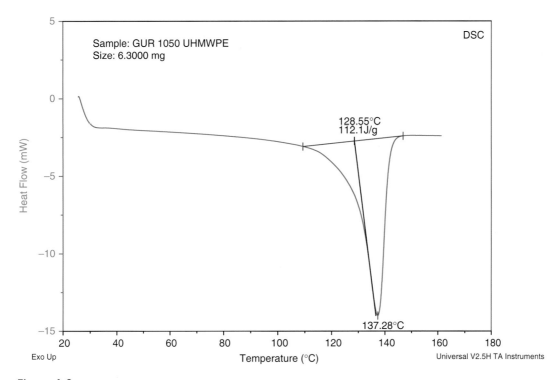

Figure 1.8
DSC trace for UHMWPE.

immense polymer chains prevents it from flowing. UHMWPE does not exhibit a flow transition for this reason.

1.6 Overview of the *Handbook*

This *Handbook* is organized into three main sections. The first section, consisting of three chapters, reviews the basic scientific and engineering foundations for UHMWPE. For example, in Chapter 2, we explain how UHMWPE must be formed into bulk components from the resin powder using extrusion or compression molding techniques. In Chapter 3, we review the techniques associated with sterilization and packaging of UHMWPE implants.

The second part of the *Handbook* is focused on the use of UHMWPE as an orthopedic biomaterial. Chapter 4 reviews the clinical origins of UHMWPE in hip replacement, which was introduced to orthopedics in 1962 by Sir John Charnley. Although several variants of UHMWPE have been investigated since the 1960s, UHMWPE remains the gold standard orthopedic bearing material still widely used today in joint replacements around the world. Chapter 5 summarizes the clinical performance of UHMWPE hip implants and discusses the patterns of wear and surface damage that occur during implantation. Chapter 6 describes alternatives to conventional UHMWPE in hip replacement. Chapters 7 and 8 relate to the development and clinical performance of UHMWPE in knee replacement. Chapter 9 is devoted to clinical applications of UHMWPE in the shoulder, and Chapter 10 covers the use of UHMWPE in the spine.

The topics outlined in this *Handbook* may be used as a resource in undergraduate, as well as graduate, courses in biomaterials and orthopedic biomechanics. Students in these disciplines can learn a great deal from exposure to the historical development of total joint replacements within the context of UHMWPE. The first two main sections of this book, which cover the fundamentals of UHMWPE and clinical applications in the spine and upper and lower extremities, are intended as a resource for undergraduate instruction.

The third section of this book, which covers more specialized topics related to UHMWPE, is intended for an audience of graduate students and orthopedic researchers. Chapter 11 covers the chemistry of UHMWPE following irradiation, which leads to oxidation and crosslinking of the material. Chapter 12 describes the characterization methods for UHMWPE in the context of regulatory submissions prior to clinical trials. In Chapter 13, we review the development of the small punch test, a miniature specimen mechanical testing technique that has recently been standardized. Chapter 14 describes the micromechanical modeling of conventional and highly crosslinked UHMWPE. The final chapter in this work, Chapter 15, is a compendium of the processing, packaging, and sterilization information for highly crosslinked and thermally treated UHMWPE materials that are currently used in hip and knee arthroplasty.

Understanding basic chemical structure and morphology is an important starting point for appreciating the unique and outstanding properties of UHMWPE. The chapters that follow and describe the processing, as well as the sterilization, of UHMWPE will continue to build on the conceptual foundation established in this introduction.

References

Bellare A., H. Schnablegger, and R.E. Cohen. 1995. A small-angle x-ray scattering study of high-density polyethylene and ultra-high molecular weight polyethylene. *Macromolecules* 17:2325–2333.

Charnley J. 1963. Tissue reaction to the polytetrafluoroethylene. *Lancet* II:1379.

Charnley J. 1979. Low friction principle. In *Low friction arthroplasty of the hip: Theory and practice*. Berlin: Springer-Verlag.

Chubberley A.H. 1965. Ultra-high molecular weight polyethylenes. In *Modern plastics encyclopaedia*. New York: McGraw-Hill.

Edidin A.A., and S.M. Kurtz. 2000. The influence of mechanical behavior on the wear of four clinically relevant polymeric biomaterials in a hip simulator. *J Arthroplasty* 15:321–331.

Kurtz S.M., O.K. Muratoglu, M. Evans, A.A. Edidin. 1999. Advances in the processing, sterilization, and crosslinking of ultra-high molecular weight polyethylene for total joint arthroplasty. *Biomaterials* 20:1659–1688.

Rodriguez F. 1989. *Principles of polymer systems.* New York: Hemisphere.

Wroblewski B.M., P.D. Siney, D. Dowson, and S.N. Collins. 1996. Prospective clinical and joint simulator studies of a new total hip arthroplasty using alumina ceramic heads and cross-linked polyethylene cups. *Journal of Bone & Joint Surgery* 78B:280–285.

Young R.J. 1983. *Introduction to polymers.* London: Chapman and Hall.

Chapter I. Reading Comprehension Questions

1.1. Let A, B, and C be monomers. What is the molecular structure of a linear homopolymer?
   a) –A-A-B-A-A-B-A-A-B-
   b) –A-B-C-A-B-C-A-B-C-
   c) -B-C-C-C-C-C-C-C-C-
   d) –B-B-B-B-B-B-B-B-B-
   e) –C-A-A-C-A-A-C-A-A-

1.2. Which of the following polymers is NOT synthesized from ethylene?
   a) LLDPE
   b) PTFE
   c) UHMWPE
   d) HDPE
   e) LDPE

1.3. What is the major difference between HDPE from UHMWPE?
   a) Molecular weight
   b) Monomer
   c) Chemical composition
   d) Color
   e) All of the above

1.4. What are the crystals in polyethylene made up of?
   a) Folded molecular chains
   b) Calcium stearate
   c) Aluminum tetrachloride
   d) Copolymer
   e) Branched chain ends

1.5. UHMWPE exhibits which of the following transition temperatures?
   a) Glass transition
   b) Melting transition
   c) Flow transition
   d) Glass and melting transitions
   e) Glass, melting, and flow transitions

1.6. What do the tie molecules in UHMWPE interconnect?
   a) Molecular chain ends
   b) Molecular chain folds
   c) Crystalline lamellae
   d) Amorphous regions
   e) Molecular branching

1.7. What happens to UHMWPE above T_m?
   a) The polymer becomes a viscous liquid
   b) The polymer becomes brittle and glassy
   c) The polymer becomes a gaseous vapor
   d) The polymer becomes opaque
   e) The polymer becomes translucent

Chapter 2

From Ethylene Gas to UHMWPE Component: The Process of Producing Orthopedic Implants

Introduction

At a conceptual level, polyethylene consists only of carbon and hydrogen, as was described in the previous chapter. However, if the discussion of polyethylene is to proceed from ideal abstractions to actual physical implants, three "real world" steps need to occur. First, the ultra-high molecular weight polyethylene (UHMWPE) must be polymerized from ethylene gas. Second, the polymerized UHMWPE, in the form of resin powder, needs to be consolidated into a sheet, rod, or near-net shaped implant (Figure 2.1). Finally, in most instances, the UHMWPE implant needs to be machined into its final shape (Figure 2.1). A small subset of implants are consolidated into their final form directly, in a process known as direct compression molding (DCM), without need of additional machining.

Each of these three principal steps produces a subtle alteration of the properties of UHMWPE. In some cases, such as machining, the change in the material may occur only in the topography and appearance of the surface. On the other hand, changes in the polymerization and conversion of the UHMWPE can impact the physical and mechanical properties of the entire implant.

Because many of the details used in the polymerization, conversion, and machining of UHMWPE are proprietary, very little is written in the public domain literature about the techniques used to make actual implants. The little that has been written about implant production can be confusing to the

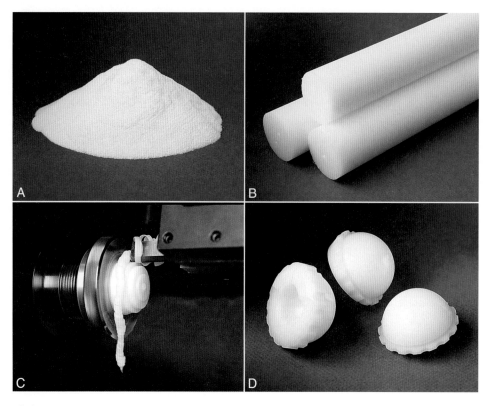

Figure 2.1

Typical processing steps in the manufacture of UHMWPE implants, starting with the resin powder (**A**). (**B**) Semifinished rods that have been consolidated from the resin powder. (**C**) Machining of the UHMWPE rods on a lathe. (**D**) UHMWPE acetabular components after machining. (Pictures provided courtesy of David Schroeder [Biomet, Warsaw, IN].)

uninitiated because of the abundance of trade names, which evolve over time. Not only do industrial trade names change over time, but the actual processing techniques have also improved. Surgeons and researchers are strongly cautioned against overgeneralizing about the properties or techniques of any single resin or process, which has likely changed over the past 40 years.

Polymerization: From Ethylene Gas to UHMWPE Powder

The polymerization of UHMWPE was commercialized by Ruhrchemie AG, based in northern Germany, during the 1950s. Ruhrchemie AG itself evolved in 1928 with the mission of developing useful chemicals from coal (carbon); its shareholders consisted of 28 coal mining companies. In 1953, chemists from the nearby Max-Plank Institute approached scientists at Ruhrchemie AG in Oberhausen, Germany, with a brown, wet (not fully dried) mass that they claimed was a new form of polyethylene, produced in a new low-pressure process. Convinced of the commercial utility of such a material (the dangers of high-pressure polymerization, such as in the production of low-density polyethylene (LDPE), were widely appreciated), development on UHMWPE began shortly thereafter at Ruhrchemie AG. In 1955, the first commercial polymerization of UHMWPE began, and during that same year, the material was first introduced at the K55, a polymer trade show.

Since the 1950s, the UHMWPE powders have been produced by Ruhrchemie (currently known as Ticona) using the Ziegler process, which has been described by Birnkraut (1991). The main ingredients for producing UHMWPE are ethylene (a reactive gas), hydrogen and titanium tetra chloride (the catalyst). The polymerization takes place in a solvent used for mass and heat transfer. The ingredients require that the polymerization be conducted in specialized production plants capable of handling these volatile and potentially dangerous chemicals. The last ingredient (catalyst) has been improved consistently since the 1950s because it is the key to producing white UHMWPE powder with reduced impurities.

The requirements for medical grade UHMWPE powder are specified in the American Society for Testing and Materials (ASTM) standard F648 and ISO standard 5834-1. In the standards, medical grade resins are described as Types 1, 2, or 3, depending on their molecular weight and producer (Table 2.1). The trace impurities of titanium, aluminum, and chlorine are residuals from the catalyst, whereas the trace levels of calcium, as well as the ash content, depend on the storage and handling of the powder after polymerization.

Table 2.1
Requirements for Medical Grade UHMWPE Powders (per ASTM F648 and ISO 5834-1)

Property	Requirements	
Resin type	Types 1–2	Type 3
Trade name	GUR 1020 & 1050	1900H
Producer	Ticona, Inc.	Basell polyolefins (now discontinued)
Ash, mg/kg, (maximum)	150	300
Titanium, ppm, (maximum)	40	150
Aluminum, ppm, (maximum)	40	100
Calcium, ppm, (maximum)	50	50
Chlorine, ppm, (maximum)	20	90

Currently Ticona (Oberhausen, Germany) produces a Type 1 and Type 2 resin with the trade names of GUR 1020 and 1050, respectively. Before 2002, Basell Polyolefins (Wilmington, DE) produced a Type 3 resin with the 1900 trade name. This resin was discontinued in January 2002 and is no longer produced. Two orthopedic manufacturers have maintained large stockpiles of this resin. Therefore, orthopedic implants will continue to be fabricated from this resin, at least in the near future.

The orthopedics literature contains numerous references to different trade names for UHMWPE, which fall into two categories: 1) GUR resins currently produced by Ticona and 2) the 1900 resins, previously produced by Basell. The UHMWPE grades currently used in the orthopedic industry are summarized in Table 2.2.

GUR Resins

Ticona (formerly known as Hoechst, and before that, as Ruhrchemie AG) currently supplies 800 tons of premium grade UHMWPE per year for orthopedic applications. This is less than 2% of Ticona's overall UHMWPE production. Resin is currently manufactured in Bishop, TX, and Oberhausen, Germany. The resin grades produced at Bishop and Oberhausen use the same catalyst technology and undergo similar resin synthesis processes. When tested in accordance with ASTM F648 and ISO 5834, the physical and mechanical properties of the Bishop and Oberhausen resins are indistinguishable. Ticona uses the designation GUR for its UHMWPE grades worldwide; the acronym GUR stands for "Granular," "UHMWPE," and "Ruhrchemie."

Hoechst changed its resin nomenclature between 1992 and 1998. Resins distributed to both the general and orthopedic marketplaces before 1992 were designated as, for example, GUR 415, and GUR 412 in the United States, or CHIRULEN® P in Europe. In October 1992, a fourth digit was added to all general and orthopedic products to allow inclusion of additional resins to the product lines. For example, GUR 415 became GUR 4150. In 1994, after the

Table 2.2

Nomenclature of Ticona and Basell UHMWPE Resins. The Average Molecular Weight is Calculated Based on Intrinsic Viscosity

Resin designation	Producer	Average molecular weight (10^6 g/mol)	Calcium stearate added?
GUR 1020	Ticona	3.5	No
GUR 1050	Ticona	5.5–6	No
1900H*	Basell	>4.9	No

*Note: Production of this resin was discontinued in 2002.

sale of the CHIRULEN trade name, Hoechst in Europe decided to exclusively designate all resins for the surgical implant market with a first digit of "1." By comparison, Hoechst in North America renamed GUR 4150 as GUR 4150HP (to designate High Purity) for this application in August 1993. In 1998, all of the nomenclature was consolidated with the availability of four grades for the worldwide orthopedic market—GUR 1150, 1050, 1120, and 1020 resins (Table 2.2).

The first digit of the grade name was originally the loose bulk density of the resin, i.e., the weight measurement of a fixed volume of loose, unconsolidated powder; the "4" corresponded to a bulk density of more than 400 g/l for standard grades. The second digit indicates the presence (1) or absence (0) of calcium stearate, whereas the third digit is correlated to the average molecular weight of the resin. The fourth digit is a Hoechst internal code designation. Most recently, in 2002, Ticona discontinued the production of GUR 1120 and 1150, which contained added calcium stearate.

1900 Resins

Although Basell, prior to 2002, produced six grades of 1900 resin, only one of these grades (1900H) is currently used in orthopedic applications (Table 2.2). Since its introduction by the Hercules Powder Company (Wilmington, DE), 1900 resin has been through transitions in nomenclature and production. During the 1960s, Hercules produced several grades of Hi-Fax 1900 UHMWPE with molecular weights ranging from 2 million to 4 million (Chubberley 1965). In 1983 Hercules formed a joint venture with Montedison in Italy to become Himont. In April 1995, Montedison and Shell Oil (Netherlands) formed Montell Polyolefins; the company was merged into BASF and Shell, forming Basell Polyolefins in 2001. In January 2002, Basell divested itself of its UHMWPE production facilities and sold its technology to Polialden, a Brazilian UHMWPE manufacturer.

The resin designation of 1900 has stayed the same through the past three decades of transitions while Montell changed the manufacturing facilities. Over this time, three different reactors were used to produce the different grades of 1900. Originally the A-line reactor was used, but in 1989–1990 Himont began producing the resin on their F-line. The latter line was a semicontinuous process as compared with the batch processing used earlier. In 1996–1997 the F-line was replaced with the G-line. In early 2002, Basell dismantled the G-line production facility. The current owner of the 1900 resin technology, Polialden, has not yet announced whether they intend to resume production for orthopedic use in Brazil.

Molecular Weight

The mechanical behavior of UHMWPE is related to its average molecular weight, which is routinely inferred from intrinsic viscosity (IV) measurements

(Eyerer, Frank, and Jin 1985, Li and Burstein 1994). There are two commonly used methods for calculating the viscosity average molecular weight (M_v) for UHMWPE based on the IV using the Mark-Houwink equation:

1. ASTM D4020-00: $M_v = 53,700\ IV^{1.37}$
2. Margolies equation, used outside North America: $M_v = 53,700\ IV^{1.49}$

In the previous equations, M_v has units of g/mol and IV has units of dL/g. Alternative methods for molecular weight characterization of UHMWPE include sequential extraction (Kusy and Whitley 1986) and gel permeation chromatography (GPC), a type of size exclusion chromatography (Wagner and Dillon 1988).

The IV of UHMWPE is related to the bulk impact strength and abrasive wear resistance after conversion to bulk form, although the relationships are nonlinear. The maximum impact strength of both Ticona and Basell is found between 16 and 20 IV, which is the equivalent to a molecular weight range of 2.4 million to 3.3 million using the ASTM calculation for molecular weight. As the IV increases, abrasive wear resistance increases, as measured by sand slurry testing, reaching a plateau for IV greater than 20.

Molecular weight also influences the static fracture response as well as the mechanical behavior of UHMWPE at large strains (Kurtz et al. 1998). For example, beyond the polymer yield point, the hardening or cold drawing portion behavior in uniaxial tension is sensitive to the molecular weight. Figure 2.2 illustrates the true-stress strain curve in uniaxial tension (room temperature, 30 mm/min) for two grades of UHMWPE, in comparison with high-density polyethylene (HDPE).

Under biaxial drawing conditions of the small punch test (Edidin and Kurtz 2001), the large-deformation mechanical behavior of polyethylene also

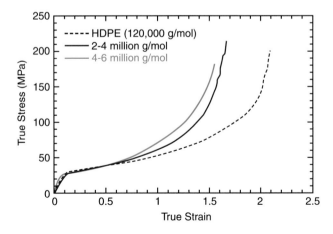

Figure 2.2

The true-stress strain behavior in uniaxial tension (room temperature, 30 mm/min) for two grades of UHMWPE, in comparison with HDPE.

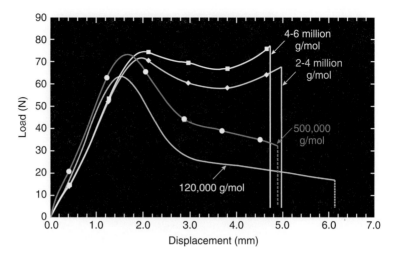

Figure 2.3
Representative small punch test data conducted at room temperature at a rate of 0.5 mm/min.

displays strong molecular weight dependence. Representative small punch test data, shown in Figure 2.3, were conducted at room temperature at a rate of 0.5 mm/min.

GUR Versus 1900 Resin

Variations between the material properties of converted GUR and 1900 resins have been explained by differences in the average resin particle size, the size distribution, and morphology of the resin particles (Gul 1997, Han et al. 1981). Ticona resins have a mean particle size of approximately 140 μm (Gul 1997, Han et al. 1981), whereas the mean particle size of 1900 resin is approximately 300 μm (Han et al. 1981). The distribution of particle sizes also varies between Ticona and Basell resins. When studied under a scanning electron microscope, GUR and 1900 resin powders consist of numerous spheroidal particles (Gul 1997, Han et al. 1981). The scanning electron micrographs seen in Figure 2.4 (provided courtesy of Rizwan Gul [1997]) illustrate the subtle differences in nascent particle morphology between Basell and Ticona resins, which are likely related to the catalyst package and polymerization conditions (Birnkraut 1991).

The most notable finding from studies of powder morphology is that the Ticona resins are characterized by a fine network of submicron-sized fibrils that interconnect the microscopic spheroids. The fibrils are illustrated in the scanning electron microscopy (SEM) micrograph in Figure 2.5 (provided courtesy of Rizwan Gul [1997]).

There is some evidence that Basell resins have a different molecular weight distribution than the Ticona resins (Gul 1997). For example, depending on the

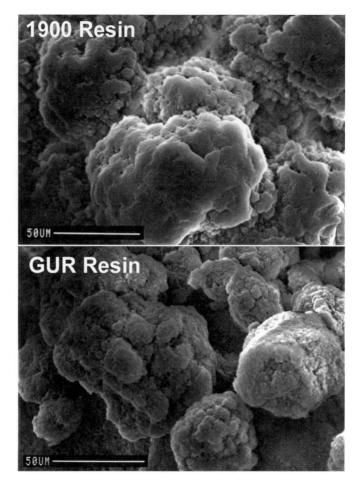

Figure 2.4
The scanning electron micrographs illustrate the subtle differences in nascent particle morphology between Basell and Ticona resins.

processing conditions, Basell 1900 exhibits spherulitic crystalline morphology (Gul 1997, Weightman and Light 1985), which is typically associated with slightly lower molecular weight polyethylenes (Voigt-Martin, Fisher, and Mandelkern 1980). Spherulites are not observed in Ticona resins, which have a lamellar crystalline morphology (Goldman et al. 1996, Gul 1997, Pruitt and Bailey 1998). Differences in the molecular weight distribution between the Basell- and Ticona-produced resins may account for the variation in impact strength observed experimentally at the same average molecular weight. However, despite clues in the literature, the differences in molecular weight distribution between Basell and Ticona resins have yet to be explicitly quantified.

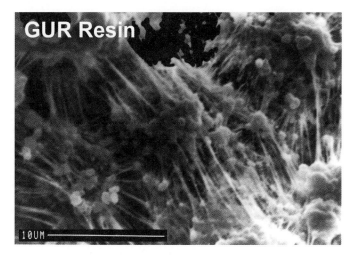

Figure 2.5
SEM image of submicron size fibrils in GUR resins.

Calcium Stearate

Calcium stearate has been used by Hoechst since 1955 and by Montell since the late 1960s. The additive acts as a scavenger for residual catalyst components that can potentially corrode conversion equipment; calcium stearate also acts as a lubricant and a release agent (Eyerer et al. 1990). The catalyst used by Ticona today has a higher activity than the one originally used for UHMWPE production. Thus, a larger quantity of UHMWPE can be synthesized with the same amount of catalyst, resulting in a lower residual catalyst concentration (Birnkraut 1991). The trace element level of calcium in UHMWPE is directly proportional to the addition of calcium stearate, which is currently certified as food grade (Stein 1997). When calcium stearate is added to any UHMWPE resin, regardless of the manufacturer or catalyst technology used, the polymer particles are surface coated by the calcium stearate.

During the 1980s, the influence of calcium stearate on the properties and performance of UHMWPE total joint replacements was considered a controversial issue (Kurtz et al. 1999). In several studies, the presence of trace calcium levels in UHMWPE has been associated with fusion defects and oxidation of UHMWPE (Biomet 1995, Blunn et al. 1997, Eyerer et al. 1990, Gsell, King, and Swarts 1997, Hamilton, Wang, and Sung 1996, Schmidt and Hamilton 1996, Swarts et al. 1996, Walker, Blunn, and Lilley 1996, Wrona et al. 1994). Using high-resolution synchrotron infrared spectroscopy, Muratoglu and colleagues (1997) resolved the molecular vibrations associated with calcium stearate in the grain boundary layers of fusion defects in GUR 4150HP; examination of converted GUR 4050 and HIMONT 1900, in contrast, showed no evidence of calcium stearate. In an accelerated aging study by Swarts and colleagues (1996),

GUR 4150HP exhibited more oxidation than reduced-calcium stearate resins (GUR 4050 and Montell 1900H).

However, the mere presence of calcium stearate does not automatically imply poor consolidation and decreased fracture resistance in UHMWPE. For example, using the J-integral method, Baldini and colleagues (1997) reported that the fracture resistance of GUR 1020 and 1120 were "comparable." Lykins and Evans (1995) have suggested that fusion defects may result from inappropriate control of processing variables (e.g., temperature, pressure, time, heating rate) during conversion of resin to stock material. Thus, it remains to be established whether calcium stearate plays a role in the mechanical behavior of well-consolidated UHMWPE except at sufficiently high enough concentrations to interfere with the sintering of the powder.

Thus, research conducted during the 1990s indicated that calcium stearate may be present at the boundaries of fusion defects in UHMWPE, and that fusion defects may in turn deleteriously affect the fatigue and fracture behavior of UHMWPE. However, in the absence of fusion defects, the deleterious effects of trace levels of calcium stearate were not conclusively established. Furthermore, polymerization and processing technology had by the late 1990s evolved to the point that the additive was no longer necessary. Consequently, orthopedic manufacturers began switching to UHMWPE resins without added calcium stearate (e.g., GUR 1020 and 1050). By 2002, demand for the calcium stearate–containing resins (GUR 1120 and 1150) dropped to the point that Ticona discontinued its production.

Conversion: From UHMWPE Powder to Consolidated Form

UHMWPE is produced as powder and must be consolidated under elevated temperatures and pressures because of its high melt viscosity. As already discussed in Chapter 1, UHMWPE does not flow like lower molecular weight polyethylenes when raised above its melt temperature. For this reason, many thermoplastic processing techniques, such as injection molding, screw extrusion, or blow molding, are not practical for UHMWPE. Instead, semifinished UHMWPE is typically produced by compression molding and ram extrusion.

The process of consolidation in UHMWPE requires the proper combination of temperature, pressure, and time. The precise combinations of these variables used to produce commercially available molded and extruded stock materials remain proprietary, but the scientific principles underlying consolidation of UHMWPE are generally well understood (Barnetson and Hornsby 1995, Bastiaansen, Meyer, and Lemstra 1990, Bellare and Cohen 1996, Chen, Ellis, and Crugnola 1981, Farrar and Brain 1997, Gul 1997, Halldin and Kamel 1977a, Halldin and Kamel 1977b, Han et al. 1981, McKenna, Crissman, and Khoury 1981, Olley et al. 1999, Shenoy and Saini 1985, Truss et al. 1980, Wang, Li, and Salovey 1988, Zachariades 1985). The governing mechanism of consolidation is self-diffusion, whereby the UHMWPE chains (or chain segments) in adjacent resin particles intermingle at a molecular level. The kinetics of intergranular diffusion are promoted by close proximity of the interfaces (at elevated pressures) and thermally activated mobility of the polymer chains (at elevated temperatures). As a diffusion-limited process, consolidation of UHMWPE requires sufficient time at elevated temperature and pressure for the molecular chains to migrate across grain boundaries.

Consolidated UHMWPE retains a memory of its prior granular structure, which is especially evident when calcium stearate is added to the resin (Bastiaansen, Meyer, and Lemstra 1990, Farrar and Brain 1997, Gul 1997, Olley et al. 1999). The ultrastructure of UHMWPE may be visualized by either optical or SEM, but special preparation methods are needed in either case. A standard technique, described in ASTM F648, involves microtoming thin films and observing them under optical microscopy, preferably under dark field conditions (Halldin and Kamel 1977a). However, in today's calcium-stearate free-resins, the presence of grain boundaries in well-consolidated material is usually difficult to detect, as illustrated in Figure 2.6, the following optical micrograph of a 100 μm–thick section of GUR 1020 (provided courtesy of Rolf Kaldeweier, Ticona, Inc.).

Grain boundaries can also be visualized by freezing an UHMWPE sample in liquid nitrogen and then fracturing the sample while frozen (Gul 1997). Etching of fracture surfaces can further highlight the intergranular regions, allowing for visualization of the UHMWPE ultrastructure (Olley et al. 1999). SEM is then used to inspect the surface of the freeze-fractured or etched UHMWPE surfaces.

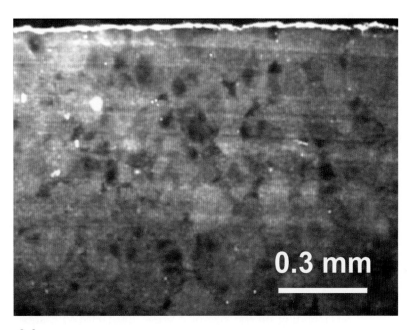

Figure 2.6
An optical micrograph of a 100 μm–thick section of GUR 1020.

In contrast with grain boundaries, which reflect the normal ultrastructure of UHMWPE, consolidation defects may arise when the proper combination of pressure, temperature, and time are not used (Lykins and Evans 1995). More typically, consolidation defects represent a single resin particle, or a highly localized region, that has not fully fused with its neighbors. A standard technique (specified in ASTM F648), involving optical microscopy of thin sections, has been developed to quantify the presence of fusion defects in UHMWPE. Interinstitutional studies are currently progressing under the auspices of ASTM to quantify the repeatability and reproducibility of the standard inspection technique in contemporary calcium-stearate–free resins. Nondestructive methods, including laser candling and ultrasound, are also used in industry for inspection of medical grade UHMWPE, but the use of these techniques has not been standardized.

Compression Molding of UHMWPE

Historically, the UHMWPE powder has been converted by compression molding since the 1950s, because the industries in the area around Ruhrchemie already had experience with this processing technique. At first, the semi-finished material was distributed under the trade name RCH 1000/Hostalen GUR 412. During the 1970s, the compression-molded UHMWPE was manufactured and distributed by Ruhrchemie/Hoechst and later distributed by the company Europlast specifically for orthopedic applications under the trade name CHIRULEN, which denoted that the material was produced in Germany by a dedicated press using established processing parameters. Thus, references in the orthopedic literature to RCH 1000 and CHIRULEN apply to compression-molded UHMWPE, which was similar to contemporary GUR 1020. Note, however, that GUR 412 would have contained calcium stearate, whereas GUR 1020 would not.

Today, compression-molded sheets of GUR 1020 and 1050 are produced commercially by two companies (Perplas Medical and Poly Hi Solidur Meditech). Ticona stopped producing compression-molded UHMWPE in 1994. Perplas's molding facility is in England, whereas Poly Hi Solidur produces medical grade UHMWPE sheets in the United States and in Germany.

An example of a compression molding press, currently installed at Poly Hi Solidur Meditech in Vreden, Germany, is shown in Figure 2.7. This particular press, originally designed by Hoechst in the 1970s for production of CHIRULEN, molds two 1 m × 2 m sheets in a single press cycle. One UHMWPE sheet is pressed between the upper and middle platens, and the second is produced between the middle and lower platens. The platens are oil heated and hydraulically actuated from below. The heating and loading systems are all computer controlled. Finally, the entire press is contained in a clean room, to reduce the introduction of extraneous matter into the sheet.

Depending on the size of a press, the UHMWPE sheet may range in size between 1 m × 2 m (shown in the photo) to 2 m × 4 m, with thicknesses of 30 mm to 80 mm. However, the facilities necessary to mold a 2 m × 4 m sheet of UHMWPE are considerably larger than that shown in the photo. The press

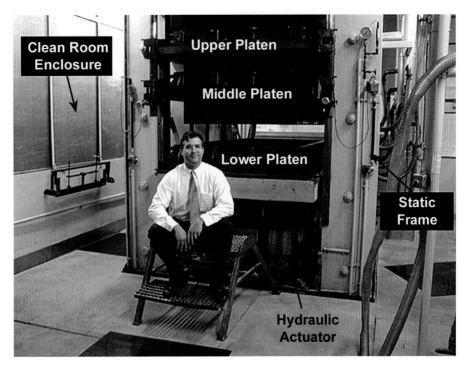

Figure 2.7
Compression molding press (along with the author, for scale) for production of 1 m × 2 m sheets of UHMWPE. This press is located at Poly Hi Solidur Meditech in Vreden, Germany, and was originally used by Ticona in the production of Chirulen sheets of UHMWPE. The press is still used today in the production of medical grade UHMWPE.

operated by Perplas Medical, for example, is over three stories tall and enclosed within its own clean room structure.

Because of the relatively low thermal conductivity of UHMWPE, the duration of the molding cycle will depend on the particular geometry of the press and the size of the sheet to be produced, but the processing time can last up to 24 hours. The long molding times are necessary to maintain the slow, uniform heating and cooling rates throughout the entire sheet during the molding process.

After molding, the sheet is typically turned into rods or other preform shapes to facilitate subsequent machining operations by orthopedic manufacturers. Thus, the final shape of UHMWPE stock material today is not necessarily dictated by the conversion method.

Ram Extrusion of UHMWPE

In contrast with compression molding, which originated in Germany in the 1950s, ram extrusion of UHMWPE was developed by converters in the

United States during the 1970s. Historically, a wide range of UHMWPE resins have been extruded since the seventies. From the early 1970s to early 1980s, ram extruded Hifax 1900 resin from Hercules Powder Company (Wilmington, DE) was commonly supplied to bulk form converters. In the early 1960s Hoechst's GUR 412 resin was made available to converters. Although GUR 412 tended to have a lower extraneous particle count than 1900, it was also more difficult to process using ram extrusion because of its lower average molecular weight and lower melt viscosity. In the mid 1980s, converters began ram extruding using the higher average molecular weight Hoechst GUR 415 resin, which also had a lower extraneous particle count relative to 1900 and was easier to extrude than GUR 412. During the 1980s and early 1990s, extruded rods of GUR 412, 415, and (more rarely) 1900 CM (a calcium-stearate–containing grade of 1900 resin) were all used in orthopedics. Thus, it is difficult to generalize about the resin types used in implants without tracing the lot numbers used by a particular manufacturer.

Today only a few converters supply medical grade, GUR 1020 and 1050 ram extruded UHMWPE to the orthopedic industry. Medical grade extrusion facilities for Poly Hi Solidur and Westlake Plastics are based in the United States, whereas Perplas Medical's medical grade extrusion is in England.

Like compression molding facilities, a medical grade extruder is typically maintained in a clean room environment to reduce the introduction of extraneous matter into the UHMWPE. The generic schematic of a ram extruder is illustrated in Figure 2.8 (the clean room is not shown).

UHMWPE powder is fed continuously into an extruder. The extruder itself consists essentially of a hopper that allows powder to enter a heated receiving chamber, a horizontal reciprocating ram, a heated die, and an outlet. Within the extruder, the UHMWPE is maintained under pressure by the ram, as well as by the back pressure of the molten UHMWPE. The back pressure is caused by frictional forces of the molten resin against the heated die wall surface as it is

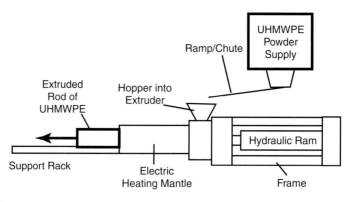

Figure 2.8
Schematic of a ram extruder.

forced horizontally through the outlet. Beyond the outlet, the rod of UHMWPE is slowly cooled in a series of electric heating mantles.

A wide range of rod diameters is achievable (up to 12 inches in diameter) using extrusion, but the rate of production depends on the rod size because of the increased cooling times needed with the larger rod diameters. Rod sizes ranging from 20 to 80 mm in diameter are most often used in orthopedic applications. Typical production rates are on the order of millimeter per minute.

Hot Isostatic Pressing of ArCom™ UHMWPE

A third conversion method, known as hot isostatic pressing (HIPing), is used by one orthopedic manufacturer (Biomet, Warsaw, IN) for conversion of resin to stock material (Gul 1997). This multistep conversion process, referred to as ArCom™ by the manufacturer, begins with the manufacture of a cylindrical compact through cold isostatic pressing, which expels most of the air (Figure 2.9). Subsequently, the compacted green rods are sintered in a HIP furnace in an argon-filled pouch to prevent degradation of the UHMWPE (Figure 2.9). The resulting ArCom rod stock is essentially isotropic as a result of the hydrostatic sintering process and may be considered a compression-molded form of the resin. Finished implants are then made from the iso-molded rod stock by either turning or milling operations (Biomet 1997), as illustrated in Figure 2.1.

Direct Compression Molding of UHMWPE

In DCM, sometimes also called net shape compression molding, the manufacturer of the polyethylene insert effectively converts the resin to a finished or semifinished part using individual molds. One advantage of DCM is the extremely smooth surface finish obtained with a complete absence of machining marks at the articulating surface. In addition, higher processing pressures may be attained, if desired, because the projected surface area of each individual part mold is relatively small compared with the area of large molds used to compression mold sheets.

DCM has been used for more than 20 years to produce tibial and acetabular inserts. Historically, the process may have been adopted because the cutting and milling machinery of the day was not numerically controlled and thus less able to accurately produce complicated curves as required to make knee inserts. The physical and mechanical properties of the finished product can be tailored to some degree by varying the DCM cycle, as detailed by Chen, Truss, and others (Han et al. 1981, Truss et al. 1980, Chen, Ellis, and Crugnola 1981). DCM can also be used to produce UHMWPE with properties indistinguishable from stock material produced by closely monitored conversion of compression-molded sheet and extruded rod.

Figure 2.9
Steps in the processing of ArCom UHMWPE (Biomet, Warsaw, IN). (**A**) Resin powder is first poured into a polymeric can with a compressible lid. (**B**) The can is inserted into a cold isostatic press for compaction of the powder. (**C**) Cold pressed ("green") rod of UHMWPE after cold compaction. (**D**) The green rod is sealed in an Argon-filled foil pouch. (**E**) The foil-wrapped green rods are loaded in a metallic mesh rack and lowered into a hot isostatic press. (**F**) The foil-wrapped rod of consolidated UHMWPE after HIPing. The foil is peeled off the rod and machined into components as shown in Figure 2.1. (Pictures provided courtesy of David Schroeder, Biomet, Inc.)

Overall, the literature suggests that the DCM process can produce a specimen or implant with mechanical properties similar to those exhibited by high-quality bulk converted UHMWPE stock. The variations in final properties seen as functions of temperature, pressure, density, and cooling rate would presumably affect the production of bulk converted sheet and rod. However, bulk conversion needs to be performed much less often to convert a given mass of resin, and thus issues of monitoring and maintaining quality need to be performed less often as well. From a manufacturability standpoint, the difficulties associated with process control of DCM has led most implant producers to obtain bulk converted sheet or rod.

ArCom

ArCom is the trade name of a proprietary conversion process, which denotes that the resin has been compression molded in the presence of argon (Biomet, Warsaw, IN). ArCom is produced by HIPing as well as by DCM.

The type of resins used with ArCom are summarized in Table 2.3. Both 1900H and GUR resins have been fabricated using the ArCom iso-molding process. Between 1993 and 2002, iso-molded ArCom was produced with 1900H resin. In 2002, after Basell discontinued the production of 1900 resin, Biomet switched production of iso-molded rod stock to GUR resins produced by Ticona (Table 2.3). On the other hand, Biomet has stockpiled sufficient 1900H resin to continue production of direct compression-molded ArCom for the foreseeable future (Table 2.3).

Properties of Extruded Versus Molded UHMWPE

As might be expected from our previous discussion of compression molding and extrusion, the conversion method could have an effect on the properties of UHMWPE. In practice, however, the differences between extrusion and

Table 2.3
Summary of Conversion Methods and Resins Used in the Production of ArCom During Both Historical and Contemporary Time Frames

ArCom	Historical	Contemporary
Conversion method	UHMWPE resin	UHMWPE resin
	(1993–2002)	(2002–Present)
Isostatically molding	1900H	GUR 1050
DCM	1900H	1900H*

*This resin continues to be used because it has been stockpiled by the manufacturer.

Table 2.4

Breakdown of 680 Individual UHMWPE Lots Compared in ASTM Survey

	GUR 1020 (Type 1)	GUR 1050 (Type 2)	Total
Ram extruded	113	218	331
Compression molded	186	163	349
Total	299	381	680

compression molding are slight in the hands of a skilled converter. In 2001, a survey was conducted by the author under the auspices of ASTM to evaluate the variation in physical and mechanical properties of UHMWPE as a function of resin and conversion method. Three commercial suppliers provided physical and mechanical property data collected during certification of 680 individual lots of medical grade UHMWPE that were produced between 1998 and 2001. The breakdown of the individual lots in the total data set, according to resin and conversion method, is summarized in Table 2.4, and indicates a relatively even distribution between resins and conversion methods.

Not surprisingly, the data reported by the converters for contemporary UHMWPE was found to exceed the requirements for Types 1 and 2 resins, as set forth in ASTM F648. For density, resin but not conversion method was found to be a significant factor (Table 2.5). For the impact strength and the tensile mechanical properties, resin as well as conversion method were both found to be significant factors (Table 2.5). Although statistically significant, differences in the density and tensile properties of UHMWPE were in general not substantial (less than 21% difference in means between the Extruded GUR 1020 and Molded GUR 1050).

Note that this study did not evaluate differences between converters, which could be another source of variation in material properties of UHMWPE. Because the sintering conditions can significantly influence the density, as well

Table 2.5

Summary of Mean (± Standard Deviation) Physical and Tensile Mechanical Properties of Extruded and Molded UHMWPE

Material	Density (kg/m³)	Tensile yield (MPa)	Ultimate tensile strength (MPa)	Elongation to failure (%)
Extruded GUR 1020	935 ± 1	22.3 ± 0.5	53.7 ± 4.4	452 ± 19
Molded GUR 1020	935 ± 1	21.9 ± 0.7	51.1 ± 7.7	440 ± 32
Extruded GUR 1050	931 ± 1	21.5 ± 0.5	50.7 ± 4.2	395 ± 23
Molded GUR 1050	930 ± 2	21.0 ± 0.7	46.8 ± 6.4	373 ± 29

as the mechanical properties of UHMWPE, one should not make conclusions of preferred resin and processing method without having all relevant data available from a particular converter.

Studies have reported subtle differences in the morphology and fatigue crack propagation behavior of extruded versus compression-molded UHMWPE. For example, investigations of UHMWPE morphology suggest that compression-molded material has an isotropic crystalline orientation, whereas the morphology in ram extruded UHMWPE varies slightly as a function of the distance from the centerline (Bellare and Cohen 1996). Similarly, crack propagation studies have found more isotropic crack propagation behavior in compression-molded sheets as opposed to ram extruded rods of UHMWPE (Pruitt and Bailey 1998).

Machining: From Consolidated Form to Implant

Orthopedic manufacturers generally machine UHMWPE components into their final form. Even components that are direct molded may be machined on the back surface to accommodate a locking mechanism. However, the actual morphology of machining marks depends on the manufacturing conditions as well as the type of UHMWPE material (e.g., conventional versus highly crosslinked) (Kurtz 2002). An example of machining marks in an as-machined (never implanted) GUR 1050 UHMWPE component is shown in Figure 2.10.

Machining of UHMWPE components consists of milling and turning operations for both roughing and finishing steps. In some cases, the resin converter may supply the stock in a shape that approximates the cross-section of the finished implant. Such preshaping or preforming offers advantages of efficiency and speed to the manufacturer. Because UHMWPE can be damaged

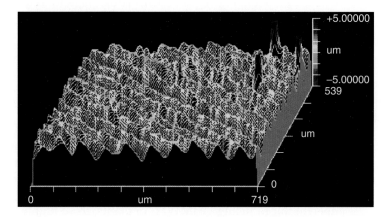

Figure 2.10

An example of machining marks in an as-machined (never implanted) GUR 1050 UHMWPE component.

by excessive heat, the feed rate, tool cutting force, and spindle speed must be closely monitored during manufacture. Close mechanical tolerances generally require the manufacturing environment temperature to be carefully controlled. Milling machine spindles in the early to mid-1980s could develop up to 4000 rpm, whereas newer machinery can develop up to 8000 rpm; the latest machines develop speeds of 12,000 rpm. Overheating of the UHMWPE is avoided by proper optimization of the feed rate and tool cutting force during machining operations.

The actual cutting speeds, tool feed rates, and depths of cut used to machine UHMWPE components are proprietary, and hence little information is available in the literature about the effect of machining parameters on the tribological properties of UHMWPE (Song et al. 1999). Song and colleagues (1999) have proposed an idealized model for the surface morphology of as-machined UHMWPE as a triangular wave, in which the peak-to-peak distance, d, is given by the following equation:

$$d = \frac{f}{2s}$$

where f is the tool feed rate and s is the cutting speed. Song's model also proposes that the base angle of the machining marks is related to the geometry of the cutting tool. Given the numerous (potentially interrelated) factors influencing the surface topology of as-manufactured components, it is difficult (based on measurements of the surface alone) to quantify the machining conditions used to produce the surface finishes of a randomly sampled component.

Cooper and colleagues (1993) have classified abrasive wear in UHMWPE as involving macroscopic or microscopic asperity contact between the sliding surfaces. Because of the initial difference in surface roughness at the UHMWPE and metal counterfaces, the initial wear rate involves the removal of the larger macroscopic asperities on the UHMWPE surface, whereas the long-term wear rate is governed by the microscopic asperity size of the metal counterface (Cooper, Dowson, and Fisher 1993). Thus, changes to the surface roughness of UHMWPE components may be expected to affect the initial wear rate, because removal of machining marks will occur within the contact zone during the first stages of wear in an orthopedic bearing (Wang, Stark, and Dumbleton 1995).

Conclusion

As will be shown in Chapter 4, the first UHMWPE hip components were produced by Charnley in his home workshop or in the machine shop at Wrightington starting in 1962. The sophistication of the implant manufacturing process has changed considerably since that time. Because of the complexities inherent in producing UHMWPE implants, three highly specialized industries have developed since the 1960s to address different stages of the production pipeline. Polymer resin producers, such as Ticona, focus on polymerization of medical grade UHMWPE resin, whereas a separate industry of converters has evolved to address consolidation of UHMWPE into semifinished stock materials, catering to the needs of orthopedic manufacturers. The task of implant design, UHMWPE selection, implant machining, packaging, sterilization, and distribution typically fall within the purview of orthopedic manufacturers.

The choice of conversion method (and converter) is at least as important a decision as the choice of resin for an UHMWPE component, because both factors introduce a subtle change in the morphology and material properties of the consolidated polymer. However, there is currently no consensus as to which resin and conversion method would be universally superior for all orthopedic applications. Consequently, it is left to each orthopedic manufacturer to determine which conversion method (and resin) is most appropriate for applications in hip, knee, shoulder, and spine implants.

References

Baldini T.H., C.M. Rimnac, and T.M. Wright. 1997. The effect of resin type and sterilization method on the static (J-integral) fracture resistance of UHMW polyethylene. *Orthopaedic Research Society* 43:780.

Barnetson A., and P.R. Hornsby. 1995. Observations on the sintering of ultra-high molecular weight polyethylene (UHMWPE) powders. *J Materials Sci Letters* 14:80–84.

Bastiaansen C.W.M., H.E.H. Meyer, and P.J. Lemstra. 1990. Memory effects in polyethylenes: influence of processing and crystallization history. *Polymer* 31: 1435–1440.

Bellare A., and R.E. Cohen. 1996. Morphology of rod stock and compression-moulded sheets of ultra-high-molecular-weight polyethylene used in orthopaedic implants. *Biomaterials* 17:2325–2333.

Biomet. 1995. Resin consolidation issues with UHMWPE. Report No. Y-BEM-069. Warsaw.

Biomet. 1997. ArCom processed polyethylene. Report No. Y-BMT-503. Warsaw.

Birnkraut H.W. 1991. Synthesis of UHMWPE. In *Ultra-high molecular weight polyethylene as a biomaterial in orthopedic surgery.* H.G. Willert, G.H. Buchhorn, and P. Eyerer. Eds. Lewiston, NY: Hogrefe & Huber.

Blunn G.W., A.B. Joshi, R.J. Minns, et al. 1997. Wear in retrieved condylar knee arthroplasties. A comparison of wear in different designs of 280 retrieved condylar knee prostheses. *J Arthroplasty* 12:281–290.

Chen K.C., E.J. Ellis, and A. Crugnola. 1981. Effects of molding cycle on the molecular structure and abrasion resistance of ultra-high molecular weight polyethylene. *ANTEC '81* 39:270–272.

Chubberley A.H. 1965. Ultra-high molecular weight polyethylenes. In *Modern plastics encyclopaedia.* New York: McGraw-Hill.

Cooper J.R., D. Dowson, and J. Fisher. 1993. Macroscopic and microscopic wear mechanisms in ultra-high molecular weight polyethylene. *Wear* 162–164, 378–384.

Edidin A.A., and S.M. Kurtz. 2001. Development and validation of the small punch test for UHMWPE used in total joint replacements. In *Functional biomaterials.* N. Katsube, W. Soboyejo, and M. Sacks. Eds. Winterthur, Switzerland: Trans Tech Publications.

Eyerer P., R. Ellwanger, H.A. Federolf, et al. 1990. Polyethylene. In *Concise encyclopaedia of medical and dental materials.* D. Williams, and R. Cahn. Eds. Oxford: Pergamon.

Eyerer P., A. Frank, and R. Jin. 1985. Characterization of ultrahigh molecular weight polyethylene (UHMWPE): Extraction and viscometry of UHMWPE. *Plastverarbeiter* 36:46–54.

Farrar D.F., and A.A. Brain. 1997. The microstructure of ultra-high molecular weight polyethylene used in total joint replacements. *Biomaterials* 18:1677–1685.

Goldman M., R. Gronsky, R. Ranganathan, and L. Pruitt. 1996. The effects of gamma radiation sterilization and aging on the structure and morphology of medical grade ultra-high molecular weight polyethylene. *Polymer* 37:2909–2913.

Gsell R., R. King, and D. Swarts. 1997. Quality indicators of high-performance UHMWPE. Warsaw, IN: Zimmer.

Gul R. 1997. Improved UHMWPE for use in total joint replacement. Ph.D. diss., Massachusetts Institute of Technology.

Halldin G.W., and I.L. Kamel. 1977a. Powder processing of ultra-high molecular weight polyethylene. I. Powder characterization and compaction. *Poly Eng Sci* 17:21–26.

Halldin G.W., and I.L. Kamel. 1977b. Powder processing of ultra-high molecular weight polyethylene. II. Sintering. *ANTEC 77* 35:298–300.

Hamilton J.V., H.C. Wang, and C. Sung. 1996. The effect of fusion defects on the mechanical properties of UHMWPE. Trans 5th World Biomater Conference 2:511.

Han K.S., J.F. Wallace, R.W. Truss, and P.H. Geil. 1981. Powder compaction, sintering, and rolling of ultra-high molecular weight polyethylene and its composites. *J Macromol Sci Phys* B19:313–349.

Kurtz S.M., L. Pruitt, C.W. Jewett, et al. 1998. The yielding, plastic flow, and fracture behavior of ultra-high molecular weight polyethylene used in total joint replacements. *Biomaterials* 19:1989–2003.

Kurtz S.M., O.K. Muratoglu, M. Evans, and A.A. Edidin. 1999. Advances in the processing, sterilization, and crosslinking of ultra-high molecular weight polyethylene for total joint arthroplasty. *Biomaterials* 20:1659–1688.

Kurtz S.M., J. Turner, M. Herr, and A.A. Edidin. 2002. Deconvolution of surface topology for quantification of initial wear in highly crosslinked acetabular components for THA. *JBMR (Applied Biomaterials)* 63:492–500.

Kusy R.P., and J.Q. Whitley. 1986. Use of a sequential extraction technique to determine the MWD of bulk UHMWPE. *J Appl Poly Sci* 32:4263–4269.

Li S., and A.H. Burstein. 1994. Ultra-high molecular weight polyethylene. The material and its use in total joint implants. *J Bone Joint Surg Am* 76:1080–1090.

Lykins M.D., and M.A. Evans. 1995. A comparison of extruded and molded UHMWPE. *Trans 21st Soc Biomater* 18:385.

McKenna G.B., J.M. Crissman, and F. Khoury. 1981. Deformation and failure of ultra-high molecular weight polyethylene. *ANTEC '81* 39:82–84.

Muratoglu O.K., M. Jasty, and W.H. Harris. 1997. High resolution synchrotron infra-red microscopy of the structure of fusion defects in UHMWPE. Transactions of the 43rd Orthopedic Research Society 22:773.

Olley R.H., I.L. Hosier, D.C. Bassett, and N.G. Smith. 1999. On morphology of consolidated UHMWPE resin in hip cups. *Biomaterials* 20:2037–2046.

Pruitt L., and L. Bailey. 1998. Factors affecting the near-threshold fatigue behavior of surgical grade ultra high molecular weight polyethylene. *Polymer* 39:1545–1553.

Schmidt M.B., and J.V. Hamilton. 1996. The effects of calcium stearate on the properties of UHMWPE. 42nd Orthopedic Research Society 21:22.

Shenoy A.V., and D.R. Saini. 1985. Compression moulding of ultra-high molecular weight polyethylene. *Plast Rubber Proc Appl* 5:313–317.

Song J., P. Liu, M. Cremens, and P. Bonutti. 1999. Effects of machining on tribological behavior of ultra-high molecular weight polyethylene (UHMWPE) under dry reciprocating sliding. *Wear* 225–229:716–723.

Stein H. Personal communication. 8 July 1997.

Swarts D., R. Gsell, R. King, et al. 1996. Aging of calcium stearate-free polyethylene. Trans 5th World Biomater Conference 2:196.

Truss R.W., K.S. Han, J.F. Wallace, and P.H. Geil. 1980. Cold compaction molding and sintering of ultra high molecular weight polyethylene. *Poly Engr Sci* 20:747–755.

Voigt-Martin I.G., E.W. Fisher, and L. Mandelkern. 1980. Morphology of melt-crystallized linear polyethylene fractions and its dependence on molecular weight and crystallization temperature. *J Poly Sci (Poly Phys)* 18:2347–2367.

Wagner H.L., and J.G. Dillon. 1988. Viscosity and molecular weight distribution of ultra-high molecular weight polyethylene. *J Appl Poly Sci* 36:567–582.

Walker P.S., G.W. Blunn, and P.A. Lilley. 1996. Wear testing of materials and surfaces for total knee replacement. *J Biomed Mater Res* 33:159–175.

Wang A., C. Stark, and J.H. Dumbleton. 1995. Role of cyclic plastic deformation in the wear of UHMWPE acetabular cups. *J Biomed Mater Res* 29:619–626.

Wang X.Y., S.Y. Li, and R. Salovey. 1988. Processing of ultra-high molecular weight polyethylene. *J Appl Poly Sci* 35:2165–2171.

Weightman B., and D. Light. 1985. A comparison of RCH 1000 and Hi-Fax 1900 ultra-high molecular weight polyethylenes. *Biomaterials* 6:177–183.

Wrona M., M.B. Mayor, J.P. Collier, and R.E. Jensen. 1994. The correlation between fusion defects and damage in tibial polyethylene bearings. *Clin Orthop* 299:92–103.

Zachariades A.E. 1985. The effect of powder particle fusion on the mechanical properties of ultra-high molecular weight polyethylene. *Poly Engr Sci* 25:747–750.

Chapter 2. Reading Comprehension Questions

2.1. When was UHMWPE first commercially introduced for industrial applications?
   a) 1940s
   b) 1950s
   c) 1960s
   d) 1970s
   e) 1980s

2.2. What trace impurities are present in UHMWPE after polymerization?
   a) Chlorine
   b) Aluminum
   c) Titanium
   d) Ash
   e) All of the above

2.3. Which of the following is a trade name for UHMWPE resin that contains calcium stearate?
   a) RCH 1000
   b) GUR 1050
   c) GUR 415
   d) Chirulen
   e) All of the above

2.4. For which of the following reasons would calcium stearate be added to UHWMPE?
   a) Scavenge residual catalyst
   b) Decrease specific gravity
   c) Improve impact strength
   d) Reduce consolidation time
   e) All of the above

2.5. Which of the following UHMWPE resins has the highest molecular weight?
   a) GUR 1020
   b) GUR 1050
   c) Teflon
   d) Chirulen
   e) RCH 1000

2.6. The molecular weight of UHMWPE is most commonly inferred from which material property?
   a) Impact strength
   b) Tensile strength
   c) Density
   d) Intrinsic viscosity
   e) Small punch test

2.7. Which of the following conversion techniques CANNOT be used to process UHMWPE resin powder?
   a) Isostatic molding
   b) Net shape molding
   c) Compression molding
   d) Injection molding
   e) Ram extrusion

2.8. Which of the following factors influence the machining of UHMWPE implants?
   a) Tool feed rate
   b) Tool cutting force
   c) Heat generation
   d) Spindle speed
   e) All of the above

Chapter 3

Packaging and Sterilization of UHMWPE

Introduction

After fabrication, either by machining or by direct compression molding, ultra-high molecular weight polyethylene (UHMWPE) components for total joint replacement (TJR) are packaged and sterilized before external distribution to the clinic. Unlike polymerization and resin conversion, which are typically under the direct control of specialized vendors, the choice of packaging and sterilization method falls within the purview of the implant designer. Although the implant designer is responsible for selecting the packaging and sterilization methods for a particular type of orthopedic component, the actual packaging and sterilization processes themselves may be amenable to outsourcing. For instance, gas plasma sterilizers are commercially distributed as stand-alone units and can be incorporated directly into a manufacturer's facility (e.g., Sterrad System: Advanced Sterilization Products, Irvine, CA), whereas gamma and ethylene oxide sterilization requires specialized facilities and is typically performed by an outside vendor (e.g., Isomedix Services, Steris Corporation, Mentor, OH).

During the 1990s, the packaging and sterilization of UHMWPE was a controversial topic. As recently as 1995, UHMWPE was typically sterilized with a nominal dose of 25 to 40 kGy of gamma radiation in the presence of air. By 1998, all of the major orthopedic manufacturers in the United States were either sterilizing UHMWPE using gamma radiation in a reduced oxygen environment or sterilizing without ionizing radiation, using ethylene oxide or gas plasma. The shift in sterilization practice was catalyzed by mounting evidence that gamma sterilization in air, followed by long-term shelf storage, promoted oxidative chain scission and degradation of desirable physical, chemical, and mechanical properties of UHMWPE.

Table 3.1

Summary of Sterilization Processes for UHMWPE Implants. Note that Gamma Air Sterilization Is Listed as a Historical Reference, for Comparison Purposes Only

Sterilization process	Packaging type	Gamma radiation dose	Contemporary method?
Gamma air	Gas permeable	25–40 kGy	No (historical)
Gamma inert	Barrier packaging, reduced oxygen atmosphere	25–40 kGy	Yes
Gas plasma	Gas permeable	None	Yes
Ethylene oxide	Gas permeable	None	Yes

A wide range of choices is currently available to the implant designer for packaging and sterilization of UHMWPE implants, as summarized in Table 3.1. Implants can be sterilized with or without ionizing radiation. When sterilized using gamma radiation, UHMWPE components are contained in a barrier package with a reduced oxygen environment. Gas plasma and ethylene oxide–sterilized implants are packaged in gas-permeable packaging to allow access of the sterilizing medium to the UHMWPE surface.

The purpose of this chapter is to review historical and contemporary packaging and sterilization methods for UHMWPE. Obviously, all of the sterilization methods currently employed by the orthopedic community fulfill their intended purpose, namely the eradication of bacterial agents, which may result in sepsis and premature revision. The diverse sterilization methods in current use reflect the lack of scientific consensus as to which of the currently favored sterilization methods provides the most advantageous long-term UHMWPE product for the ultimate user, namely the patient.

Gamma Sterilization in Air

Starting in the 1960s, UHMWPE components for joint replacement have been stored in air-permeable packaging and gamma sterilized with a nominal dose of 25 kGy (2.5 Mrad) (Isaac, Dowson, and Wroblewski 1996). Although the packaging of orthopedic implants has evolved since then, gamma sterilization of UHMWPE components remains an industry standard today (Kurtz et al. 1999). Historically, UHMWPE components have received a dosage ranging between 25 and 40 kGy during sterilization. An example of air-permeable packaging, shown in Figure 3.1, consists of a box, two nested, polymeric packages, and an inner foam insert (the UHMWPE component is not shown).

The outer box also typically contains a booklet of information for the surgeon, as well as stickers identifying the catalog and lot numbers of the

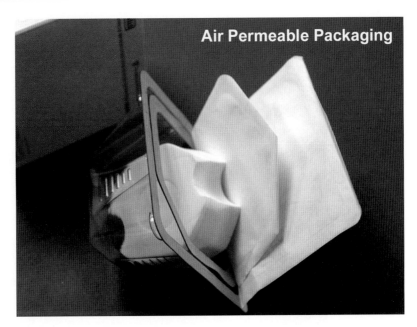

Figure 3.1
An example of historical air-permeable packaging used with gamma sterilization, consisting of a box, two nested, polymeric packages, and an inner foam insert (Osteonics, Allendale, NJ). The UHMWPE component is not shown.

implant, for affixing to the patient's medical records (not shown in Figure 3.1). More examples of air-permeable packaging from different manufacturers are included in Figures 3.2 and 3.3.

Air-permeable packaging was replaced by barrier packaging with a low oxygen environment starting during the mid 1990s. The motivation for changing packaging techniques was to prevent oxidation of free radicals in the UHMWPE, which persist for years after irradiation (Jahan and Wang 1991), and can be replenished by the cascade of chemical reactions that follows oxidation (Premnath et al. 1996). During shelf storage, UHMWPE components that were gamma sterilized in air-permeable packaging undergo oxidative degradation, resulting in an increase in density and crystallinity (Bostrom et al. 1994, Kurth et al. 1988, Kurtz, Bartel, and Rimnac 1998, Kurtz, Rimnac, and Bartel 1997, Rimnac et al. 1994) and, more importantly, in a loss of mechanical properties, associated with progressive embrittlement (Collier et al. 1996, Edidin et al. 2000, Sutula et al. 1995). The loss of mechanical properties during long-term shelf aging in air often manifests most severely in a subsurface band, located 1–2 mm below the articulating surface. The development of this subsurface embrittled region has been associated with fatigue damage, including delamination, of tricompartmental and unicondylar knee replacements (Kennedy et al. 2000, McGovern et al. 2002, Sutula et al. 1995).

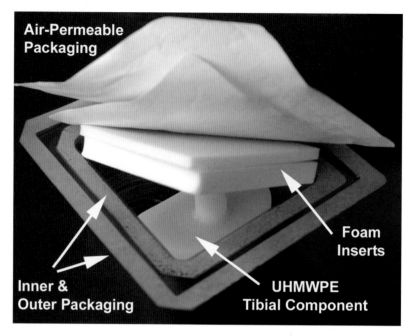

Figure 3.2
Historical air-permeable packaging used with gamma sterilization (Wright Medical, Arlington, TN).

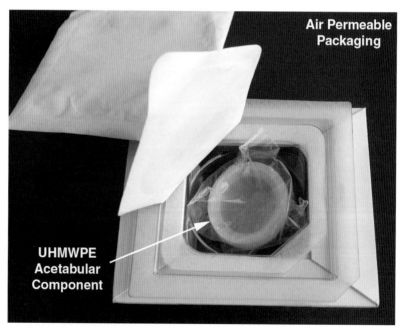

Figure 3.3
Historical air-permeable packaging used with gamma sterilization (DePuy, Warsaw, IN).

Although currently discontinued by major orthopedic manufacturers, gamma sterilization of UHMWPE components in the presence of air will likely continue as a clinically relevant issue well into the 21st century. If we assume that the frequency of total hip and knee procedures performed annually in the United States increased at a rate of 10% in the 1980s and at a rate of 5% in the 1990s, reaching an estimated 500,000 annual procedures in 1995 (Orthopaedic Products 1996), then at least 2 million U.S. patients may have been implanted with an UHMWPE component that was sterilized in air during the period of 1980–1989, when gamma irradiation in air was the standard sterilization practice. From the same calculation, an additional 2 million patients are estimated to have been implanted with air-sterilized UHMWPE components in the United States between 1990 and 1995. Therefore, based on the large population of patients already implanted with air-sterilized UHMWPE components during the 1980s and 1990s, clinical interest in the long-term *in vivo* effects of oxidative degradation on UHMWPE will likely continue for at least another decade.

Gamma Sterilization in Barrier Packaging

There are many types of barrier packaging currently in use by the orthopedic industry (Table 3.2). The details about current packaging systems for orthopedic components are proprietary, but basically consist of evacuating the air from the packaging and backfilling with an inert gas (e.g., nitrogen or argon).

Table 3.2

Summary of Contemporary Barrier Packaging Techniques Used for Gamma Sterilization of UHMWPE Among Major Orthopedic Manufacturers (as of January 2003)

Company	Package environment	Sterilization/packaging trade name
Biomet	Argon flushed, near-vacuum sealed	ArCom
DePuy, Inc.	Near vacuum	GVF ("gamma vacuum foil")
Stryker Howmedica Osteonics	Nitrogen	N_2-Vac; Duration
Centerpulse	Near vacuum with oxygen scavenger (U.S.); Nitrogen (Europe)	Oxygenless, O_2 less (USA)
Zimmer, Inc.	Nitrogen	

The contents of this table were verified by personal communications between the author and the respective manufacturers in January 2003.

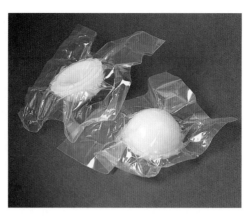

Figure 3.4
Contemporary barrier packaging used with the fabrication of ArCom™ (Biomet, Inc., Warsaw, IN). ArCom packaging currently uses a glass film interposed between polymer sheets for an oxygen barrier. The packaging is argon flushed and then vacuum sealed.

The barrier in the package consists of polymer laminates or metallic foils to block gas diffusion (Figure 3.4). Three further examples of contemporary barrier packaging for UHMWPE components from different manufacturers are shown in Figures 3.5, 3.6, and 3.7.

The goal of barrier packaging is to minimize oxidative degradation during long-term shelf storage, and a series of recent studies would suggest that contemporary packaging techniques have been effective in this regard (Edidin et al. 2000, Lu, Buchanan, and Orr 2002). Furthermore, accelerated aging in an

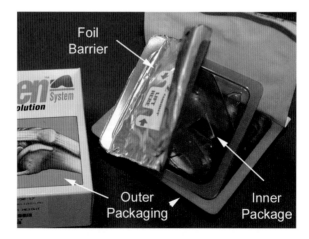

Figure 3.5
Contemporary nitrogen-filled barrier packaging for gamma sterilization of UHMWPE components used by Zimmer, Inc. (Warsaw, IN).

Figure 3.6

Contemporary N$_2$-Vac barrier packaging for gamma sterilization of UHMWPE components used by Stryker Howmedica Osteonics (Mahwah, NJ).

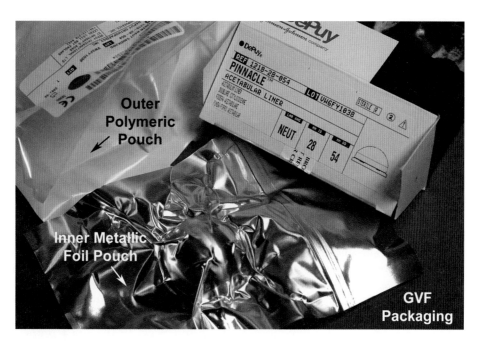

Figure 3.7

Contemporary Gamma Vacuum Foil (GVF) barrier packaging for gamma sterilization of UHMWPE components, used by DePuy Orthopedics (Warsaw, IN).

inert environment (e.g., nitrogen) has demonstrated that the exclusion of oxygen prevents the degradation of UHMWPE, even if it was originally sterilized in the presence of air (Edidin et al. 2002). Thus, currently available data indicates that barrier packaging addresses the risk of oxidative degradation of UHMWPE during shelf storage as was observed in the past.

Concern about degradation during shelf aging prompted some orthopedic manufacturers to adopt sterilization of conventional UHMWPE using gas plasma or ethylene oxide. These sterilization methods admittedly generate no free radicals that can subsequently oxidize during shelf storage. However, UHMWPE sterilized in this manner also does not receive a tribological benefit associated with radiation-induced crosslinking. McKellop and colleagues (1999) reported on the wear performance of UHMWPE in a contemporary hip simulator following gamma irradiation in air, gamma irradiation in an inert gas, ethylene oxide, or gas plasma. Between 2 and 5 million cycles (each million of cycles corresponds to about a year of use *in vivo* for an average patient), the wear rate of the gamma sterilized UHMWPE was significantly lower than UHMWPE sterilized by either gas plasma or ethylene oxide. For example, the wear rate from 3.5 to 5 million cycles for ethylene oxide–sterilized UHMWPE was reported as 40 ± 0.6 mm^3/million cycles; in contrast, the wear rate for UHMWPE that was gamma irradiated in an air-permeable package was found to be 18.5 ± 0.9 mm^3/million cycles (McKellop et al. 1999). A similar trend has been reported by Wang and associates (1998), who observed a more than 50% drop in the hip simulator wear rate following a single 25 kGy dose of gamma sterilization. When oxygen is excluded from the package during sterilization, further crosslinking, and additional improvement in wear performance, may be achieved relative to gamma sterilization in air (McKellop et al. 1999). As shown in Table 3.2, many major orthopedic manufacturers continue to distribute UHMWPE components that are barrier packaged and gamma sterilized in an inert environment.

Ethylene Oxide Gas Sterilization

Ethylene oxide gas (EtO) has been a commercially available sterilization method since the 1980s (Bruck and Mueller 1988). Highly toxic, EtO neutralizes bacteria, spores, and viruses. UHMWPE is a good candidate for EtO sterilization because it contains no constituents that will react with or bind to the toxic gas. The efficacy of EtO sterilization depends on stringent control of process conditions, including humidity, duration, and temperature (Bruck and Mueller 1988, Ries, Weaver, and Beals 1996). When properly performed, EtO has been shown to effectively sterilize the interface of assembled modular components, such as metal-backed tibial and patellar devices (Ries, Weaver, and Beals 1996). However, because of its toxicity and hazardous residues, ethylene oxide sterilization is conducted in accordance with domestic and international standards (Page and Cyr 1998, Ries, Weaver, and Beals 1996). Laboratory studies suggest that sterilization using EtO does not substantially influence the physical, chemical, and mechanical properties of UHMWPE (Collier et al. 1996, Goldman et al. 1997, Ries et al. 1996, Wang et al. 1997).

Sterilization of UHMWPE is accomplished by diffusion of EtO into the near-surface regions during several hours (Bruck and Mueller 1988, Ries, Weaver, and Beals 1996). After sterilization, residual EtO is then allowed to diffuse out of the UHMWPE to reduce the likelihood of an adverse topical reaction (Bruck and Mueller 1988). In a study by Ries and colleagues (1996), which provided details of a validated EtO sterilization cycle for UHMWPE, the following protocol was employed: 18 hours of preconditioning at 65% relative humidity, followed by 5 hours of exposure to 100% EtO at 0.04 MPa, followed by 18 hours of forced air aeration. Thus, the entire sterilization cycle with EtO took a total of 41 hours. Preconditioning, exposure, and aeration were conducted at 46°C.

Based on a limited number of retrieval studies, the clinical experience with EtO-sterilized UHMWPE components has thus far been favorable (Sutula et al. 1995, White et al. 1996). In a study by White and associates (1996), researchers compared the surface damage and physical and mechanical properties of 26 retrieved UHMWPE tibial components of identical design that had been sterilized with either EtO or gamma radiation (presumably in air). Components that were sterilized using EtO showed significantly less surface damage and delamination than gamma radiation sterilized components. The radiation sterilized components, on the other hand, had decreased ductility (elongation to failure), decreased toughness, and increased crystallinity. In a study of 150 retrieved acetabular components, Sutula and colleagues observed no evidence of rim cracking or delamination in any of the 17 ethylene oxide sterilized components (1995). In contrast, rim cracking was found in 19% of the acetabular components that were gamma radiation sterilized in air. Although the cited retrieval studies are necessarily limited by their retrospective nature, EtO sterilization appears not to induce certain modes of surface damage, which may be exacerbated by oxidative embrittlement of UHMWPE following gamma radiation sterilization in air.

Today, ethylene oxide continues to be used as a contemporary sterilization method for UHMWPE by major orthopedic manufacturers. Smith & Nephew, Inc. (Memphis, TN) and Wright Medical Technology, Inc. (Arlington, TN) have employed ethylene oxide since the 1990s (Kurtz et al. 1999). More recently, however, Centerpulse Orthopaedics (Austin, TX) has chosen ethylene oxide for sterilization of its highly crosslinked UHMWPE (Durasul) hip and knee products (Muratoglu and Kurtz 2002) (Figure 3.8).

Gas Plasma Sterilization

Low-temperature gas plasma is a relatively new commercially available sterilization method that was applied to UHMWPE in the 1990s (Collier et al. 1996, Goldman and Pruitt 1998, McKellop et al. 1999). Gas plasma is a surface sterilization method that relies on ionized gas for deactivation of biological

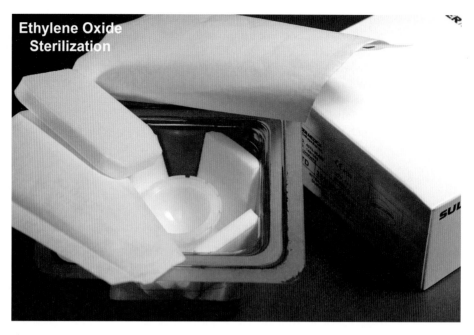

Figure 3.8
Contemporary gas-permeable packaging for ethylene sterilization of Durasul highly crosslinked UHMWPE components, used by Centerpulse, Inc. (Austin, TX).

organisms (Bruck and Mueller 1988). Two examples of commercially available gas plasma sterilization methods that have been evaluated for compatibility with UHMWPE include Plazlyte (Abtox, Inc., Mundelein, IL), which involves low-temperature peracetic acid gas plasma, and Sterrad (Advanced Sterilization Products, Irvine, CA), which uses low-temperature hydrogen peroxide gas plasma. Gas plasma sterilization is accomplished at temperatures below 50°C (Feldman and Hui 1997, Kyi, Holton, and Ridgway 1995). Recent laboratory investigations suggest that low-temperature gas plasma does not substantially affect the physical, chemical, or mechanical properties of UHMWPE (Charlebois, Daniels, and Lewis 2003, Collier et al. 1996, Goldman and Pruitt 1998, McKellop et al. 1999, McKellop et al. 2000, Reeves et al. 2000).

Gas plasma is an attractive sterilization method because it does not leave toxic residues or involve environmentally hazardous byproducts (Feldman and Hui 1997, Kyi, Holton, and Ridgway 1995). Based on product literature available from the manufacturer, the sterilization cycle time for the Plazlyte system is 3 to 4 hours; the Sterrad system has an even shorter sterilization cycle of 75 minutes (Feldman and Hui 1997, Kyi, Holton, and Ridgway 1995). Because no lengthy aeration period is required after gas plasma sterilization, it potentially offers substantial time and cost savings over EtO sterilization (Feldman and Hui 1997). Because of its recent introduction, retrieval data from *in vivo* gas plasma sterilized UHMWPE components are not yet available.

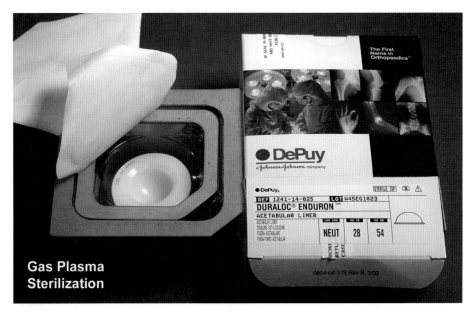

Figure 3.9
Contemporary packaging for gamma sterilization of Enduron™ conventional UHMWPE
components, used by DePuy Orthopedics (Warsaw, IN).

Gas plasma has gained increased acceptance as a method for sterilizing
UHMWPE components for TJR. Gas plasma (Plazlyte system) has been used by
DePuy Orthopaedics, Inc., (Warsaw, IN) since the 1990s to routinely sterilize
UHMWPE components (Kurtz et al. 1999) (Figure 3.9). More recently, Zimmer,
Inc. (Warsaw, IN) has chosen gas plasma for sterilization of their highly
crosslinked UHMWPE components (e.g., Longevity, Prolong) (Muratoglu and
Kurtz 2002).

Shelf Life of UHMWPE Components for Total Joint Replacement

Until recently, a consensus practice of 5-year shelf life was adopted in Europe
for medical implants so that sterility can be ensured. However, within the past
few years, the European Committee for Standardization (CEN) has established
standards that limit the shelf life of UHMWPE components to 5 years. Because
implants are distributed multinationally, packaging for components sold in the
United States may stipulate a 5-year shelf life. However, it should be noted that,
even today, there is no U.S. standard for shelf life of UHMWPE components
after sterilization by gamma radiation, ethylene oxide, or gas plasma.

Accelerated aging methods have been developed for evaluating the
integrity of medical packaging (ASTM F 1980-02 2002). Nevertheless, because

of proprietary variations in industry packaging practices, regulatory agencies such as the U.S. Food and Drug Administration (FDA) currently require orthopedic manufacturers to submit real-time shelf aging data to validate accelerated aging tests and to establish the maximum shelf life for a particular UHMWPE product. Efforts are under way at the American Society for Testing and Materials (ASTM) to develop protocols for establishing the shelf life of UHMWPE components used in TJR.

Overview of Current Trends

Ten years ago, gamma sterilization of UHMWPE in air-permeable packaging was the norm in orthopedics. Today, a broad range of sterilization and packaging methods are employed by orthopedic companies for UHMWPE components. For gamma sterilization, barrier packaging methods have been widely adopted that are recognized to satisfactorily address the historical problem of oxidation during shelf aging. Gas-permeable packaging continues to be used, however, for ethylene oxide and gas plasma, which have emerged as industrially viable, alternative methods for sterilizing UHMWPE. The proliferation of sterilization and proprietary packaging techniques has thus far made it difficult to establish broad industry standards in this area. Standards organizations, such as ASTM, have begun to harmonize the methods used to establish the shelf life of contemporary UHMWPE packaging and sterilization techniques.

Acknowledgments

Special thanks to Janet Krevolin (Centerpulse Orthopaedics, Inc.), Ray Gsell (Zimmer, Inc.), Shi-Shen Yao and Paul Serekian (Howmedica Osteonics Corp.), Jorge Ochoa and Mark Haynes (DePuy Orthopaedics, Inc.), and David Schroeder (Biomet, Inc.) for their helpful discussions and editorial assistance with this chapter.

References

ASTM F 1980-02. 2002. Standard guide for accelerated aging of sterile medical device packages. West Conshohocken, PA: American Society for Testing and Materials.

Bostrom M.P., A.P. Bennett, C.M. Rimnac, and T.M. Wright. 1994. The natural history of ultra high molecular weight polyethylene. *Clin Orthop* 309:20–28.

Bruck S.D., and E.P. Mueller. 1988. Radiation sterilization of polymeric implant materials. *J Biomed Mater Res* 22:133–144.

Charlebois S.J., A.U. Daniels, G. Lewis. 2003. Isothermal microcalorimetry: an analytical technique for assessing the dynamic chemical stability of UHMWPE. *Biomaterials* 24:291–296.

Collier J.P., D.K. Sperling, J.H. Currier, et al. 1996. Impact of gamma sterilization on clinical performance of polyethylene in the knee. *J Arthroplasty* 11:377–389.

Collier J.P., L.C. Sutula, B.H. Currier, et al. 1996. Overview of polyethylene as a bearing material: Comparison of sterilization methods. *Clin Orthop* 333:76–86.

Edidin A.A., C.W. Jewett, K. Kwarteng, et al. 2000. Degradation of mechanical behavior in UHMWPE after natural and accelerated aging. *Biomaterials* 21:1451–1460.

Edidin A.A., J. Muth, S. Spiegelberg, and S.R. Schaffner. 2000. Sterilization of UHMWPE in nitrogen prevents oxidative degradation for more than ten years. Transactions of the 46th Orthopaedic Research Society 25:1.

Edidin A.A., M.P. Herr, M.L. Villarraga, et al. 2002. Accelerated aging studies of UHMWPE. I. Effect of resin, processing, and radiation environment on resistance to mechanical degradation. *J Biomed Mater Res* 61:312–322.

Feldman L.A., and H.K. Hui. 1997. Compatibility of medical devices and materials with low-temperature hydrogen peroxide gas plasma. *Med Dev Diag Indust* 19:57–62.

Goldman M., M. Lee, R. Gronsky, and L. Pruitt. 1997. Oxidation of ultrahigh molecular weight polyethylene characterized by Fourier Transform Infrared Spectrometry. *J Biomed Mater Res* 37:43–50.

Goldman M., and L. Pruitt. 1998. A comparison of the effects of gamma radiation and plasma sterilization on the molecular structure, fatigue resistance, and wear behavior of UHMWPE. *J Biomed Mat Res* 40:378–384.

Isaac G.H., D. Dowson, and B.M. Wroblewski. 1996. An investigation into the origins of time-dependent variation in penetration rates with Charnley acetabular cups—wear, creep or degradation? *Proc Inst Mech Eng [H]* 210:209–216.

Jahan M.S., and C. Wang. 1991. Combined chemical and mechanical effects on free radicals in UHMWPE joints during implantation. *J Biomed Mat Res* 25:1005–1017.

Kennedy F.E., J.H. Currier, S. Plumet, et al. 2000. Contact fatigue failure of ultra-high molecular weight polyethylene bearing components of knee prostheses. *J Tribology* 122:332–339.

Kurth M., P. Eyerer, R. Ascherl, et al. 1988. An evaluation of retrieved UHMWPE hip joint cups. *J Biomater Appl* 3:33–51.

Kurtz S.M., C.M. Rimnac, and D.L. Bartel. 1997. Degradation rate of ultra-high molecular weight polyethylene. *J Orthop Res* 15:57–61.

Kurtz S.M., D.L. Bartel, and C.M. Rimnac. 1998. Post-irradiation aging affects the stresses and strains in UHMWPE components for total joint replacement. *Clin Orthop* 350:209–220.

Kurtz S.M., O.K. Muratoglu, M. Evans, A.A. Edidin. 1999. Advances in the processing, sterilization, and crosslinking of ultra-high molecular weight polyethylene for total joint arthroplasty. *Biomaterials* 20:1659–1688.

Kyi M.S., J. Holton, and G.L. Ridgway. 1995. Assessment of the efficacy of a low temperature hydrogen peroxide gas plasma sterilization system. *J Hosp Infect* 31:275–284.

Lu S., F.J. Buchanan, and J.F. Orr. 2002. Analysis of variables influencing the accelerated ageing behaviour of ultra-high molecular weight polyethylene (UHMWPE). *Polymer Testing* 21:623–631.

McGovern T.F., D.J. Ammeen, J.P. Collier, et al. 2002. Rapid polyethylene failure of unicondylar tibial components sterilized with gamma irradiation in air and implanted after a long shelf life. *J Bone Joint Surg Am* 84-A:901–906.

McKellop H.A., F.W. Shen, P. Campbell, and T. Ota. 1999. Effect of molecular weight, calcium stearate, and sterilization methods on the wear of ultra-high molecular weight polyethylene acetabular cups in a hip simulator. *J Orthop Res* 17:329–339.

McKellop H., F.W. Shen, B. Lu, et al. 2000. Effect of sterilization method and other modifications on the wear resistance of acetabular cups made of ultra-high molecular weight polyethylene. A hip-simulator study. *J Bone Joint Surg Am* 82-A:1708–1725.

Muratoglu O.K., and S.M. Kurtz. 2002. Alternative bearing surfaces in hip replacement. In *Hip replacement: current trends and controversies.* R. Sinha, Ed. New York: Marcel Dekker.

Orthopaedic Products. 1996. In *Medical & healthcare marketplace guide.* R.C. Smith, M.A. Geier, J. Reno, and J. Sarasohn-Kahn, Eds. New York: IDD Enterprises, L.P., 1996.

Page B.F.J., and H. Cyr. 1998. A guide to AAMI's TIR for EtO-sterilized medical devices. *Med Dev Diag Indust* 120:73–78.

Premnath V., W.H. Harris, M. Jasty, and E.W. Merrill. 1996. Gamma sterilization of UHMWPE articular implants: an analysis of the oxidation problem. *Biomaterials* 17:1741–1753.

Reeves E.A., D.C. Barton, D.P. FitzPatrick, and J. Fisher. 2000. Comparison of gas plasma and gamma irradiation in air sterilization on the delamination wear of the ultra-high molecular weight polyethylene used in knee replacements. *Proc Inst Mech Eng* [H] 214:249–255.

Ries M.D., K. Weaver, and N. Beals. 1996. Safety and efficacy of ethylene oxide sterilized polyethylene in total knee arthroplasty. *Clin Orthop* 331:159–163.

Ries M.D., K. Weaver, R.M. Rose, et al. 1996. Fatigue strength of polyethylene after sterilization by gamma irradiation or ethylene oxide. *Clin Orthop* 333:87–95.

Rimnac C.M., R.W. Klein, F. Betts, and T.M. Wright. 1994. Post-irradiation aging of ultra-high molecular weight polyethylene. *J Bone Joint Surg* 76A:1052–1056.

Sutula L.C., J.P. Collier, K.A. Saum, et al. 1995. Impact of gamma sterilization on clinical performance of polyethylene in the hip. *Clin Orthop* 319:28–40.

Wang A., D.C. Sun, S-S Yau, et al. 1997. Orientation softening in the deformation and wear of ultra-high molecular weight polyethylene. *Wear* 203–204, 230–241.

Wang A., A. Essner, V.K. Polineni, et al. 1998. Lubrication and wear of ultra-high molecular weight polyethylene in total joint replacements. *Tribology International* 31:17–33.

White S.E., R.D. Paxson, M.G. Tanner, and L.A. Whiteside. 1996. Effects of sterilization on wear in total knee arthroplasty. *Clin Orthop* 331:164–171.

Chapter 3. Reading Comprehension Questions

3.1. Which of the following methods is no longer used to sterilize UHMWPE implants?
a) Gamma irradiation in inert (low oxygen) packaging
b) Gamma irradiation in air-permeable packaging
c) Gas plasma in inert (low oxygen) packaging
d) Ethylene oxide in air-permeable packaging
e) All of the above

3.2. Which of the following radiation doses is sufficient for gamma sterilization?
a) 0 kGy
b) 10 kGy
c) 15 kGy
d) 20 kGy
e) 30 kGy

3.3. Which of the following radiation doses is sufficient for gas plasma sterilization?
   a) 0 kGy
   b) 10 kGy
   c) 15 kGy
   d) 20 kGy
   e) 30 kGy

3.4. Which of the following statements about ethylene oxide sterilization are true?
   a) A maximum radiation dose of 40 kGy is required
   b) Free radicals are produced during sterilization
   c) Barrier packaging is required for sterilization
   d) Implants must be quarantined after sterilization for complete degassing
   e) All of the above

3.5. When was gas plasma first used by the orthopedic community to sterilize UHMWPE implants?
   a) 1990s
   b) 1980s
   c) 1970s
   d) 1960s
   e) 1950s

3.6. What is the maximum shelf life currently permitted for sterilized UHMWPE implants in Europe?
   a) 1 week
   b) There are no required shelf life limits for UHMWPE implants in Europe
   c) 1 year
   d) 2 years
   e) 5 years

3.7. What is the maximum shelf life currently permitted for sterilized UHMWPE implants in the United States?
   a) 1 week
   b) There are no required shelf life limits for UHMWPE implants in the United States
   c) 1 year
   d) 2 years
   e) 5 years

3.8. Orthopedic manufacturers are responsible for which of the following decisions during UHMWPE implant production?
   a) Resin selection
   b) Conversion method selection
   c) Sterilization method selection
   d) Packaging design and selection
   e) All of the above

Chapter 4

The Origins of UHMWPE in Total Hip Arthroplasty

Introduction and Timeline

Introduced clinically in November 1962 by Sir John Charnley, UHMWPE articulating against a metallic femoral head remains the gold standard bearing surface combination for total hip arthroplasty. Considering how rapidly technology can change in the field of orthopedics, the long-term role that ultra-high molecular weight polyethylene (UHMWPE) has played in joint arthroplasty since the 1960s is fairly remarkable. Starting in the 1970s, researchers have attempted to improve UHMWPE for orthopedic applications, starting with the introduction of a carbon-fiber reinforced material (Poly II) (Kurtz et al. 1999). However, this composite UHMWPE was not found to exhibit consistent and improved clinical results relative to the conventional UHMWPE introduced by Charnley.

In Japan during the 1970s, two important technological advancements occurred: one was the introduction of alumina ceramic as an alternative bearing surface for UHMWPE (Shikata, Oonishi, and Hashimato 1977). The second advancement involved the clinical introduction of a highly crosslinked UHMWPE by more than 1000 kGy of gamma irradiation in air (Oonishi 1995). A similar advancement in highly crosslinked UHMWPE also occurred in South Africa during the 1970s, where researchers in Praetoria clinically introduced a UHMWPE that was gamma irradiated with up to 700 kGy in the presence of acetylene (Grobbelaar, Du Plessis, and Marais 1978).

During the 1980s, two other noteworthy developments occurred relative to polyethylene in joint replacements. In the early 1980s, Thackray Ltd. of Leeds began development on an injection molded high-density polyethylene (HDPE) that could be crosslinked by silane chemistry (the same type of crosslinking used in silicone polymers) (Atkinson and Cicek 1983, Atkinson and Cicek 1984). Only 22 of these implants were produced and implanted by

Dr. Wroblewski starting in 1986 (1996). After an initial bedding-in period, these crosslinked HDPE components have been found to exhibit very low clinical wear rates.

In the late 1980s, a joint venture between DePuy Orthopedics and DuPont developed a highly crystalline form of UHMWPE distributed under the trade name of Hylamer (Champion et al. 1994, Li and Burstein 1994). The clinical history of Hylamer, which unfolded during the 1990s, has been mixed and therefore controversial (Kurtz et al. 1999). Although several orthopedic centers have reported worse clinical performance using Hylamer than with conventional UHMWPE, other surgeons have experienced satisfactory or even improved performance.

Thus, it would be misleading to think that the use of UHMWPE in the field of orthopedics has stood still for the past four decades. As summarized in Table 4.1, researchers and clinicians have tried on several occasions to modify UHMWPE for use in joint replacements. Nevertheless, the UHMWPE introduced by Charnley remains the gold standard for artificial hips and now other artificial joints, including the knee and shoulder.

Table 4.1

Timeline of UHMWPE Development for Joint Replacement

Date	Comment
1958	Charnley develops the technique of low friction arthroplasty (LFA). Using polytetrafluoroethylene (PTFE) as the bearing material (Charnley 1961) implants were fabricated either by Charnley in his home workshop or in the machine shop at Wrightington and chemically sterilized.
1962	Charnley adopts UHMWPE for use in his LFA. Components were chemically sterilized.
1968	Start of Leeds production of the Charnley LFA by Chas F. Thackray, Ltd., of Leeds. The UHMWPE was gamma irradiated.
1969	General commercial release of the Charnley LFA by Chas F. Thackray, Ltd., of Leeds. UHMWPE was marketed as gamma irradiated (in air) with a minimum dose of 2.5 Mrad.
1970s	Commercial release of the poly II—carbon fiber reinforced UHMWPE for THA/TKA* by Zimmer, Inc.
1972	Use of alumina ceramic heads articulating against UHMWPE in Japan.
1980-84	Co-development of Silane-crosslinked HDPE by University of Leeds, Wrightington Hospital, and Thackray.
1980s	Commercial release of Hylamer (extended chain recrystalized UHMWPE) for THA/TKA/TSA* by DePuy Orthopedics.

*THA, total hip arthroplasty; TKA, total knee arthroplasty; TSA, total shoulder arthroplasty.

This chapter focuses on the historical development of UHMWPE for use in hip replacements by John Charnley. The main sources of this chapter have been Charnley's journal publications and books, as well as an outstanding biography of Charnley written by William Waugh (1990). Preparation of this chapter also entailed interviews and the review of archives and implants at Wrightington Hospital, DePuy International (Leeds), and the Thackray Museum of Leeds.

This chapter addresses the following questions: 1) How was UHMWPE introduced into hip arthroplasty? 2) What methods were used to sterilize the first UHMWPE implants? 3) What methods were used to test and evaluate UHMWPE (as well as other candidate biomaterials) prior to implantation? The goal of this chapter is to review the historical origins of UHMWPE as an orthopedic biomaterial.

The Origins of a Gold Standard (1958–1982)

The rationale for John Charnley's design of an artificial joint started with a series of frictional experiments on animal and human joints in the 1950s (Charnley 1959, Waugh 1990). From these experiments, Charnley concluded that natural joints functioned well because of their low coefficient of friction, which results from the unique properties of cartilage tissue that promote lubrication. When the natural joint is compressed, the cartilage tissue expels water between the contacting surfaces, which become separated, at least in part, by a thin film of pressurized synovial fluid. In addition to water, the synovial fluid contains proteins and other biological constituents that facilitate lubrication. The pressurized synovial fluid film carries the joint force that protects cartilage tissue from wear during walking or other load-bearing activities. Because of arthritis and other joint diseases, cartilage can lose its unique lubricious characteristics.

Charnley realized that an artificial joint fabricated from synthetic materials like metal or plastic could not operate with purely hydrodynamic lubrication. In 1959, Charnley wrote that "attempts to lubricate any artificial joint must be based on the idea of using boundary lubrication. The substance which seems ideally suited for this purpose is polytetrafluoroethylene (PTFE) because not only has this a low coefficient of friction (0.04–0.05) but it is a substance which is readily tolerated by animal tissues by virtue of its chemical inertness" (Charnley 1959). Therefore, friction, not wear, was Charnley's primary reason for selecting PTFE. From the onset, Charnley's artificial joint design was based on the principles of boundary lubrication, during which pressure in the synovial fluid is not sufficient to fully separate the joint surfaces for load bearing. Thus, the artificial joint surfaces are expected to contact each other during boundary lubrication. This regime of partial lubrication is in contrast with a situation (common for some industrial applications) in which bearing surfaces must articulate dry, or without any lubrication to assist with reducing friction.

Charnley's First Hip Arthroplasty Design with PTFE (1958)

Charnley's design of an artificial hip joint was the product of an evolutionary process between 1958 and 1960 (Charnley 1979). Five design iterations occurred. Charnley's initial "double cup" design in 1958 mimicked the natural joint. The acetabulum was replaced with a thin shell of PTFE and femoral head surface was replaced with a PTFE ball, as shown in Figure 4.1, taken of the collection at the Charnley Museum at Wrightington.

For this initial design, the femoral head and the shell were press fit into the bone, with the idea that the friction between the bone and the PTFE would prevent relative motion. Unfortunately, the tolerances between the head and shell were such that the articulation tended to occur mainly between the PFTE shell and the acetabulum, resulting in abrasion of the shell and destruction of the underlying bone. Wear also occurred of the PTFE femoral head prosthesis, but more concerning from a clinical standpoint was the concomitant loss of blood supply to the femoral bone inside the PTFE cavity, which lead to necrosis (Charnley 1961).

Implant Fixation with Pink Dental Acrylic Cement (1958–1966)

In subsequent designs of the hip arthroplasty, Charnley made use of pink dental acrylic cement, circulated under the trade name of Nu-Life (Waugh

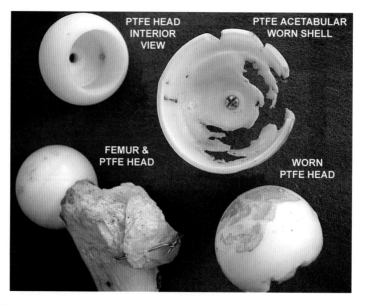

Figure 4.1

Acetabulum was replaced with a thin shell of PTFE and femoral head surface was replaced with a PTFE ball.

1990), to fix the stem to the bone. After the fixation failures of the PTFE shell were observed, Charnley also started to fix the PTFE acetabular cups as well as the stems. Starting in 1958, the cement used by Charnley consisted of acrylic powder that was mixed with liquid monomer. The acrylic powder was chemically sterilized by formaldehyde vapor (Waugh 1990), but it was not necessary to sterilize the monomer because the liquid (methyl methacrylate) itself is bactericidal. As shown in Figure 4.2, the cross-sectional view of a worn implant, retrieved with its cement mantle, the cup was designed with grooves in the back surface to facilitate interlocking at the PTFE-cement interface.

Although Charnley popularized the use of bone cement for hip arthroplasty, Haboush previously reported the use of acrylic cement to fix a femoral component in hip arthroplasty in 1953. In Haboush's previous experience, however, the fixation failed because the femoral prosthesis did not extend into the femur and was attached to the cut end of the bone. In the case of Charnley, the femoral stems extended deep into the femur canal. The bone cement acted to effectively transfer the compressive forces and twisting (i.e., moments) that are imparted to the artificial hip during regular daily activities from Charnley's prosthesis to the bone. Like the grout that holds bathroom tiles to a shower wall, bone cement acted as a void filler and an interlocking agent between the implant and the bone. Bone cement is a poor adhesive to bone or PTFE. According to Charnley's biography (Waugh 1990) and records at Wrightington, Charnley used the pink Nu-Life dental cement until 1966, when he switched to the "Calculated Molecular Weight" (CMW) formulation.

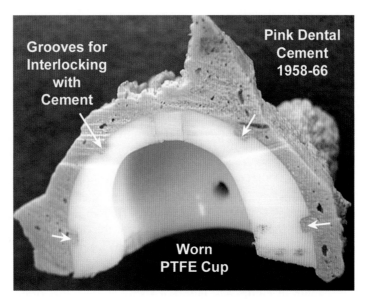

Figure 4.2
The cross-sectional view of a worn PTFE implant, retrieved with its cement mantle.

Interim Hip Arthroplasty Designs with PTFE (1958–1960)

Charnley's second, interim design for a total hip replacement included an acetabular component of PTFE articulating against a metallic femoral head from a cemented Austin Moore or Thompson prosthesis (Waugh 1990). The femoral head was 41.5 mm (1 5/8 inches) in diameter. He initially employed the largest feasible femoral head, which would result in the lowest contact stress. However, his engineering colleagues soon persuaded Charnley to reduce the femoral head diameter, based on the rationale that it would decrease the frictional torque. In other words, Charnley theorized that the frictional resistance during twisting of the joint would be reduced by a smaller femoral head. The concern was that the friction imparted to the joint during normal walking could lead to loosening of the cup from the acetabulum. This theory was judged to be valid because the joint surfaces were not intended to be lubricated hydronamically. In a hydrodynamic bearing, having a large femoral head would be an asset, because the surface-sliding speeds would be greater, facilitating the development of a fluid film to separate the articulating surfaces. On the other hand, in an artificial joint in which the surfaces would always be in contact, reducing the frictional resistance of the joint was a major concern for Charnley. Consequently, third and fourth iterations of this design, between 1958 and 1960 (Charnley, Kamangar, and Longfield 1969), employed cemented femoral components with diameters of 28.5 mm and 25.25 mm, respectively.

These three interim implant designs, preserved in the collection at Wrightington Hospital, are shown in Figure 4.3. The PTFE components in Figure 4.3 were retrieved at revision surgery and are severely worn. The wear is most evident in the sectioned 25.3 mm diameter acetabular component. The femoral components, on the other hand, appear pristine. The femoral heads are polished to a mirror finish (note the reflection of my hands holding the camera).

Final Hip Arthroplasty Design with PTFE (1960–1962)

The fifth and final design iteration, which was clinically introduced in January 1960, included an acetabular component of PTFE, articulating against a cemented femoral component having a 22.225 mm (7/8 inch) diameter head (Charnley 1961). In May 1961, Charnley published the early results of his LFA in the *Lancet* (1961). The early results of the new operation were extremely encouraging, with patients, previously handicapped by joint disease, returning to pain-free mobility within weeks of the operation. He wrote that "negligible wear" was observed on "close scrutiny" of radiographs for patients implanted for 10 months (Charnley 1961).

An example of a radiograph from a short-term implanted PTFE LFA is shown in Figure 4.4. Figure 4.4 shows the initial orientation of the PTFE cup

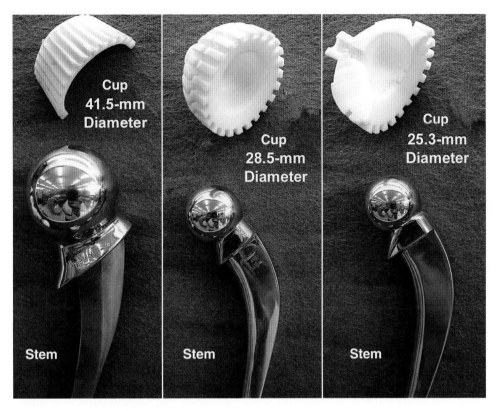

Figure 4.3
Three interim PTFE implant designs by Charnley.

with respect to the pelvis. Initially, the femoral head is centered in the acetabular cup.

Within only a few years, patients began reporting to Wrightington with pain and inflammation associated with their PTFE artificial joints. Radiographic examination revealed severe wear, and upon revision, the joints were found surrounded by 100–200 ml of caseous, purulent tissue (examples of which were thoughtfully saved and can still be viewed, preserved at the Charnley museum at Wrightington). Although the joint articulation was successful in the first one to two years after surgery, more than 99% had to be revised within two to three years of implantation because of severe wear and the inflammatory response provoked by the PTFE wear debris.

The radiograph in Figure 4.5 shows the PTFE artificial hip joint prior to revision (same patient as in Figure 4.4). Note that the femoral head has now translated vertically with respect to the center of acetabular cup, indicative of severe wear. The photograph to the right (not the same component as in the radiographs) shows how the femoral head tunnels into the PTFE acetabular cup during wear. The cup has been sectioned for illustrative purposes.

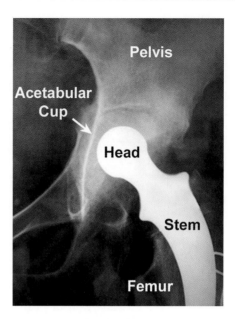

Figure 4.4
The initial orientation of the PTFE cup with respect to the pelvis.

Charnley was faced with an ethical dilemma. On one hand, he had many elderly patients demanding relief from their debilitating pain, and some were willing to receive a PTFE component even if it would require revision in a few years. On the other, the short-term benefit was balanced by the risks associated with revision surgery and loss of bone stock caused by the PTFE debris.

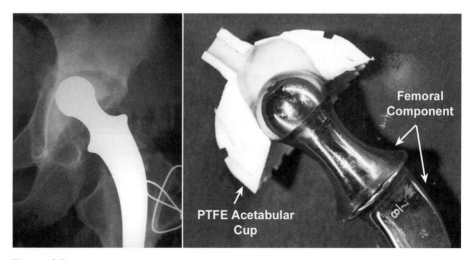

Figure 4.5
PTFE artificial hip joint before revision (same patient as in Figure 4.4).

Implant Fabrication at Wrightington

Up until 1963, the acetabular cups used at Wrightington were fabricated either by Charnley himself at his home workshop or by technicians at the hospital workshop in Wrightington (Craven 2002, Waugh 1990) (Figure 4.6). Charnley's home workshop was used to develop implant prototypes, as well as instruments. His 1/2 hp lathe, purchased in 1946, is now on display at the Thackray museum in Leeds. Charnley's biography recounts that "throughout his life, he was able to 'mock up' any surgical device in his workshop, and try it out himself before having it manufactured" (Waugh 1990).

Charnley's 1961 publication indicates that "the research committee of the Manchester Regional Hospital Board has built and equipped a research workshop in the hospital and has provided the salary for a fitter and turner who will make the surgical implants." Charnley's research workshop and laboratory still remain at Wrightington Hospital, and they have been partly converted into a museum.

Between 1958 and 1966, H. Craven was Charnley's technician and responsible for machining PTFE as well as UHMWPE cups for use at Wrightington (Craven 2002). He remembers ordering the PTFE in billets, which were predrilled by the supplier with a hole to improve the dimensional stability of the polymer. Because the cups were turned on a lathe, the center hole originally present in the billet ended up along the central axis of the cup, as shown in the section of the PTFE cup shown in Figure 4.7. The spigot was a design feature, which was used to orient the cup with respect to a locator hole drilled in the acetabulum.

Figure 4.6
The first UHMWPE hip components were fabricated either at the green lathe in his home workshop (the photo behind the lathe was taken of Charnley), or at the machine shop in Wrightington Hospital.

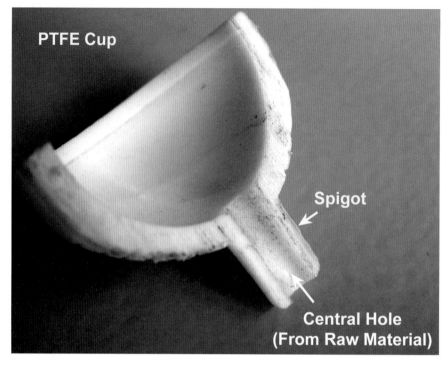

Figure 4.7
Center hole originally present in the billet ended up along the central axis of the cup.

The cups originally took about 45 minutes to machine. However as the number of procedures increased, Craven developed a machine in 1962 that could produce a UHMWPE cup in less than 5 minutes. Craven's machine was sold to Thackray in 1963 after it was decided that the cups would be produced commercially (Craven 2002).

The First Wear Tester

Craven built the first wear testing rig at Wrightington for Charnley between 1959 and 1960. With little money available to support research, the stainless steel parts of the wear tester were scrounged from the local scrap yard, where the proprietor was accustomed to reserving raw materials for the hospital. The rig, shown in Figure 4.8 on display in Wrightington Hospital, consisted of four stations.

Within each test chamber, a stationary quarter-inch diameter polymer pin was mounted superiorly with respect to a sliding, stainless steel plate. The height of each pin was recorded continuously during the test by a height gauge,

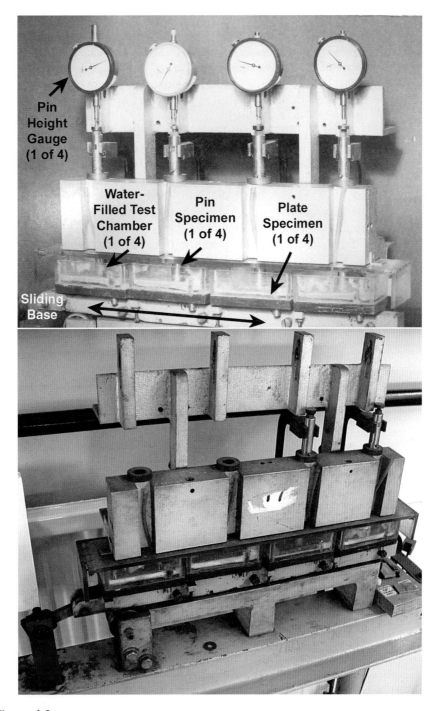

Figure 4.8
Wear testing rig built at Wrightington, constructed by Harry Craven, which was used to test pins of PTFE and the first UHMWPE.

each having a sensitivity of 0.001 inch. The test chambers were mounted on a reciprocating, sliding base. The sliding distance was approximately 1 inch, and the contact pressure was 700 psi (5 MPa). In this rig, 1 full day of testing PTFE (24 hours) was considered by Craven to be equivalent to 18 months of *in vivo* wear. Thus, Craven's testing rig subjected polymer pins of candidate biomaterials to unidirectional sliding conditions in an aqueous (saline) environment, using PTFE as the control.

Charnley was certainly aware of the effects of testing joint surfaces in synovial fluid as opposed to aqueous environment. His animal and human joint frictional testing, for example, which were conducted in the 1950s, included a range of lubricating fluids (Charnley 1959, Waugh 1990). However, for wear testing of artificial joint materials, Charnley advocated saline as the medium because he felt that synovial fluid would not be a harsh enough test medium for the joint materials.

Still on display in Wrightington, the rig is only partially assembled today. Inspection reveals a very sturdily constructed apparatus. Craven's pride in his craftsmanship is evident because his name is stamped into the top of the rig.

In the 1970s, more complex testing rigs were built at Wrightington, which would subject polymer pins to multidirectional sliding (still using saline as a lubricant). However, from a historical perspective, the first test rig built by Craven is the most significant of the wear testers at Wrightington because it was used by Charnley to screen materials for the artificial hip after PTFE began to show evidence of severe wear *in vivo*. As we shall see, data collected using Craven's wear tester ultimately established the superiority of UHMWPE over other polymeric materials for use in joint replacement.

Searching to Replace PTFE

After the failure of PTFE cups *in vivo*, Charnley experimented with filled PTFE (Charnley 1979). Charnley tested and ultimately implanted glass-filled PTFE as well as a proprietary-filled PTFE produced by Polypenco under the trade name Fluorosint (Figure 4.9).

In the wear tester, Craven showed that both materials showed evidence of scratching the stainless steel counterface (presumably because of the glass fibers). However, the wear rates were improved. For instance, the wear tester predicted that the wear rate of Fluorosint would be 10 times lower than PTFE. Charnley ordered that cups be fashioned and implanted immediately.

In the body, both types of filled PTFE exhibited severe wear, similar to PTFE. With the Fluorosint components, the wear debris presented an added complication of darkened tissue around the joint, which needed to be excised. Figure 4.10, a retrieved Fluorosint component (encased in pink acrylic dental cement) at Wrightington, illustrates the severe wear exhibited by such devices.

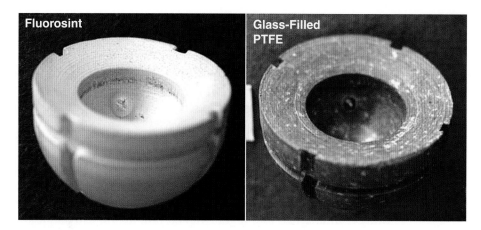

Figure 4.9
Proprietary-filled PTFE with trade name Fluorosint and glass-filled PTFE.

After the clinical failures of filled PTFE, Charnley abandoned polymers entirely in early 1962 and started to implant Thompson prostheses. However, the gloom that pervaded Wrightington at that time was short lived, because a promising new material soon arrived at the hospital through a series of most fortunate events.

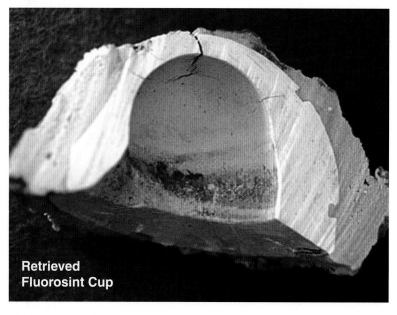

Figure 4.10
Retrieved Fluorosint component (encased in pink acrylic dental cement).

UHMWPE Arrives at Wrightington

The story of how UHMWPE was introduced to orthopedics is recounted in considerable detail by Waugh in Charnley's biography (Charnley 1963). Interviews with Craven, who was 32 years old at the time, as well as Charnley's writings, provide the basis for what is known about the circumstances at Wrightington in 1962.

In May of that year, a salesman (Mr. V.C. Binns) arrived at Wrightington and was introduced to Craven by the hospital's supply officer. Binn's products, small gears and bushings, were fabricated from UHMWPE and he provided Craven with a sample of the material. Charnley's first reaction to UHMWPE is described in his biography as follows: "When shown the material, Charnley dug his thumb-nail into it and walked out, telling Craven he was wasting his time" (Waugh 1990).

Despite Charnley's negative initial remarks, Craven decided to test UHMWPE on his wear testing rig anyway while the surgeon was away in Copenhagen for a meeting. When Charnley returned to Wrightington in June, Craven showed him the 3 weeks of continuous testing data he had collected in his absence. Charnley then wrote (Waugh 1990):

> After running day and night for three weeks, this new material, which very few people even in engineering circles had heard about at that time, had not worn as much as PTFE would have worn in 24 hours under the same conditions. There was no doubt about it, "we were on".

Although the wear results were encouraging, Charnley did not implant UHMWPE into patients until he was convinced of its biocompatibility. He first wrote to the producers of the UHMWPE, which was marketed under the trade name of RCH 1000. The German company that produced UHMWPE was known as Ruhrchemie, although it was later merged into Hoechst. From his correspondence, Charnley concluded that "by its chemistry, polyethylene had a very good chance of resisting attack by body fluids" (Waugh 1990).

Before implanting cups in patients, however, he implanted the UHMWPE in his own thigh, both in the bulk and finely divided (particulate) state. He also implanted PTFE in his thigh, to serve as a historical control. The results of this personal biocompatibility test, published in the *Lancet* in 1963 (Charnley 1963), convinced Charnley that UHMWPE-wear debris was biocompatible, whereas the particulate PTFE was not. Armed with this information, Charnley began implanting UHMWPE in patients during November 1962.

Implant Sterilization Procedures at Wrightington

Because the sterilization of UHMWPE continues to be a controversial topic, even today, it is helpful to review the early methods in use at Wrightington for joint replacement components. The stainless steel femoral components were produced by Thackray (Waugh 1990). Before 1968, only the final polishing of the femoral heads would be performed at Wrightington. After 1968, the stems were fabricated entirely by Thackray. The femoral components were provided nonsterile by Thackray and sterilized at the hospital by autoclaving. This sterilization method was not appropriate for UHMWPE, which would distort after prolonged steam exposure.

According to Dr. Wroblewski, who started his service at Wrightington in 1967, the UHMWPE cups were chemically sterilized by wrapping them in gauze and soaking them in Cidex (glutaraldehyde) overnight. Although UHMWPE cups produced commercially by Thackray were gamma irradiated starting in 1968, it remains unclear based on the documents available today, when, before 1968, gamma irradiation was initiated for cups that were implanted at Wrightington (Isaac, Dowson, and Wroblewski 1996). Figure 4.11 shows a Chas F. Thackray Ltd., catalog page from the 1969 product release brochure of the Charnley acetabular cup, indicating that the components were provided after sterilization with 2.5 Mrad of gamma radiation.

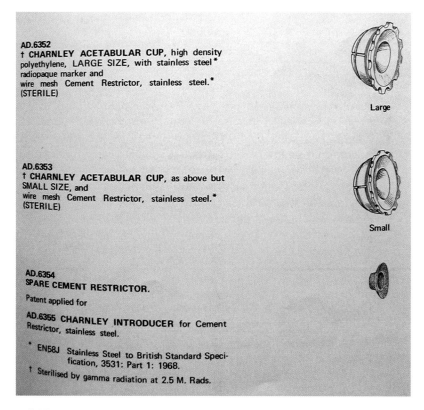

Figure 4.11

Chas F. Thackray Ltd., catalog page from the 1969 product release brochure.

Overview

In summary, UHMWPE was subjected to wear testing and biocompatibility testing prior to its first use in patients, during November 1962. After its clinical introduction, Charnley continued his research on the wear properties of polymers, but he never found a material better suited for joint replacements than UHMWPE. Although it is clear that he evaluated different polymers, including Hi-Fax UHMWPE material from the United States, his cups were always fabricated from the RCH 1000 compression molded UHMWPE material produced in Germany by Ruhrchemie (now Ticona).

Understanding the origins of UHMWPE in orthopedics is didactic in several respects. The first lesson is that the clinical introduction of UHMWPE was made possible by the close collaboration of surgical and engineering talent. That is not to suggest that the collaboration was necessarily harmonious, as the available evidence would suggest that Charnley and Craven were both stubborn and opinionated. However, there was clearly a respect for the talents of the surgeon for the engineer, and vice versa. Mutual respect, collaboration, as well as genuine talent, were the necessary foundations to achieve the breakthrough for joint arthroplasty that was desperately needed. The remarkable clinical performance of UHMWPE after its introduction is recounted in Chapter 5.

Acknowledgments

This chapter would not have been possible without the assistance of many supportive colleagues in England. Professor John Fisher (University of Leeds), Dr. Mike Wroblewski (Wrightington Hospital), and Harry Craven (Lancashire) were all instrumental in providing details regarding the undocumented history of Wrightington during the time of Charnley. Drs. Ken Brummit and Graham Isaac (DePuy International) and Alan Humphries (Thackray Museum) were extremely helpful in tracing the historical background of Thackray during the early development period of the artificial hip.

References

Atkinson J.R., and R.Z. Cicek. 1983. Silane cross-linked polyethylene for prosthetic applications. Part I. Certain physical and mechanical properties related to the nature of the material. *Biomaterials* 4:267–275.

Atkinson J.R., and R.Z. Cicek. 1984. Silane cross-linked polyethylene for prosthetic applications. Part II. Creep and wear behavior and a preliminary moulding test. *Biomaterials* 5:326–335.

Champion A.R., S. Li, K. Saum, E. Howard, and W. Simmons. 1994. The effect of crystallinity on the physical properties of UHMWPE. *Transactions of the 40th Orthopedic Research Society* 19:585.

Charnley J. 1959. The lubrication of animal joints. *Institution of Mechanical Engineers: Symposium on Biomechanics* 17:12–22.

Charnley J. 1961. Arthroplasty of the hip: A new operation. *Lancet* I:1129–1132.

Charnley J. 1963. Tissue reaction to the polytetrafluoroethylene. *Lancet* II:1379.

Charnley J., A. Kamangar, and M.D. Longfield. 1969. The optimum size of prosthetic heads in relation to the wear of plastic sockets in total replacement of the hip. *Med Biol Eng* 7:31–39.

Charnley J. 1979. Low Friction Principle. In *Low friction arthroplasty of the hip: Theory and practice*. Berlin: Springer-Verlag.

Craven H. July, 2002. Personal Communication.

Grobbelaar C.J., T.A. Du Plessis, and F. Marais. 1978. The radiation improvement of polyethylene prostheses: A preliminary study. *J Bone Joint Surg* 60–B:370–374.

Isaac G.H., D. Dowson, and B.M. Wroblewski. 1996. An investigation into the origins of time-dependent variation in penetration rates with Charnley acetabular cups—wear, creep or degradation? *Proc Inst Mech Eng [H]* 210:209–216.

Kurtz S.M., O.K. Muratoglu, M. Evans, and A.A. Edidin. 1999. Advances in the processing, sterilization, and crosslinking of ultra-high molecular weight polyethylene for total joint arthroplasty. *Biomaterials* 20:1659–1688.

Li S., and A.H. Burstein. 1994. Ultra-high molecular weight polyethylene: The material and its use in total joint implants. *J Bone Joint Surg Am* 76:1080–1090.

Oonishi H. 1995. Long-term clinical results of THR. Clinical results of THR of an alumina head with a cross-linked UHMWPE cup. *Orthopaedic Surgery and Traumatology* 38:1255–1264.

Shikata T., H. Oonishi, Y. Hashimato, et al. 1977. Wear resistance of irradiated UHMW polyethylenes to Al2O3 ceramics in total hip prostheses. *Transactions of the 3rd Annual Meeting of the Society for Biomaterials* 118.

Waugh W. 1990. *John Charnley: The man and the hip*. London: Springer-Verlag.

Wroblewski B.M., P.D. Siney, D. Dowson, and S.N. Collins. 1996. Prospective clinical and joint simulator studies of a new total hip arthroplasty using alumina ceramic heads and cross-linked polyethylene cups. *Journal of Bone & Joint Surgery* 78 B: 280–285.

Chapter 4. Reading Comprehension Questions

4.1. Which polymer did Charnley choose for his initial design of total hip replacements?
   a) Polytetrafluoroethylene
   b) High density polyethylene
   c) High density polyurethane
   d) Ultra-high molecular weight polyethylene
   e) None of the above

4.2. Charnley's initial choice of implant polymer was based on which material property?
   a) Wear rate
   b) Coefficient of friction
   c) Ultimate Strength
   d) Density
   e) Small punch test–Ultimate load

4.3. For which properties did Charnley choose UHMWPE over other polymers as a bearing material for total hip replacements?
 a) Wear rate
 b) Coefficient of friction
 c) Impact strength
 d) Biocompatibility
 e) All of the above

4.4. UHMWPE was first used in hip replacements during which year?
 a) 1970
 b) 1968
 c) 1964
 d) 1962
 e) 1958

4.5. The trade name for the UHMWPE first used by Charnley was
 a) Fluon
 b) Chirulen
 c) RCH 1000
 d) Hifax 1900
 e) None of the above

4.6. When he developed hip replacements, Charnley practiced medicine in
 a) Oberhausen, Germany
 b) Paris
 c) New York City
 d) London
 e) Wrightington, England

4.7. Which method was first used by Charnley to sterilize UHMWPE for joint replacements?
 a) Steam autoclaving
 b) Glutaraldehyde immersion
 c) Gamma irradiation in air
 d) Gas Plasma
 e) Ethylene oxide

4.8. Which of the following components was NOT included in Charnley's hip replacement design?
 a) Femoral component with head and stem
 b) UHMWPE acetabular cup
 c) Metal backing
 d) Acrylic (PMMA) bone cement
 e) All of the above

Chapter 5

The Clinical Performance of UHMWPE in Hip Replacements

Introduction

Although ultra-high molecular weight polyethylene (UHMWPE) has a clinical track record spanning more than 40 years, the first decade of clinical implementation occurred at Wrightington under the exclusive direction of John Charnley. Following the polytetrafluoroethylene (PTFE) debacle, Charnley withheld from publishing his experience with UHMWPE low friction arthroplasties (LFAs) until the 1970s. Thus, in the 1960s, the proliferation of total joint replacements (TJRs) beyond Wrightington was strictly controlled by Charnley (Waugh 1990). However, copies of Charnley's designs began to appear in the United States during the mid-1970s, and after the details of the operation became widely known, total hip replacement (THR) underwent a period of explosive growth, which lasted through the 1980s. Starting in the 1990s, joint arthroplasty entered a period of more steady, predictable growth, which is expected to last throughout the next three decades. The American Academy of Orthopedic Surgeons (AAOS) has proclaimed the first decade of our century (2002–2011) as The United States Bone and Joint Decade.

Prevailing optimism regarding joint replacement is founded not only on the historically successful performance of these surgical procedures in the United States, but also upon the demographics of our gentrifying population. The chart in Figure 5.1 shows the age distribution receiving a primary hip or knee replacement in 2000, along with the projections for 2030 calculated by the AAOS.

Thus, the elderly population (ages 65 years and older) receiving a hip or knee replacement is expected to increase by 101% over the next 30 years, whereas the population of young patients (younger than 65 years) is projected to increase

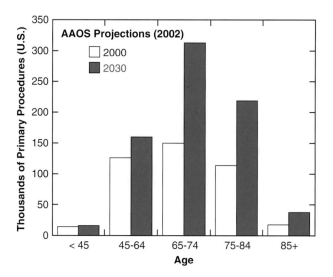

Figure 5.1
Age distribution receiving a primary hip or knee replacement in 2000, along with the projections for 2030 calculated by the AAOS.

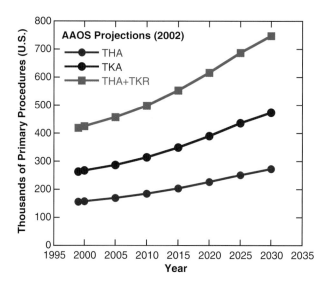

Figure 5.2
Projected growth of primary hip and knee procedures in the United States over the next 30 years.

by only 26%. The graph in Figure 5.2 charts the projected growth of primary hip and knee procedures in the United States over the next 30 years.

The data plotted in Figure 5.2 indicate that approximately 36 to 37% of primary total joint procedures are performed for the hip and opposed to the knee, and this ratio is not expected to change substantially over the next 30 years. Although total hip arthroplasty (THA) is performed less often than total knee arthroplasty, the clinical performance of the hip has been studied to a much greater extent than that of the knee. Consequently, the focus of this chapter is on the clinical performance of THA.

Because of the role of wear in promoting osteolysis, aseptic loosening, and ultimately implant revision, this chapter reviews the historical development of concepts related to quantifying clinical wear performance of UHMWPE in hip replacements. The foundations of our current understanding of this topic were established in the publications of Charnley, published in the 1970s. Additional background material about the biological, design, and implant factors related to implant wear may be found in the review by Schmalzreid and colleagues (1998), as well as in the comprehensive book *Implant Wear,* recently published by the AAOS in conjunction with the National Institutes of Health (2001). Several excellent book chapters and review articles also have been published that describe the range of factors contributing to wear of UHMWPE materials in the laboratory (Clarke and Kabo 1991, McKellop 1998, Sauer and Anthony 1998, Wang et al. 1998).

Joint Replacements Do Not Last Forever

Joint replacements are highly successful, especially during their first decade of use. Nevertheless, joint replacements may still be revised at a rate of about 1% per year during the first 10 years of implantation for a variety of implant-, patient-, and surgeon-related factors (Barrack 1996, Fitzpatrick et al. 1998, Malchau et al. 2000, Malchau et al. 2002). Outcome studies for THA suggest that the long-term survivorship decreases markedly after 10 years of implantation, especially for patients younger than 55 years (Malchau et al. 2002).

The revision rate for TJRs remains a significant burden to the health care economies of Western countries, and varies from 10–20% depending on the nationality (Malchau et al. 2002). In the United States, for instance, Medicare data (for patients aged 65 years and older) suggests that revision procedures occur at a rate of about 18% relative to the number of primary surgeries (Malchau et al. 2002). According to data for 11,543 revisions in Sweden from 1979 to 1998, 75.7% of hip revision surgeries occurred as a result of aseptic loosening (Malchau et al. 2000).

Wear of UHMWPE is currently recognized as the primary culprit responsible for aseptic loosening and late revision of hip replacements. Researchers have estimated that for each day of patient activity, around 100 million microscopic UHMWPE wear particles are released into the tissues surrounding the

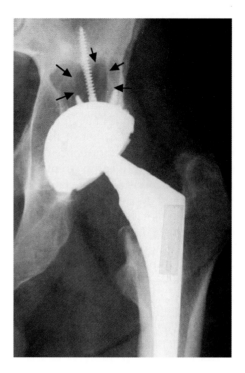

Figure 5.3
Example of an osteolytic lesion in the pelvis, located superior to the metal-backed acetabular component. (Courtesy of Av Edidin, Ph.D., Drexel University.)

hip joint (Muratoglu and Kurtz 2002). This particulate wear debris can initiate a cascade of adverse tissue response leading to osteolysis (bone death) and ultimately aseptic loosening of the components (Goldring et al. 1986, Jasty et al. 1986, Willert 1977, Willert, Bertram, and Buchhorn 1990). Figure 5.3 shows an example of an osteolytic lesion in the pelvis, located superior to the metal-backed acetabular component.

Based on a review of the literature, Dumbleton and colleagues (2002) suggest that radiographic wear rates of less than 0.05 mm/year are below an "osteolysis threshold," below which patients are not expected to be at risk of developing osteolysis. Osteolysis, in turn, may be associated with the need for revision, depending on the location (e.g., in the pelvis or femur) and rate of progression. As noted by Hozack and associates (1996), "Polyethylene wear alone (stage I) is an indication of impending failure, and when symptoms develop (stage IIA), revision should be undertaken. The development of radiographic lysis is a critical event, and as soon as osteolysis develops (stage IIB or III), revision should be undertaken immediately. From the perspective of the revision surgeon, there is great value in the early intervention for polyethylene wear and pelvic osteolysis." Therefore, understanding the natural history of UHMWPE wear in a clinical setting, starting with the pioneering work of Charnley, is an important first step to improving the longevity of THA.

Range of Clinical Wear Performance in Cemented Acetabular Components

Charnley and coworkers first developed radiographic techniques for evaluating the wear rate of UHMWPE acetabular components in patients. In 1973, Charnley and Cupic reported on the long-term wear performance in the first cohort of patients to receive a UHMWPE component between November 1962 and December 1963. During this time period, 170 patients received a cemented LFA with an UHMWPE component; a total of 185 acetabular cups were implanted. Because of the elderly population originally implanted with the components, many had died or were too infirm to travel to the clinic for follow-up examination (more than two-thirds of the patients were older than 60 years of age at the time of implantation). Thus, only 106 out of the original 185 UHMWPE cups could still be evaluated after 9 or 10 years of implantation. The complications for this series included a 4–6% rate of infection, 1–2% rate of mechanical loosening, and a 2% incidence of late dislocation.

Early in 1963, Charnley introduced the use of a semicircular wire marker on the back of the acetabular component to assist with the measurement of wear using radiographs. The photograph in Figure 5.4, taken of an unused Charnley cup from the collection at the Thackray Museum at Leeds, shows the wire marker inserted into the cement groove. Although the implant shown was produced between 1968 and 1975, the configuration of the wire marker was similar to that used in Charnley and Cupic's study.

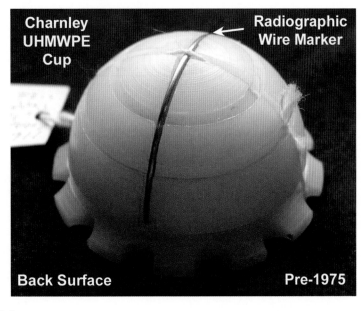

Figure 5.4

Produced between 1968 and 1975, the configuration of the wire marker was similar to that used in Charnley and Cupic's study.

Charnley and Cupic (1973) calculated wear from the radiographs by measuring the narrowest distance between the femoral head and the back of the liner in the weight-bearing and nonweight-bearing regions of the cup. The difference in thickness in the weight-bearing and nonweight-bearing regions was then divided by two to estimate the radial penetration of the head, which was attributed to wear of the cup. These measurements had a precision of only 0.5 mm, so that wear of less than 0.5 mm could not be detected using this technique. In contrast, contemporary computer-based radiographic wear analysis techniques have a precision of 0.1 mm or less (Martell and Berdia 1997).

A total of 72 components out of Charnley's 9- to 10-year series were measured for wear. The other hips in the series either did not have a wire marker or did not have radiographs. Charnley reported the average wear rate was 0.15 mm/year, but analysis of his data indicates that this average excluded patients with undetectable (0 mm) wear. The distribution of wear rates in Charnley's 72 patients is shown in Figure 5.5 and includes patients with undetectable (0 mm) wear.

The average (mean) wear, including those patients with undetectable wear, was 1.18 mm (range: 0 to 4 mm). A standard normal (Gaussian) distribution is also superimposed on this data, showing that the distribution is skewed to the left with a greater number of patients exhibiting wear below the mean than above it. Also, there is a small number of patients exhibiting wear much greater than the mean. This skewed distribution in UHMWPE wear performance in a patient population was well recognized by Charnley (Charnley and Cupic 1973, Griffith et al. 1978) and has been observed in subsequent studies using contemporary modular hip designs (AAOS 2001). For this reason, it is sometimes more useful to characterize clinical wear performance using nonparametric statistical methods, which make no assumptions about

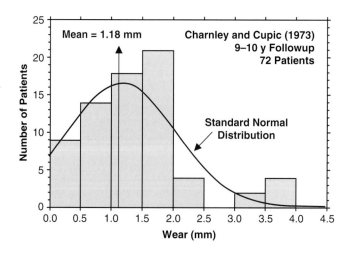

Figure 5.5

Histogram showing the distribution of wear measured radiographically in Charnley and Cupic's study (1973).

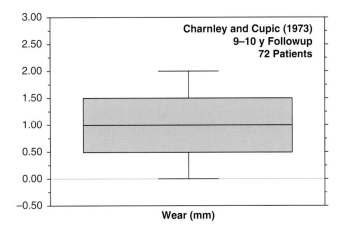

Figure 5.6
This box and whiskers plot, illustrating the same data shown in Figure 5.5, is helpful to summarize the distribution of wear results.

the underlying distribution of the data. A "box and whiskers" plot (or more simply, a box plot), shown in Figure 5.6 for the same wear data as in the histogram derived from Charnley and Cupic's study (1973), is helpful to summarize the distribution of the wear data.

For readers who may not be familiar with a box plot, it can be interpreted as follows. The top and bottom faces of the box correspond to the 25th and 75th percentiles of the data (e.g., 0.5 mm and 1.5 mm of wear, respectively). The median (50th percentile) is the horizontal line inside the box (at 1.0 mm of wear). Thus, the box itself represents the interquartile range, or the central tendency of the distribution corresponding to the middle half of the population. The median is the center of the distribution, with half the population lying to either side. Note that the median for the wear data (1 mm) is a bit lower than the arithmetic mean or average (1.18 mm). Finally, the whiskers of the plot correspond to the 10% and 90% of the distribution (e.g., 0.0 mm and 2.0 mm of wear, respectively), providing an indication of breadth of the central 80% of the population. Some statisticians like to plot the data outside the 10th and 90th percentiles as individual data points or outliers, but I have not adopted that convention here.

Wear Versus Wear Rate of Hip Replacements

The absolute amount of wear observed in a group of patients is going to depend on the time interval between radiographic examinations. Consequently, clinical researchers starting with Charnley have been interested in the linear wear rate (LWR), or the change in apparent femoral head penetration (P) over time (t):

$$LWR = \frac{\Delta P}{\Delta t} \qquad (1)$$

Charnley also suggested that the volumetric wear rate, as opposed to the LWR, may also be a clinically relevant metric for wear, as the biological stimulus may be related to the volume of wear debris (Charnley, Kamangar, and Longfield 1969). If the femoral head penetrates the cup following a linear trajectory (an assumption that was verified by Charnley's clinical experience with PTFE), the wear volume will be approximated as a cylinder having the projected circular area (A) of the femoral head and a height equal to the depth of penetration. Under this assumption, the volumetric wear rate (VWR) can be calculated as follows:

$$VWR = A \times LWR = \frac{\pi}{4}D^2 \times \frac{\Delta P}{\Delta t} \qquad (2)$$

In a study published in 1975, Charnley and Halley (1978) calculated the wear rate for the 72 patients with 9 to 10 years of follow-up by analyzing the serial radiographs. Charnley and Halley found that after an initial bedding-in period during the first several years, the rate of wear progressed linearly with time, as shown in Figure 5.7.

Figure 5.7 plots the *average* radiographic wear (again, for patients with measurable wear) for 72 patients, so one must understand that some variability would be superimposed on this picture if all of the individual data points had been plotted. Overall, the average wear for the entire series was observed to decrease over time. In the first 5 years, the wear rate was 0.18 mm/year for all 72 patients, and during the 5- to 10-year period, the average wear dropped to 0.10 mm/year.

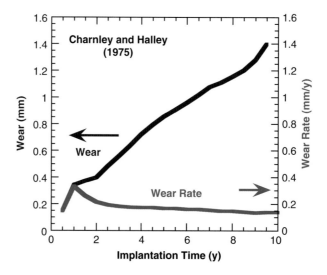

Figure 5.7

After an initial bedding-in period during the first several years, the rate of wear progressed quasilinearly with time. The overall wear rate, however, appeared to slowly decrease over time. (Adapted from Charnley and Halley 1978.)

Comparing Wear Rates Between Different Clinical Studies

If two studies have different follow-up periods, it should be possible to compare the wear rates. In practice, however, differences between patient groups, surgeon groups, and implant systems greatly complicate the task of reconciling differences in wear behavior observed in clinical studies. For example, consider studies published by Charnley and coworkers from Wrightington that describe the wear behavior of UHMWPE in two cohorts. The first cohort, which we have already discussed, consisted of 72 arthroplasties that were implanted between 1962 and 1963 and followed for 9 to 10 years (Charnley and Cupic 1973). The second cohort, described by Griffith and coworkers, consisted of 493 arthroplasties implanted between 1967 and 1968 and followed for 7 to 9 years (mean, 8.3 years) (Griffith et al. 1978).

Both cohorts of patients were implanted at Wrightington using the same cemented design using an UHMWPE cup fabricated from RCH 1000 (Ruhrchemie, Germany). Presumably, these two studies should be the easiest to compare, because they were directed by the same surgeons and drew patients from the same elderly populations. However, there are significant differences in the resulting wear rates for both studies: the average wear rate for patients in Charnley's study was 0.15 mm/year, whereas the wear rate in Griffith's study was 0.07 mm/year. The average wear rates reported by both studies did not include patients exhibiting undetectable wear, which is an important omission if we consider that 110 of Griffith's 491 hips (i.e., 22%) were measured with 0 mm of wear after 8.3 years of implantation.

When we include the patients with zero wear, the distribution of wear rates in both studies is summarized as a box plot in Figure 5.8, taking into account the calculation as described in Equation (1).

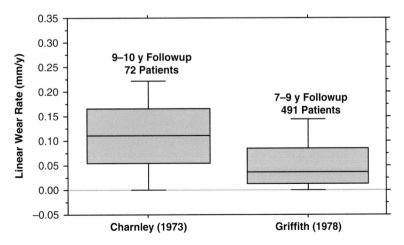

Figure 5.8

Distribution of THA wear rates observed in two studies from Wrightington Hospital.

Table 5.1

Linear wear rate (mm/year) based on radiographs for Charnley components implanted between 1962 and 1968

Percentiles	Wear Rate (mm/year) Charnley (1973)	Wear Rate (mm/year) Griffith (1978)
10	0.00	0.00
25	0.06	0.01
50	0.11	0.04
75	0.17	0.09
90	0.22	0.15

The data indicates that the distribution of wear rates is more skewed by low-wearing patients in Griffith's study (1978) than in Charnley's study (1973). The distribution of linear wear data for the two studies, in terms of percentiles, is shown in Table 5.1.

Thus, more than 60% of the patients in both studies exhibited wear lower than the average values of 0.15 and 0.07 mm/year reported by Charnley and colleagues and Griffith and associates, respectively. Also, in both studies, 10% of the patients (i.e., in the 90th percentile) exhibited wear at 2 to 3 times the median rate.

Of particular interest to Griffith, Charnley, and coworkers was to explain the variability in clinical wear performance, not only within the context of a single study, but also between the two different studies. For example, Griffith and colleagues (1978) examined whether age, gender, postoperative activity level, and the involvement of other joints with hip disease influenced the magnitude of wear. The researchers observed that male patients younger than 60 years of age at the time of implantation appeared to be associated with the incidence of heavy wear (defined as greater than 1.4 mm). Griffith and associates (1978) also tried to explain the difference between their results and Charnley and Cupic's previous study. The researchers suggest that an improvement in surface finish of the femoral head may have contributed to the apparent improvement clinical wear performance. They also suggest that the UHMWPE used in Charnley and Cupic's earlier study may have been "of inferior quality" (Griffith et al. 1978).

Over the years, the difference in wear rates observed between the two studies has continued to intrigue researchers (Isaac, Dowson, and Wroblewski 1996). Griffith's paper strongly suggests that differences in manufacturing of the implants, as opposed to the patients, is most likely responsible for the difference in wear rates as compared with Charnley and Cupic's earlier study. Unfortunately, it is no longer possible to obtain written documentation describing the manufacture of components in the early 1960s at Wrightington. Many of the details about implant manufacturing during that time period have been provided by Harry Craven, who served as Charnley's engineer and technician between 1958 and 1966 when he was between 30 and 36 years old (Craven 2002).

On one hand, commercial production of the Charnley prosthesis by Thackray Ltd. is reported to have started between 1968 and 1969. The production date has been confirmed from several sources (Isaac, Dowson, and Wroblewski 1996, Brummit 2002), including a commercial release pamphlet from 1969 produced by Thackray that is available at the Thackray Museum in Leeds (see Figure 4-11). Thackray's documentation clearly indicates that the UHMWPE cups were gamma irradiated by a dose of 2.5 Mrad. It has been reported that no sterilization records from that time survive today (Brummit 2002).

On the other hand, documents obtained at Wrightington suggest that 1968 to 1969 may have applied only to distribution of Charnley implants to the outside world. According to an internal Wrightington publication (Charnley 1966):

> The total hip prosthesis which I call "low friction arthroplasty," in which a plastic socket is an integral part of the design, is being released (November 1966) under pressure from surgical colleagues to a restricted number of surgeons (in particular those who have worked with me over six months) prior to considering a general release in January 1968.

The release described in Charnley's internal report is at least consistent with Griffith's statement that Wrightington shifted to commercial production of femoral heads in 1966 (Griffith et al. 1978). Craven, who was himself responsible for manufacturing some of the implants and instruments for Charnley between 1958 and 1966, recalls that Thackray was producing components for Wrightington as early as 1963 (Craven 2002). Specifically, Craven recalls fabricating a specialized machine for stamping out the scalloping for the rim in the standard cup and selling the machine to Thackray between 1963 and 1964 for the sum of 360 £. Taken together, the available evidence suggests that Thackray was producing implants for Wrightington earlier than 1969. Because Thackray sterilized UHMWPE using only gamma irradiation, the inference is that prior to 1969, perhaps as early as 1963, UHMWPE components were used at Wrightington that had been gamma irradiated. Therefore, there is convincing evidence to suggest that the components in Griffith's study, which were implanted at Wrightington between 1967 and 1968, were also gamma irradiated.

The crosslinking produced by a single dose of gamma radiation (even in air) has the beneficial result of increasing the resistance to adhesive/abrasive wear and could explain, at least in part, the lower wear rates in Griffith's study. According to wear testing by Wang and associates using a contemporary multidirectional hip simulator (1997), changing from 0 to 2.5 Mrads of irradiation (in air) drops the wear rate from 140 to 90 mm^3/million cycles (using 32 mm diameter heads), corresponding to a reduction of about 36%.

A substantial increase in the molecular weight of the UHMWPE could also explain the improved wear resistance observed in Griffith's study. The wear resistance of UHMWPE is imparted by its molecular weight, which ranges today, on average, between 2 and 6 million. Griffith and coworkers suggest that the quality of the UHMWPE may have been inferior in 1961 to 1962 (1978), although the source of the material remained the same (Ruhrchemie AG, currently known as Ticona, distributed locally in Lancashire by High Density Plastics [Craven 2002]). In wear testing of UHMWPE and high-density polyethylene (molecular weight 200,000) using a contemporary multidirectional hip simulator (Edidin and Kurtz 2000), differences in the molecular weight of the polyethylene can affect the wear rate by a factor of 4.

The surface finish of the heads, suggested by Griffith (1978), is a possible, but unlikely reason for the change in clinical wear rates. A wide range in surface roughness values were observed in Isaac's analysis of retrieved Charnley cups (Isaac, Dowson, and Wroblewski 1996, Isaac et al. 1992), with no association to the clinical wear rate. Similarly, the retrieval work of Hall and colleagues also showed no significant relationship between surface roughness of the femoral head and clinical wear rate in a group of retrieved metal-backed acetabular components that were implanted without cement (1997). Thus, in light of recently published studies, the roughness of the femoral head seems an unlikely explanation for the change in wear behavior for the entire cohort of Griffith's patients.

In summary, the following three factors, either alone or in combination, could explain the differences in wear rates between Charnley's and Griffith's studies:

1. Radiation-induced crosslinking
2. Molecular weight of the UHMWPE
3. Surface finish of the femoral head

Our inability to conclusively identify the explanation for the difference in wear rates between these two studies underscores the difficulty in comparing two retrospective series of patients, even when performed at the same institution and by the same group of investigators.

Comparison of Wear Rates in Clinical and Retrieval Studies

Two additional clinical studies of interest related to the Charnley prosthesis were performed by Isaac and associates (1992, 1996) on a group of 100 explanted UHMWPE cups that were retrieved at revision surgery after an average of 9 years of implantation, with an implantation time of up to 17.5 years. Unlike previous studies conducted by Charnley and his research colleagues, who focused on the wear behavior of UHMWPE functioning arthroplasties, the retrieved components represented a group of patients in which the surgery had failed. Thus, the study of retrieved implants provides insight into factors leading to the revision (i.e., clinical failure) of total joint arthroplasty. In the retrieval study of retrieved Charnley components, wear of the cups was measured radiographically, as well as directly from the actual explants, in 86 out of 100 cases (in 14 components the wear was too severe to be accurately measured). The average wear rate for this series of explanted cups was 0.2 mm/year, but again a wide variation in rates was observed. The wear rates for individual cups are plotted in Figure 5.9 as a function of implantation time.

Current Methods for Measuring Clinical Wear in Total Hip Arthroplasty

Wear can be measured from radiographs using manual or computer-assisted techniques. The Charnley method, developed in the 1970s for measuring radiographic wear was reviewed earlier in this chapter. Livermore and colleagues (1990) later improved on the Charnley wear measurement technique with the use of circular templates. Despite the widespread use of Charnley and Livermore techniques throughout Europe and the United States, these manual methods have poor intraobserver and interobserver repeatability and are unsuitable for measuring wear of less than 0.3 to 0.5 mm (Dumbleton, Manley, and Edidin 2002, Martell and Berdia 1997), as would presumably be the case for newly developed highly crosslinked UHMWPE materials for hip replacement.

For detecting low levels of initial wear in UHMWPE hip components, researchers have developed both two-dimensional and three-dimensional computer-assisted radiographic wear measurement methods, as summarized in a recent review (Dumbleton, Manley, and Edidin 2002). In addition, radiostereometric analysis (RSA) in conjunction with digital radiography promises to have improved accuracy for measuring initial wear in UHMWPE (Bragdon et al. 2002). However, a number of practical limitations preclude the use of RSA and three-dimensional computer-assisted techniques in a generalized clinical setting. For instance, the use of RSA requires the addition of tantalum

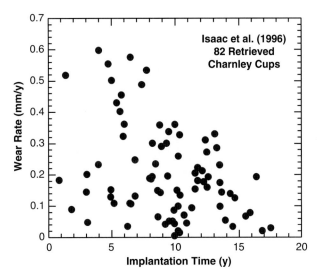

Figure 5.9
Wear rate versus implantation time. (Redrawn from Isaac et al. 1996.)

markers to the acetabular and femoral components, and is limited to only a few, specialized radiographic suites worldwide. Another advanced three-dimensional wear-assessment technique, developed by Devane and colleagues (1995), requires that an manufacturer provide a computer-aided design (CAD) model of an implant design in order to reconstruct wear from multiple radiographic views.

In contrast, a two-dimensional computer-assisted technique using edge detection and computer vision, developed by Martell and Berdia, improves intraobserver and interobserver repeatability tenfold over manual techniques (Martell and Berdia 1997). Figure 5.10 illustrates the Martell technique for computer-assisted measurement of radiographic wear.

The Martell technique has been adapted to allow three-dimensional wear assessment. Comparing two-dimensional and three-dimensional analysis on a large clinical population, Martell has shown that 87–90% of wear is detected using two-dimensional on anterior-posterior (A-P) pelvis radiographs alone, with two-dimensional and three-dimensional wear values being highly correlated ($r^2 = 0.993$). Additionally, as a result of poor lateral pelvic radiographic quality, repeatability of the three-dimensional analysis is up to four times worse than the two-dimensional technique (Bragdon et al. 2002, Martell, Berkson, and Jacobs 2000). Two-dimensional analysis is therefore the preferred method for wear detection in THA using the Martell technique. Using a validated dynamic phantom total hip wear model, the accuracy of the Martell two-dimensional analysis has been shown to be ±0.073 mm whereas the precision is ±0.023 mm (Bragdon et al. 2001). Finally, the Martell technique is implant independent, requiring only digitized A-P radiographs and knowledge of the head size.

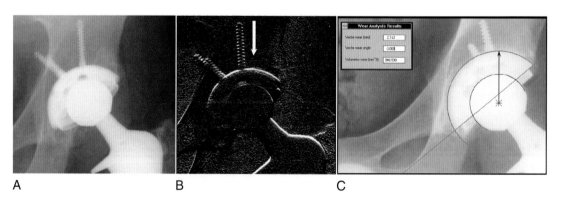

A   B   C

Figure 5.10
Computer-assisted radiographic wear measurement using the method of Martell and Berdia (1997). (**A**) Digital radiographic image prior to wear analysis. (**B**) The results of processing the digital image using edge detectors in the direction of the white arrow. Each white dot represents a potential point on the edge of the prosthesis. (**C**) The computer has chosen the best circle fits for the acetabular shell and femoral head using the edge points created in the previous image. The change in position of the femoral head with respect to the acetabular center over time is reported as wear, and displayed on the screen. (Images provided courtesy of J. Martell, M.D.)

Table 5.2

Linear Wear Rate (mm/year) Based on Radiographs for the Modular Harris-Gallante I Acetabular Component as a Function of Implantation Time

Percentiles	Implanted 1–6 years N = 35 patients		Implanted 7–10 years N = 43 patients		Implanted 11+ years N = 29 patients	
	2D	3D	2D	3D	2D	3D
10	-0.070	0.080	0.040	0.058	0.054	0.064
25	0.080	0.122	0.060	0.070	0.077	0.090
50	0.150	0.200	0.090	0.110	0.120	0.130
75	0.228	0.287	0.170	0.210	0.182	0.185
90	0.340	0.480	0.272	0.272	0.218	0.232

2D, two-dimension; 3D, three-dimension.
Previously published data provided courtesy of Martell 2000. See also Figure 5.11.

Range of Clinical Wear Performance in Modular Acetabular Components

Qualitatively, the trends observed for cemented components also appear to be applicable to modular acetabular components, evaluated using current computer-assisted wear measurement techniques (Tables 5.2 and 5.3, and Figure 5.11). The box and whisker plots shown in Figure 5.11 compare the linear

Table 5.3

Volumetric Wear Rate (mm^3/year) Based on Radiographs for the Modular HG I Acetabular Component as a Function of Implantation Time

Percentiles	Implanted 1–6 years N = 35 patients		Implanted 7–10 years N = 43 patients		Implanted 11+ years N = 29 patients	
	2D	3D	2D	3D	2D	3D
10	2.2	26.6	17.4	20.9	28.9	31.4
25	44.3	48.9	28.0	31.1	37.3	41.6
50	75.0	88.3	42.2	51.7	60.2	56.1
75	127.5	149.3	90.1	91.8	82.0	80.9
90	182.8	226.4	135.2	132.3	113.7	108.0

2D, two-dimension; 3D, three-dimension.
Previously published data provided courtesy of Martell 2000. See also Figure 5.11.

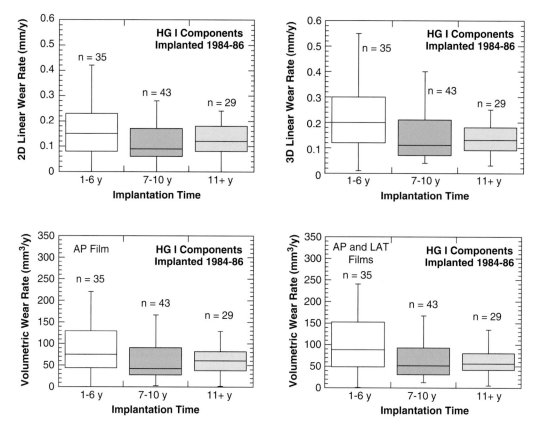

Figure 5.11

Distribution of LWRs and VWRs estimated in a first-generation modular acetabular component design (Harris-Gallante: Zimmer: Warsaw, IN). These graphs show the tendency for the wear rates to decrease with implantation time. Results are compared based on analysis of A-P radiographs alone (two-dimensional), as well as based on the combined analysis of A-P and lateral radiographs (three-dimensional). (Based on data provided courtesy of J. Martell, M.D.)

wear (penetration) of UHMWPE for a first-generation modular acetabular component design. Data previously published by Martell and colleagues (2000) has compared radiographs using planar radiographs alone, as compared with a three-dimensional analysis that includes planar and lateral radiographs (Figure 5.11). Note that, as already indicated, the lateral radiographs capture about 90% of the total wear of the hip joint.

Conclusion

UHMWPE has a long successful clinical track record in THRs dating back to the 1960s. However, the *in vivo* wear rate of conventional UHMWPE for some THA patients falls above the threshold for osteolysis and may ultimately lead to the need for long-term revision. Although accurate computer-assisted methods have now been developed to track the progression of *in vivo* wear of UHMWPE, these techniques reveal a range of performance within the human body. Even when implant- and surgeon-related factors are held constant, patient factors (e.g., activity level, obesity, etc.) play major roles in the clinical wear rate (McClung et al. 2000, Schmalzried, Dorey, and McKellop 1998, Schmalzried et al. 2000, Zahiri et al. 1998). Because of the importance of reducing the broad distribution of wear rates observed *in vivo*, major research efforts have been directed to developing improved UHMWPE materials for hip replacements, as is described in Chapter 6.

Acknowledgments

The author is indebted to John Martell, M.D., from the University of Chicago, and to Av Edidin, Ph.D., Drexel University, for their helpful comments and editorial assistance with this chapter.

References

AAOS. 2001. How should wear-related implant surveillance be carried out and what methods are indicated to diagnose wear-related problems. In *Implant wear*. T.M. Wright and S.B. Goodman, Eds. Rosemont, IL: American Academy of Orthopedic Surgeons.

Barrack R.L. 1996. Concerns with cementless modular acetabular components. *Orthopedics* 19:741–743.

Bragdon C.R., X. Yuan, R. Perinchief, et al. 2001. Precision and reproducibility of radiostereometric analysis (RSA) to determine polyethylene wear in a total hip replacement model. *Trans ORS* 47:1005.

Bragdon C.R., H. Malchau, S.L. Larson, et al. 2002. Validation of digital radiography, for use with radiostereometric analysis (RSA) using a dynamic phantom wear model. *Transactions of the 48th Orthopedic Research Society* 27:1020.

Brummit K. July 17, 2002. Personal communication.

Charnley J. 1966. Total prosthetic replacement of the hip joint using a socket of high density polyethylene. Report No. 1. Wigan: Center for Hip Surgery, Wrightington Hospital.

Charnley J., A. Kamangar, and M.D. Longfield. 1969. The optimum size of prosthetic heads in relation to the wear of plastic sockets in total replacement of the hip. *Med Biol Eng* 7:31–39.

Charnley J., and Z. Cupic. 1973. The nine and ten year results of the low-friction arthroplasty of the hip. *Clin Orthop Rel Res* 95:9–25.

Charnley J., and D.K. Halley. 1978. Rate of wear in total hip replacement. *Clin Orthop* 112:170–179.

Clarke I.C., and J.M. Kabo. 1991. Wear in total hip replacement. In *Hip arthroplasty*. H.C. Amstutz, Ed. New York: Churchill Livingstone.

Craven H. July 31, 2002. Personal communication.

Devane P.A., R.B. Bourne, C.H. Rorabeck, et al. 1995. Measurement of polyethylene wear in metal-backed acetabular cups. I. Three-dimensional technique. *Clin Orthop* 319:303–316.

Dumbleton J.H., M.T. Manley, and A.A. Edidin. 2002. A literature review of the association between wear rate and osteolysis in total hip arthroplasty. *J Arthroplasty*; 17:649–661.

Edidin A.A., and S.M. Kurtz. 2000. The influence of mechanical behavior on the wear of four clinically relevant polymeric biomaterials in a hip simulator. *J Arthroplasty* 15:321–331.

Fitzpatrick R., E. Shortall, M. Sculpher, et al. 1998. Primary total hip replacement surgery: A systematic review of outcomes and modelling of cost-effectiveness associated with different prostheses. *Health Technol Assess* 2:1–64.

Goldring S.R., M. Jasty, C.M. Roueke, et al. 1986. Formation of a synovial-like membrane at the bone-cement interface. Its role in bone resorption and implant loosening after total hip replacement. *Arthritis and Rheumatism* 29:836–842.

Griffith M.J., M.K. Seidenstein, D. Williams, and J. Charnley. 1978. Socket wear in Charnley low friction arthroplasty of the hip. *Clin Orthop* 137:37–47.

Hall R.M., P. Siney, A. Unsworth, and B.M. Wroblewski. 1997. The effect of surface topography of retrieved femoral heads on the wear of UHMWPE sockets. *Med Eng Phys* 19:711–719.

Hozack W.J., J.J. Mesa, C. Carey, and R.H. Rothman. 1996. Relationship between polyethylene wear, pelvic osteolysis, and clinical symptomatology in patients with cementless acetabular components. A framework for decision making. *J Arthroplasty* 11:769–772.

Isaac G.H., B.M. Wroblewski, J.R. Atkinson, and D. Dowson. 1992. A tribological study of retrieved hip prostheses. *Clin Orthop* 276:115–125.

Isaac G.H., D. Dowson, and B.M. Wroblewski. 1996. An investigation into the origins of time-dependent variation in penetration rates with Charnley acetabular cups—wear, creep or degradation? *Proc Inst Mech Eng [H]* 210:209–216.

Jasty M., W.E.I. Floyd, A.L. Schiller, et al. 1986. Localized osteolysis in stable, non-septic total hip replacement. *J Bone Joint Surg* 68A:912–919.

Livermore J., D. Ilstrup, and B. Morrey. 1990. Effect of femoral head size on wear of the polyethylene acetabular component. *J Bone Joint Surg* 72:518–528.

Malchau H., P. Herberts, P. Söderman, and A. Odén. 2000. Prognosis of total hip replacement: Update and validation of results from the Swedish National Hip Arthroplasty Registry, 1979–1998. 67th Annual Meeting of the American Academy of Orthopaedic Surgeons, Scientific Exhibition, Orlando, FL.

Malchau H., P. Herberts, P. Söderman, and A. Odén. 2002. Prognosis of total hip replacement: Update of results and risk-ratio analysis for revision and re-revision from the Swedish National Hip Arthroplasty Registry, 1979–2000. 69th Annual Meeting of the American Academy of Orthopaedic Surgeons, Scientific Exhibition, Dallas, TX.

Martell J.M., and S. Berdia. 1997. Determination of polyethylene wear in total hip replacements with use of digital radiographs. *J Bone Joint Surg* 79:1635–1641.

Martell J., E. Berkson, and J.J. Jacobs. 2000. The performance of 2D vs. 3D computerized wear analysis in the Harris Galante acetabular component. *Orthopaedic Transactions* 25:564.

McClung C.D., C.A. Zahiri, J.K. Higa, et al. 2000. Relationship between body mass index and activity in hip or knee arthroplasty patients. *J Orthop Res* 18:35–39.

McKellop H.A. 1998. Wear assessment. In *The adult hip*. J.J. Callaghan, A.G. Rosenberg, and H.E. Rubash, Eds. Philadelphia: Lippincott-Raven.

Muratoglu O.K., and S.M. Kurtz. 2002. Alternative bearing surfaces in hip replacement. In *Hip replacement: Current trends and controversies*. R. Sinha, Ed. New York: Marcel Dekker.

Sauer W.L., and M.E. Anthony. 1998. Predicting the clinical wear performance of orthopaedic bearing surfaces. In *Alternative bearing surfaces in total joint replacement*. J.J. Jacobs and T.L. Craig, Eds. West Conshohoken: American Society for Testing and Materials.

Schmalzried T.P., E.F. Shepherd, F.J. Dorey, et al. 2000. The John Charnley award. Wear is a function of use, not time. *Clin Orthop* 381:36–46.

Schmalzried T.P., F.J. Dorey, and H. McKellop. 1998. The multifactorial nature of polyethylene wear in vivo. *J Bone Joint Surg Am* 80:1234–1242, discussion 1242–1243.

Wang A., A. Essner, V.K. Polineni, et al. 1997. Wear mechanisms and wear testing of ultra-high molecular weight polyethylene in total joint replacements. *Polyethylene Wear in Orthopaedic Implants Workshop, Society for Biomaterials* 22:4–18.

Wang A., A. Essner, V.K. Polineni, et al. 1998. Lubrication and wear of ultra-high molecular weight polyethylene in total joint replacements. *Tribology International* 31:17–33.

Waugh W. 1990. The plan fulfilled 1959–1969. In *John Charnley: The man and the hip*. London: Springer-Verlag.

Willert H.G. 1977. Reactions of the articular capsule to wear products of artificial joint prostheses. *J Biomed Mater Res* 11:157–164.

Willert H.G., H. Bertram, and G.H. Buchhorn. 1990. Osteolysis in alloarthroplasty of the hip. The role of ultra-high molecular weight polyethylene wear particles. *Clin Orthop* 258:95–107.

Zahiri C.A., T.P. Schmalzried, E.S. Szuszczewicz, and H.C. Amstutz. 1998. Assessing activity in joint replacement patients. *J Arthroplasty* 13:890–895.

Chapter 5. Reading Comprehension Questions

5.1. Which of the following complications has not been associated with the production UHMWPE wear debris in hip replacements?
a) Aseptic loosening
b) Revison surgery
c) Osteolysis
d) Infection
e) All of the above

5.2. What is the osteolysis threshold?
a) Wear debris size above which osteolysis rarely occurs
b) Time interval during which osteolysis rarely occurs
c) Wear rate below which osteolysis rarely occurs
d) Number of wear particles above which osteolysis rarely occurs
e) All of the above

5.3. What design feature allowed Charnley to detect in vivo wear from radiographs?
a) Grooves in the backside of the cup
b) Metallic wire marker in the backside of the cup
c) Acrylic bone cement
d) Metallic femoral head
e) All of the above

Table 5.4.

Clinical data to be used for answering Questions 5.4 to 5.7.

Patient Study Number	Date of Index Radiograph	Date of Follow-up Radiograph	Femoral Head Penetration from Index Radiograph (mm)	Femoral Head Penetration at Follow-up Radiograph (mm)
1	2/17/2000	2/16/2001	0.03	0.3
2	2/19/2000	2/18/2002	0.00	0.25
3	2/26/2000	2/25/2001	0.05	0.38
4	3/4/2000	3/4/2003	0.05	0.4
5	3/12/2000	9/10/2003	0.00	0.5
6	3/18/2000	3/18/2003	0.03	0.44
7	3/19/2000	5/31/2003	0.05	0.34
8	3/25/2000	8/18/2003	0.03	0.2
9	3/26/2000	3/26/2002	0.00	0.4
10	3/31/2000	5/6/2001	0.03	0.33

5.4. You are a clinical researcher studying radiographic wear of UHMWPE in a group of 10 patients. The relevant clinical data is listed in Table 5.4. Which patient has the longest time interval between radiographs?
a) Patient #5
b) Patient #10
c) Patient #3
d) Patient #6
e) Patient #2

5.5. Which of the patients listed in Table 5.4 has the greatest change in femoral head penetration during the follow-up period?
   a) Patient #1
   b) Patient #2
   c) Patient #4
   d) Patient #6
   e) Patient #8

5.6. Which of the patients listed in Table 5.4 has the highest radiographic wear rate?
   a) Patient #1
   b) Patient #3
   c) Patient #5
   d) Patient #7
   e) Patient #9

5.7. What is the median wear rate for the group of patients listed in Table 5.4?
   a) 0.05 mm/y
   b) 0.07 mm/y
   c) 0.14 mm/y
   d) 0.21 mm/y
   e) 0.28 mm/y

5.8. Three-dimensional wear rate is expressed in which units of measurement?
   a) mm
   b) mm/y
   c) mm^2/y
   d) mm^3/y
   e) mm^4/y

5.9. In addition to digitized radiographs, the wear measurement method of Martell and Berdia (1997) requires the prior knowledge of which variable to calculate a patient's wear rate?
   a) Femoral head size
   b) Acetabular shell diameter
   c) Femoral neck diameter
   d) Patient weight
   e) All of the above

Chapter 6

Alternatives to Conventional UHMWPE for Hip Arthroplasty

Introduction

During the 1980s and early 1990s, aseptic loosening and osteolysis emerged as major problems in orthopedics that were perceived to limit the longevity of joint replacements (NIH Consensus Statement 1994). As discussed in Chapter 5, it was not until the early 1990s that the production of ultra-high molecular weight polyethylene (UHMWPE) debris at the articulating surface of joint replacements was widely recognized to play a central role in initiating osteolysis (Harris 1991, Harris 1994, NIH Consensus Statement 1994, Peters et al. 1992, Schmalzried et al. 1994, Willert, Bertram, and Buchhorn 1990). Since that time, orthopedic research efforts have focused increasingly on improving UHMWPE for joint replacements, with the goals of reducing wear and, by implication, improving implant survival, especially for young active patients. Throughout this chapter, we will refer to UHMWPE that has been irradiated with the historical sterilization dose of 25 to 40 kGy as "conventional" material, whereas UHMWPE that has been irradiated with a dosage higher than 40 kGy will be termed "highly crosslinked."

It has been estimated that more than 90% of total hip replacements (THRs) implanted worldwide since the 1990s have incorporated either a conventional or highly crosslinked UHMWPE insert (Muratoglu and Kurtz 2002). In addition to highly crosslinked UHMWPE, other alternative bearing solutions incorporating metal-on-metal (MOM) or ceramic-on-ceramic (COC) articulations have received increased attention in the past 10 years because of their ultra-low wear rates in hip simulator studies (Clarke et al. 2000), even when clinical factors such as subluxation are taken into account. In the laboratory, current MOM and COC bearings have been reported to reduce the production of wear debris by *two to three orders of magnitude* relative to conventional UHMWPE.

The first generation of alternative bearing technologies has been available since the 1950s and 1970s, but MOM and COC joints were eventually superceded in orthopedics by Charnley's low-friction metal-on-UHMWPE design. Today, however, MOM and COC have received increased attention as a result of the potential for current designs to substantially reduce the wear rate of arthroplasties, especially for young active patients. Nevertheless, the proliferation of current MOM and COC alternative bearing technologies for hip replacement has been limited due, in part, to stringent manufacturing requirements, contributing to substantially higher cost relative to designs incorporating UHMWPE. The survivorship of MOM and COC designs is especially sensitive to implantation technique, and thus these alternatives are perceived as less forgiving than UHMWPE bearings for an orthopedic surgeon who may perform only a few total hip arthroplasty (THA) procedures per year. Finally, widespread clinical adoption of MOM and COC has also been limited because of additional unique risks related to long-term toxicity (in MOM bearings) and implant fracture (in COC bearings), which are not encountered with hip replacements incorporating UHMWPE. Viewed primarily for young and highly active patients, alternative MOM and COC bearings are not expected to replace UHMWPE bearings in the near future. The reader is referred to a recent review for additional information about the materials, designs, and risks associated with first- and second-generation MOM and COC designs (Muratoglu and Kurtz 2002).

This chapter first summarizes MOM and COC alternative bearing designs and some of the unique risks associated with their use. The history of MOM bearings is particularly noteworthy, because it predates the use of UHMWPE in artificial hip joints. We also review the use of ceramics as a counter face in articulations with UHMWPE. For all practical purposes, however, highly crosslinked UHMWPE remains the most widely used alternative to conventional UHMWPE in orthopedics today. Thus, this chapter also summarizes the development of highly crosslinked and thermally stabilized UHMWPE and describes the characteristics of the most prevalent alternative to conventional UHMWPE in joint arthroplasty.

Metal-on-Metal Alternative Hip Bearings

Although a variety of metallic biomaterials have been employed in joint arthroplasty, alloys of cobalt, 28% chromium, and 6% molybdenum (CoCr) are viewed as the gold standard for use in MOM bearings. CoCr alloy is also considered the gold standard as a femoral head material for articulations against conventional as well as highly crosslinked UHMWPE (Muratoglu and Kurtz 2002, Sauer and Anthony 1998). CoCr alloys (e.g., Vitallium®) have been used for hip replacements since 1938, when the biomaterial was employed in the Smith-Petersen mold arthroplasty (Smith-Petersen 1948).

Interestingly, earlier (unsuccessful) iterations of the mold arthroplasty, dating from 1923 to 1937, employed glass, Pyrex, viscaloid, and bakelite (Smith-Petersen

1948), prior to the use of CoCr alloy. An example of a metallic Smith-Petersen mold arthroplasty (with its original packaging) is shown in Figure 6.1 (photographed at the Thackray Medical Museum, Leeds, England).

The Smith-Petersen implant is an early example of a hemiarthroplasty, in which only the femoral side of the hip joint is replaced. By contrast, in a total hip joint replacement, both the femoral and acetabular surfaces are replaced. The radiograph, shown in Figure 6.1, was obtained at the Charnley Museum at Wrightington Hospital (Wigan, England) and illustrates this early design.

The production of CoCr for use in implants has changed considerably since the days of the Smith-Petersen. CoCr implants were originally produced solely by investment casting, whereas higher strength and ductility CoCr can be achieved today by forging or by use of wrought CoCr (Table 6.1). In particular, wrought CoCr alloys have superior hardness, yield strength, and ultimate strength relative to cast CoCr because of the more uniform carbide microstructure (Muratoglu and Kurtz 2002).

Today, MOM implants are fabricated from wrought CoCr alloys with either a high-carbon (>0.20%) or low-carbon (<0.05%) composition. The specifications for CoCr medical grade alloys (cast, wrought, and forged) are covered in American Society for Testing and Materials (ASTM) standards F75, F799, and F1537 (Table 6.1). For more details about the composition and microstructure of CoCr alloys used in orthopedics, the reader is referred to several recent reviews and book chapters (Dearnley 1999, Lemons 1991, Muratoglu and Kurtz 2002, Park 1995, Varano et al. 1998).

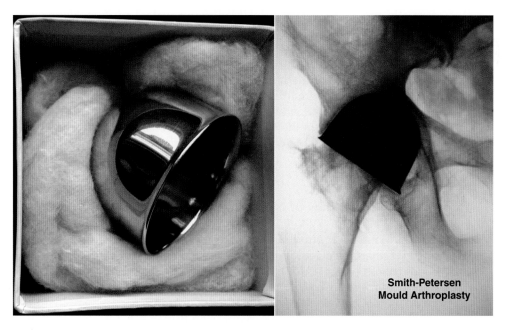

Smith-Petersen
Mould Arthroplasty

Figure 6.1

An example of a metallic Smith-Petersen mold arthroplasty (with its original packaging).

Table 6.1

Minimum American Society for Testing and Materials Specifications for Mechanical Properties of Cast, Forged, and Wrought CoCr Alloys for Use in Implants

Condition	ASTM standard	Yield strength (MPa)	Ultimate tensile strength (MPa)	Ultimate elongation (%)	Hardness, HRC, typical
Casting	F75	450	655	8	NA
Forged	F799	827	1172	12	35
Wrought and annealed	F1537	517	897	20	25
Wrought and hot worked	F1537	700	1000	12	28
Wrought and warm worked	F1537	827	1172	12	35

MOM bearings have been documented with very low long-term clinical wear rates. At a consensus development meeting in 1996, orthopedic researchers generally agreed that MOM joints wear clinically at a rate of 1–5 mm^3/year (Amstutz et al. 1996), which is 10 to 100 times lower than the typical wear rate for conventional CoCr/UHMWPE bearing couple. Generally speaking, the *in vivo* wear rates are too low to be detected using the radiographic techniques currently employed for CoCr/UHMWPE hip replacements, as were described in Chapter 5. Instead, clinical wear rates for MOM joints have generally been obtained from geometric measurements of retrieved historical and contemporary components, performed using high-precision coordinate measurement machines (CMM) (Kothari, Bartel, and Booker 1996, McKellop et al. 1996, Nevelos et al. 1999, Sieber, Rieker, and Kottig 1999, Willert and Buchhorn 1999).

Further details about first-generation (historical) and contemporary MOM designs are provided in the following two sections. In a third section, we also address the potential biological risks associated with long-term implantation of MOM hip replacements.

Historical Overview of Metal-on-Metal

In 1938, Wiles is reported to have performed the first hip arthroplasty, consisting of stainless steel femoral and acetabular components that were fixed to the bone without cement (1957). Although many clinical records of Wiles' patients were lost in World War II, some radiographs of Wiles' prosthesis can still be found at the Charnley Museum, as shown in Figure 6.2.

Between the 1950s and 1970s, pioneering British surgeons like McKee (1966) and Ring (1968) developed MOM joint replacements fabricated from CoCr alloy.

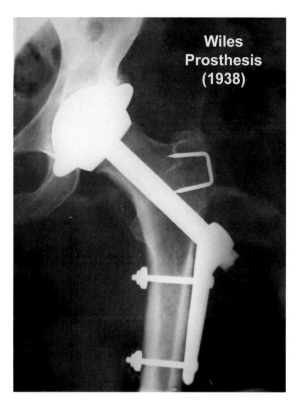

Figure 6.2
Radiograph of Wiles' prosthesis found at the Charnley Museum at Wrightington Hospital (Wigan, England).

McKee's first designs of a MOM joint from the 1950s employed screw fixation. Figure 6.3 shows an early McKee design from 1957.

Later versions of McKee's design, referred to as the McKee-Farrar prosthesis, were clinically introduced in the 1960s. The McKee-Farrar prosthesis employed cement fixation, whereas the Ring prosthesis, developed in the 1960s, employed screw fixation. The photographs of the McKee-Farrar and the Ring prosthesis, shown in Figure 6.4, were taken from implants in the collection at the Thackray Medical Museum.

In the case of first-generation MOM designs, problems with design tolerances and manufacturing of the articulations have been cited as contributing to bearing seizure and early loosening (Kothari, Bartel, and Booker 1996). Surgical factors may also have contributed to poor short-term survivorship of first-generation MOM designs (Zahiri et al. 1999). Issues related to the potential for metallic hypersensitivity, as well as the (unknown) carcigenicity associated with the long-term exposure to metal wear debris, has been consistently raised with early MOM joint replacements (Jacobs et al. 1996, Tharani, Dorey, and Schmalzried 2001, Willert 1977, Willert et al. 1996).

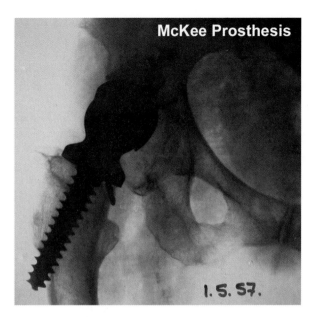

Figure 6.3

Radiograph is an early McKee design from 1957 obtained at the Charnley Museum at Wrightington Hospital (Wigan, England).

Contemporary (Second-Generation) Metal-on-Metal Hip Designs

Starting in the 1980s, members of the clinical community in Europe became intrigued by observations of long-term successful survivorship among some first-generation MOM designs (Muller 1995). Studies in England and Scandinavia, published in the 1980s and 1990s, suggested that the long-term survivorship of McKee-Farrar prostheses is comparable to the Charnley designs (August, Aldam, and Pynsent 1986, Jacobsson, Djerf, and Wahlstrom 1990, Jacobsson, Djerf, and Wahlstrom 1996).

Second-generation MOM designs were clinically introduced during 1988 by Sulzer Orthopedics (currently Centerpulse Orthopedics, Winterthur, Switzerland) (Muller 1995, Wagner and Wagner 1996, Wagner and Wagner 2000, Weber 1999). Sulzer's design was approved for marketing in the United States by the Food and Drug Administration (FDA) in August 1999. Between 1988 and 2000, it is estimated that 125,000 of these second-generation MOM components have been implanted worldwide (Muratoglu and Kurtz 2002).

Sulzer's second-generation MOM designs incorporated a CoCr articulating surface, but the acetabular component consisted of a modular shell and an UHMWPE liner embedded with a CoCr insert. A cross-section of the contemporary "sandwich" type design, distributed under the trade name METASUL (Centerpulse Orthopedics, Winterthur, Switzerland) is illustrated in Figure 6.5.

UHMWPE continues to be used in second-generation MOM and, as we shall see, in certain COC designs as well, primarily as a means for achieving

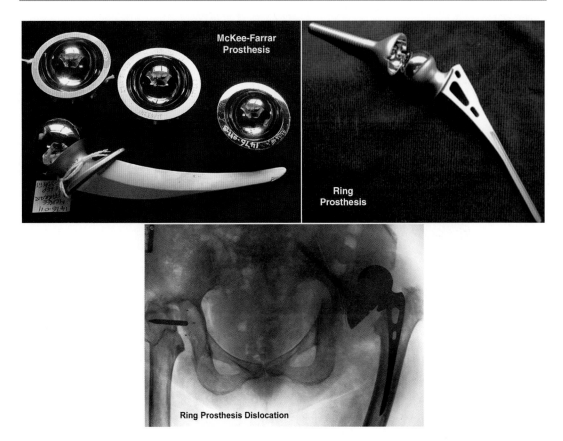

Figure 6.4
McKee-Farrar and the Ring prosthesis taken from implants in the collection at the Thackray Medical Museum, along with a radiograph of a dislocated Ring prosthesis obtained from the Charnley Museum at Wrightington Hospital (Wigan, England).

implant fixation. However, UHMWPE is also used in these alternative bearing designs for the objective of preserving intraoperative modularity (Schmidt, Weber, and Schon 1996). According to Rieker and colleagues, "This embedded solution was chosen to assure complete compatibility with the shells already commercially available (same operative technique and instruments)" (1999). Also, by incorporating UHMWPE into the bearing design, the same acetabular shell could be used for a wider range of liner designs, both conventional and alternative.

Other contemporary MOM designs, employing a modular taper-fit connection between the CoCr insert and the metal shell, have also been clinically introduced by companies such as Biomet (Warsaw, IN), Wright Medical (Arlington, TN), and Smith & Nephew (Memphis, TN). Unlike the METASUL design, these other taper-lock modular MOM designs do not incorporate an interpositional UHMWPE layer.

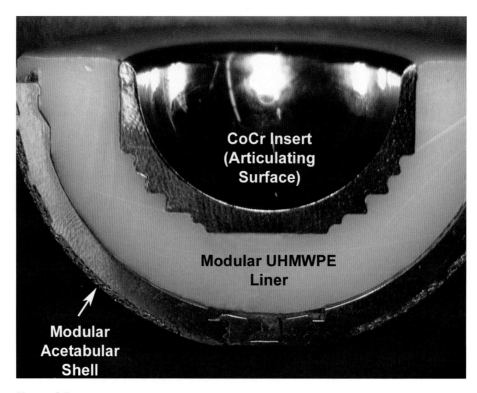

Figure 6.5
A cross-section of the contemporary "sandwich" type design, distributed under the trade name METASUL (Centerpulse Orthopedics, Winterthur, Switzerland).

Potential Biological Risks Associated with Metal-on-Metal Joints

Despite the ultra-low wear rates afforded by second-generation MOM hip implants, concerns remain about the potential health risks associated with long-term metal ion exposure. The wear particles in MOM articulations range between 6 nm and 5 mm (Doorn et al. 1998, Doorn, Campbell, and Amstutz 1999, Firkins et al. 2001, Fisher et al. 1999). Because of their smaller wear particle size, MOM hip implants have been estimated to release about 100 times the number of wear particles than conventional UHMWPE hip implants (Doorn, Campbell, and Amstutz 1999, Firkins et al. 2001). In particular, the nanometer-sized metallic wear particles are more easily digested by cells, bound into proteins, and/or dissolved into body fluids than the larger UHMWPE wear particles (Brodner et al. 1997, Fisher et al. 1999, Jacobs et al. 1996, Jacobs et al. 1998, Jacobs et al. 1999).

The wear products of MOM joint articulation are transported systemically and are manifested in elevated chromium and cobalt levels in a patient's serum and urine, raising the potential risk for carcinogenesis (Jacobs et al. 1996, Jacobs et al. 1998, Jacobs et al. 1999). However, epidemiological studies of cancer risk in patients with MOM remain inconclusive, because of the relatively small patient populations evaluated, the Scandinavian basis of the studies, and the typically rare incidence of the disease (Tharani, Dorey, and Schmalzried 2001, Visuri et al. 1996). There have also been reports of metal hypersensitivity associated with the implantation of MOM prostheses, but the incidence of this complication is reported to be extremely rare (Willert, Buchhorn, and Fayyazi 2003). In summary, for the reasons outlined earlier, the orthopedics community continues to study the biological and carcinogenic implications of metallic wear debris, which are not fully understood at the present time. Because of ongoing clinical concern, researchers are continuing to monitor the long-term health effects associated with MOM alternate bearings.

Ceramics in Hip Arthroplasty

Ceramic materials have been used successfully, as both femoral head and acetabular components, in hip replacements since the 1970s (Boutin 1971, Boutin 1972, Shikata et al. 1977). During the past decade, a new generation of high performance ceramic materials—as well as improved modular acetabular component designs—have been developed to address the historical limitations of ceramic-on-UHMWPE and COC alternative bearings. Most importantly, from the perspective of reducing wear debris–induced osteolysis, modern ceramic bearings are associated with ultra-low wear rates in contemporary hip simulator studies (Clarke et al. 2000).

In contrast with the debris produced from the articulation of MOM bearings, the particles produced by ceramic components are considered to be biologically inert. Thus, COC bearings do not share the same potential biological limitations associated with MOM articulations for hip replacement. In contrast, the principal limitation of current COC implants is perceived to be the risk of *in vivo* component fracture.

In this section, we outline the history of ceramics in orthopedics, and provide an introductory overview of ceramic materials relevant to hip replacement. This section also discusses the use of ceramic femoral heads as bearing surfaces with UHMWPE, and covers current designs of COC alternative bearings. The final part of this section contains an overview of ceramic fracture risk in historical, as well as in current, ceramic materials.

Historical Overview of Ceramics in Total Hip Arthroplasty

The application of ceramic materials in hip arthroplasty has its origins in Europe and Japan. Pierre Boutin, in collaboration with Ceraver, Inc., from France, first reported on the clinical results of COC hip arthroplasty in 1971 and 1972. In 1977, Shikata in Japan introduced the concept of using alumina femoral heads with UHMWPE acetabular components.

The first-generation of COC components consisted of monolithic acetabular component, fabricated entirely from alumina ceramic (Al_2O_3), articulating against an alumina ceramic femoral head, fitted using a tapered interlock to the metallic femoral stem. COC designs developed by Sedel in Paris (Sedel et al. 1990), and by Mittelmeier from Germany (1984), also employed Al_2O_3/Al_2O_3 bearing surfaces. These early designs incorporated tapered screw threads and/or spikes on the backside of the acetabular component for initial fixation to the pelvis. These early COC implants were at first implanted without cement. Although a wide range of COC designs were investigated in Europe during the 1970s and 1980s, only the Mittelmeier design was clinically released (as the Autophor design) in significant numbers within the United States during the 1980s.

Monolithic alumina acetabular components, including those of the Autophor or Mittelmeier design, were generally associated with unacceptable levels of loosening (Mahoney and Dimon 1990, Winter et al. 1992). Because of the bio-inertness of the ceramic material, it proved to be more difficult to achieve long-term stable fixation with uncemented monolithic ceramic cups, as compared with metal-backed UHMWPE acetabular component designs. Migration or tilting of the monolithic ceramic acetabular component in some cases con-tributed to edge loading and a variety of undesirable late complications, includ-ing edge loading, accelerated wear, and impingement (Winter et al. 1992). *In vivo* fracture of the ceramic components was also a significant problem associated with first-generation COC components (Winter et al. 1992).

However, those early COC components that survived the hurdles of fracture, migration, and loosening exhibited extremely low clinical wear rates (Dorlot, Christel, and Meunier 1989, Jazrawi et al. 1999). The *in vivo* wear rates for first-generation COC components range between 3–9 µm/year for linear wear (Bizot et al. 2000, Dorlot, Christel, and Meunier 1989) and 1–5 mm^3/year for volumetric wear (Nevelos et al. 1993). As with MOM designs, the radiographic wear assessments for COC components are complicated by the very low wear rates, and thus direct measurements from retrievals has generally been reported in the literature (Dorlot, Christel, and Meunier 1989, Nevelos et al. 1993).

Alumina and Zirconia Ceramic Materials

The potential for extremely low clinical wear rates, necessary to reduce the risk of osteolysis, has led to renewed interest in developing new COC designs for hip arthroplasty during the 1990s (Boehler, Plenk, and Salzer 2000). In addition, the desire to reduce the fracture risk has led to continuous improvement of ceramic materials for orthopedic load bearing applications over the past since the early 1980s (Table 6.2).

The ceramic materials used in orthopedics are formed by fusing or sintering microscopic grains of alumina (Al_2O_3) and/or zirconia (ZrO_2) ceramic powder into a consolidated product. The processing of ceramic parts for hip replace-ment is now performed in accordance with a wide range of international quality standards and stringent FDA regulatory requirements (Dobbs 2003). The manufacturing stages for ceramic components generally include an isostatic pressing step into near net (or "green") shape, a firing step in a furnace for

Table 6.2

Various Properties of Alumina and Zirconia Ceramics Used in the Total Hip Reconstruction

Property	Alumina in 1970s	Alumina in 1980s	Alumina in 1990s	Zirconia	Alumina composite
Bending strength (MPa)	>450	>500	>550	>900	>1000
Fracture toughness $(MPa.m^{1/2})$	4	4	4	8	5.7
Vickers hardness (0.1)	1800	1900	2000	1250	1975
Grain size (micron)	4.5	3.2	1.8	<0.5	<1.5 (alumina matrix)
Young's modulus (GPa)	380	380	380	210	350

From Merkert 2003, Willmann 1998, Willmann 2000.

final sintering of the powder, and several high-precision machining and/or polishing steps to achieve the desired surface finish (Griesmayr et al. 2003). The strength of the ceramic component depends on the purity and size of the granular powder particles, as well as upon the powder composition (i.e., alumina versus zirconia). Figure 6.6 shows an example of sintered alumina microstructure, in which the average grain size is 2 μm or less.

Ceramics are brittle, polycrystalline solids. For this family of materials, the fracture resistance is intimately related to the size and distribution of internal flaws or defects, which may occur near grain boundaries. Thus, reducing the grain size of alumina ceramics over the past 30 years (Table 6.2) has resulted in an overall improvement in its fracture strength.

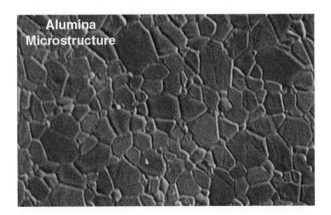

Figure 6.6

The micrograph (provided courtesy of CeramTec AG, Plochingen, Germany) shows an example of sintered alumina microstructure, in which the average grain size is 2 μm or smaller.

Despite brittle characteristics, ceramic components enjoy several outstanding tribological properties, including their hardness (Table 6.2), which contributes to wear resistance and scratch resistance. Ceramic surfaces are also more hydrophilic than the CoCr surfaces of a femoral head, as illustrated in Figure 6.7 by the water droplets. The improved wettability of ceramics contributes to lower friction than CoCr when articulated against UHMWPE under physiologic loading and lubrication conditions (Morlock et al. 2002).

There are three types of ceramics that are relevant to THR, including alumina, zirconia, and alumina matrix composites. Alumina has the longest history of successful use in articulations with UHMWPE and against itself in COC bearings. For detailed reviews of alumina materials used in orthopedics, the reader is referred to recent review articles (Willmann 1998). Currently, CeramTec AG (Plochingen, Germany) is the world's largest supplier of medical grade alumina ceramic, which is distributed under the trade name BIOLOX® Forte. CeramTec has reported that 1.3 million BIOLOX Forte femoral heads have been sold for hip arthroplasty between 1995 and 2002 (Willmann 2003). An additional 350,000 ceramic inserts for COC components have been sold by CeramTec during the same period (Willmann 2003).

Zirconia ceramic has been used in orthopedic applications since 1985 (Piconi and Maccauro 1999), due largely to its higher strength relative to alumina (Table 6.2). However, zirconia is considerably more complex than alumina. The reader is referred to several excellent reviews summarizing the properties of zirconia ceramic biomaterials (Cales 2000, Piconi and Maccauro 1999). Zirconia is a metastable ceramic, consisting of monoclinic, tetragonal, and cubic phases (Cales 2000). Under a combination of temperature, humidity, and stress, zirconia can undergo a phase transformation from tetragonal to monoclinic, which results in a volume change (Cales 2000, Willmann 1998). This phase transformation can have desirable consequences, such as the generation of a compressive stress field at the tip of a propagating crack, resulting in crack growth resistance.

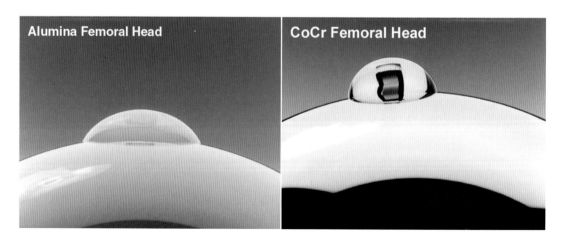

Figure 6.7

Ceramic surfaces are also more hydrophilic than the CoCr surfaces of a femoral head, shown by the water droplets in the pictures (provided courtesy of CeramTec AG, Plochingen, Germany).

This material characteristic is referred to as "phase transformation toughening" (Cales 2000). However, the phase-induced volumetric transformation can also have disastrous consequences if not properly controlled. For this reason, the phase transformation property of zirconia is typically stabilized with the addition of magnesia or yttria (Willmann 1998). The most common type of zirconia used in orthopedics is termed Y-TZP, corresponding to yttria stabilized-tetragonal phase, polycrystalline zirconia (Willmann 1998). The chemical composition of Y-TZP is approximately 5.1% yttria (Y_2O_3) and 93–94% zirconia (ZrO_2) (Cales 2000).

Until recently, St. Gobain Desmarquest (Vincennes, France) was the world's largest supplier of medical grade stabilized zirconia ceramic. Between 1985 and 2001, the company reported that it has sold a total of 500,000 components fabricated from this material under the trade name Prozyr® (Saint-Gobain Céramiques Avancées Desmarquest 2002a). Starting in 2000, Desmarquest received reports of an unusually large number of fractures from components that were fabricated in early 1998, when the company implanted a change in manufacturing processes from use of a batch furnace to a continuous kiln or belt furnace (Saint-Gobain Céramiques Avancées Desmarquest 2002b). These zirconia fractures were particularly troubling, because they involved components that complied with applicable technical standards. During August 2001, Desmarquest announced a voluntary recall of nine batches of femoral heads, all fabricated with a belt furnace during the late 1990s (Saint-Gobain Céramiques Avancées Desmarquest 2002b). The recall had global implications because the femoral heads were distributed to more than 51 companies worldwide (United States Food and Drug Administration 2001). Recall announcements soon followed from national regulatory agencies. On September 13, 2001, the FDA warned that "surgeons should not implant artificial hips with the St. Gobain zirconia ceramic heads manufactured since the processing change in 1998." The company suspended distribution of zirconia ceramics for orthopedic applications in August 2001 (Saint-Gobain Céramiques Avancées Desmarquest 2002b). Since the recall, zirconia has fallen out of favor as an orthopedic bearing biomaterial, regardless of the manufacturer.

The 2001 recall has not, however, dampened the general enthusiasm for ceramic materials in orthopedics. Alumina is currently the ceramic material of choice for orthopedic applications, either for articulations with UHMWPE or for use in COC alternate bearings. Starting at the end of 2002, a new alumina composite material (BIOLOX Forte: CeramTec, Plochingen, Germany) has been available as a femoral head material (Merkert 2003). This ceramic composite, consisting of 75% alumina matrix, is reinforced by 25% zirconia. The improved strength of this new ceramic composite, in comparison with alumina and zirconia, is summarized in Table 6.2. Clinical studies are still needed to determine the effectiveness and reliability of this new biomaterial.

Ceramic on UHMWPE

As summarized in the previous section, laboratory tests of alumina and zirconia ceramics have demonstrated substantial advantages over CoCr for articulation against UHMWPE. The harder ceramic surfaces should theoretically be more scratch resistant than CoCr femoral heads (Cuckler, Bearcroft, and Asgian 1995, Lancaster et al. 1997). Unfortunately, it has thus far proven difficult to measure the clinical benefit of ceramic femoral heads in patients.

Sugano and colleagues (1995) reported on the use of first-generation alumina ceramic femoral heads (Bioceram: Kyocera Corporation, Kyoto, Japan) in a series of 57 hips in 50 patients, who were implanted between 1981 and 1983. After 10 years of follow-up, they observed an average wear rate of 0.1 mm/year. Because a control group was not included in this study, the authors did not compare with the wear observed using CoCr heads in a comparable patient population. A similar limitation has been noted in previous retrospective clinical studies reporting wear rates for zirconia against UHMWPE, which show an overall average wear rate of 0.1 mm/year (Cales 2000). However, a wear rate of 0.1 mm/year is typically observed in the literature as an average clinical wear rate for CoCr when used with UHMWPE (Dumbleton, Manley, and Edidin 2002).

Sychterz and associates (2000) compared the *in vivo* wear of a cohort of 81 patients implanted with alumina femoral heads, with a well-matched control group of 43 patients implanted CoCr femoral heads. At 7 years follow-up, the radiographic wear rate for patients with alumina was 0.09 ± 0.07 mm/year; the wear rate in the control group was 0.07 ± 0.04 mm/year. The authors concluded that the wear rates of UHMWPE hip replacements using alumina and CoCr femoral heads were similar.

Hendrich and colleagues (2003) reported on the largest matched series of patients with 28 mm diameter alumina (n = 100) and CoCr (n = 109) femoral heads using the Harris-Galante acetabular component design. The alumina components in this case were identified as BIOLOX (CeramTec). Despite the identical acetabular component designs and femoral head sizes, there were significant differences in follow-up period, which averaged 8 years for the CoCr group and 5 years for the alumina group. The average wear rates were 0.14 ± 0.11 mm/year for components with CoCr femoral heads, and 0.13 ± 0.08 mm/year for components with the alumina femoral heads. This difference was not statistically significant (p = 0.46). Furthermore, a power analysis showed that the clinical study was powered sufficiently to detect a 0.038 mm/year difference (28%) in wear rates with 80% power.

Although the clinical studies reported thus far in the literature have been retrospective in nature, overall the research would suggest that ceramic heads do not substantially improve the wear rate of UHMWPE acetabular components. It may be that prospective, randomized trials with large numbers of patients will be necessary to detect the relatively small difference in wear rates with ceramic and CoCr femoral heads in THA.

Contemporary Ceramic-on-Ceramic Hip Implants

Today's manufacturers of second-generation COC alternative bearings claim to have addressed the issue of implant migration in contemporary COC designs with the use of modular acetabular components incorporating a metal backing having some type of biocompatible coating (Bohler et al. 2000). Short-term studies of migration in contemporary COC designs have yielded mixed results.

Bohler and colleagues (2000) reported on a series of 73 modular hip arthroplasties with identical stems and metal-backed acetabular components, except for a ceramic or UHMWPE insert. Radiographic analysis of these patients during the first 3 years of implantation revealed significantly higher vertical migration of the acetabular components with ceramic inlays in older (>60 years) and osteoporotic patients. However, a more recent randomized study with 53 patients used radiostereometric analysis (RSA) techniques to measure changes in acetabular lilt and migration after 2 years of follow-up (Schwämmlein et al. 2002). The researchers concluded that there was "no marked difference in outcomes" between acetabular components with an UHMWPE or ceramic liner. Thus, the potential for migration in current COC designs remains a somewhat controversial topic, and longer-term studies are still needed to address this issue.

Modern COC implants are modular, and the articulating surfaces have been fabricated from high-purity alumina (e.g., BIOLOX Forte). The acetabular insert may be fabricated entirely from alumina, and fitted to the metal shell with a taper junction. Alternatively, the alumina insert may be embedded within an UHMWPE liner. Currently, the majority of COC bearings are fabricated with a tapered insert design, rather than using the sandwich design (Willmann 2003). Both types of COC designs are depicted in Figure 6.8.

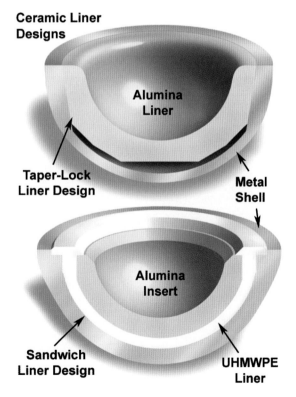

Figure 6.8
Contemporary COC designs (provided courtesy of CeramTec AG, Plochingen, Germany).

COC hip designs are clinically available throughout Europe and Asia, but have only recently been approved for marketing in the United States because of restrictions by the FDA. Starting during the 1990s, the FDA required implant manufacturers to conduct clinical studies to demonstrate the safety of their second-generation COC designs. Recently, the FDA required CeramTec (currently the sole supplier of ceramic components to the U.S. market) to maintain the same quality systems as orthopedic manufacturers (Dobbs 2003), due in part to the manufacturing issues that were associated with the 2001 recall of zirconia femoral heads. In February 2003, two orthopedic companies (Howmedica Osteonics, Mahwah, NJ; Wright Medical Technology, Arlington, TN) were granted permission by the FDA to market COC designs in the United States. Thus, COC implants have undergone particularly rigorous regulatory scrutiny during the past decade before their release by the FDA and the general clinical introduction to the American orthopedic community.

In Vivo Fracture Risk of Ceramic Components for Total Hip Replacement

The defining issue with ceramic components for hip replacement remains their *in vivo* fracture risk. The *in vivo* fracture of a ceramic component is a serious complication requiring immediate revision. The revision of a fractured femoral head may be complicated. As shown in Figure 6.9, a ceramic femoral head

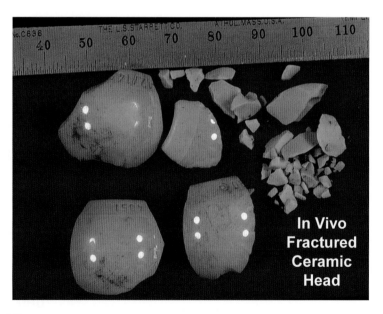

Figure 6.9
A ceramic femoral head typically fractures into multiple fragments.

typically fractures into multiple fragments, which may be difficult for the revising surgeon to completely clean from the surrounding tissues.

Furthermore, the ceramic fragments typically damage the Morse taper of the femoral component, requiring the use of a CoCr head for revision. In some cases, the taper is sufficiently damaged by *in vivo* abrasion by ceramic fracture fragments that the entire stem must be revised. Because of the potential for trunnion damage, surgeons are cautioned not to implant a new ceramic head during a revision if the stem is not also going to be revised (Krikler and Schatzker 1995).

The early clinical experience with COC in Europe was associated with high catastrophic fracture rates. The historical fracture rates ranged between 0% and 13% in some studies (Willmann 2003). However, the *in vivo* fracture rates within the overall patient population for contemporary alumina hip components have been reported by CeramTec to range between 1 and 3% in 10,000 (Willmann 2003). The risk of cement fracture is thus anticipated to be much lower relative to other clinical risks, which have been associated with all joint replacements, including loosening and infection. Other clinical complications of THA typically occur at a rate of 1% or higher.

Similar to alumina, zirconia also historically enjoyed a low overall failure rate, until the events of 2001 forced St. Gobain Desmarquest to suspend its international sales of orthopedic ceramic products. For example, in one of the groups of recalled zirconia femoral heads, a total of 227 *in vivo* fractures were reported for a series of 683 manufactured heads, corresponding to a fracture rate of 33%, according to the data made publicly available by St. Gobain (2003). Due to the possibility that an unforeseen manufacturing change may significantly influence the clinical performance of ceramics, the FDA has taken the unprecedented step of requiring alumina ceramic component suppliers to comply with the strict manufacturing and design controls that historically have been applied only to orthopedic companies. Thus far, only one ceramic manufacturer (CeramTec) has produced COC components in compliance with the stringent quality requirements of the FDA.

Highly Crosslinked and Thermally Stabilized UHMWPE

Since 1998, highly crosslinked and thermally stabilized UHMWPE has become the most widely used alternative to conventional UHMWPE for hip arthroplasty. During the late 1990s, it was realized that irradiation above the typical sterilization dose range of 25 to 40 kGy could substantially improve the wear performance of UHMWPE (Kurtz et al. 1999, McKellop et al. 1999, Muratoglu et al. 2001, Muratoglu and Kurtz 2002, Wang et al. 1998). Hip joint simulator studies indicated more than 90% reduction in wear for a dose of 100 kGy. Doses higher than 100 kGy offered diminished returns in terms of additional wear reduction. The wear reductions were the result of the increased crosslinking resulting from the higher radiation dose. On this basis, highly crosslinked

UHMWPE was introduced as a total hip bearing surface. The high wear resistance of highly crosslinked UHMWPE offers the possibility of avoiding osteolysis related to the accumulation of wear particles.

In the next section, we first briefly review the historical experience of highly crosslinked UHMWPE for hip replacement and summarize the general characteristics of contemporary materials in current clinical use. The next section also describes the effect of thermal treatment on the properties of this family of materials. In the final part of this section, we summarize the latest short-term clinical results using highly crosslinked and thermally treated UHMWPE. For more detailed information about specific highly crosslinked formulations, see Chapter 15.

Historical Clinical Experience with Highly Crosslinked UHMWPE

During the 1970s, Oonishi and colleagues in Japan (Oonishi 1995, Oonishi, Takayama, and Tsuji 1992, Oonishi, Takayama, and Tsuji 1995) and Grobbelaar and colleagues in South Africa (1978) implanted highly irradiated UHMWPE hip cups. These formulations of UHMWPE were produced using doses of up to 1000 kGy in air, by Oonishi, and up to 100 kGy in the presence of acetylene, by Grobbelaar. During the 1980s, a small clinical study of 22 patients was conducted by Wroblewski and associates using an experimental, chemically crosslinked high-density polyethylene (HDPE) (1996). Because of the small number of patients, lack of matched controls, and a substantial number of lost to follow-up, these three clinical studies of highly crosslinked polyethylene provide only anecdotal confirmation of the benefits of increased crosslinking.

Contemporary Highly Crosslinked and Thermally Treated UHMWPEs

Starting in 1998, orthopedic manufacturers introduced highly crosslinked UHMWPEs for THR. These materials are processed with a total dose ranging from 50 to 105 kGy, depending on the manufacturer. Besides choice of dosage, each manufacturer adopted a different route for production that includes a proprietary combination of three important factors: 1) an irradiation step, 2) a postirradiation thermal processing step, and 3) a sterilization step (Figure 6.10).

The latest generation of UHMWPE for THRs has thus been thermally stabilized to inhibit oxidation and highly crosslinked to reduce wear. Although barrier packaging techniques have been developed to reduce oxidation of UHMWPE, recent studies now strongly suggest that thermal treatment of irradiated UHMWPE can also substantially reduce the concentration of free radicals resulting from ionizing radiation, thereby resulting in a polymer that is resistant to long-term oxidative degradation. Radiation crosslinking of UHMWPE has been performed using gamma and electron beam radiation with the polymer in the solid or molten state.

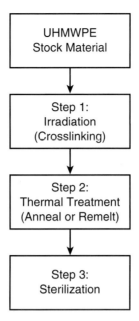

Figure 6.10
Route for production that includes a proprietary combination of three important steps.

Effect of Thermal Treatment on the Properties of Highly Crosslinked UHMWPE

Three important processing steps are necessary to produce highly crosslinked polyethylene for hip bearings. These steps are an irradiation step to promote crosslinking, an intra or postirradiation thermal processing step to increase the level of crosslinking and remove residual stress, and a sterilization step. In the irradiation step, gamma and electron beam radiation produce free radicals (unpaired electrons) in the polyethylene, which in secondary chemical reactions leads to a combination of crosslinking and chain scission. Crosslinking is beneficial for reducing wear. Chain scission produces a decrease in molecular weight, with concomitant reduction of wear resistance and mechanical properties. When irradiation is conducted in the presence of oxygen, scission predominates over crosslinking. However, when conducted in an inert environment, such as nitrogen, crosslinking predominates over scission. Regardless of whether irradiation is conducted in air or in an inert environment, some of the free radicals will remain entrapped within the crystalline phase of the UHMWPE. Over time, these entrapped free radicals can migrate to the surface of crystals. If irradiation is done in air, these free radicals react with available oxygen, causing further time-dependent chemical degradation.

Increased crosslinking improves the wear performance of UHMWPE compared with conventional material. However, the presence of the crosslinks adversely affects uniaxial ductility (Kurtz et al., 2002), and the uniaxial failure strain of UHMWPE decreases linearly with increasing radiation dosage (Kurtz et al., 2002). During irradiation, the loss of ductility depends on the crystalline microstructure of the UHMWPE, because crosslinking occurs primarily in the amorphous phase, where the molecular chains are in sufficient proximity such that a covalent bond can be created between adjacent polymer molecules by the applied energy (Muratoglu and Kurtz 2002). Unirradiated UHMWPE typically has a crystallinity in the region of 50% (Kurtz et al. 1999), so some 50% of the material is amorphous content that may be crosslinked during irradiation. If the temperature of the UHMWPE changes during the crosslinking process, this can influence the distribution of crosslinking in the polymer and, hence, influence its ability to accommodate large strains prior to failure.

The first choice an implant designer has to make is the method of crosslinking (e.g., gamma versus electron beam). If irradiation is to be carried out using electron beam irradiation, the designer must consider the additional factor of irradiation temperature, because the rate of energy dissipation increases the temperature above the melting temperature. Of the six orthopedic manufacturers currently producing highly crosslinked UHMWPE implants, two have chosen electron beam irradiation, whereas the other four use gamma radiation crosslinking. In this review, we will restrict our attention to gamma radiation crosslinking at room temperature, because it is the most widely used crosslinking modality. For more information about the differences between electron beam and gamma irradiation of UHMWPE, the reader is referred to a recent review (Muratoglu and Kurtz 2002).

In the production of a highly crosslinked UHMWPE, the material is subjected to a thermal treatment step to reduce the level of free radicals via further crosslinking reactions. At higher temperatures the polymer molecules have increased mobility, thereby increasing the probability of free radicals on adjacent chains reacting to form crosslinks. For the thermal treatment to be effective at eliminating all free radicals, it must be conducted at 150°C, above the melt temperature of the material. Heating above the melting temperature destroys the crystalline regions of the material thus making the free radicals that were in the crystals available for crosslinking. The disadvantage of melting is the reduction crystal size and in material yield and the ultimate strength that ensues. A compromise solution is to heat the material to just below the melting temperature. This solution preserves the original crystal structure, retains mechanical properties, and makes more free radicals available for crosslinking than would be available without thermal treatment while still retaining some free radicals in the crystal domains. When thermal treatment is conducted below the melt transition of 135°C, it is referred to as "annealing," and above the melt transition, it is called "remelting." Typically, annealing is carried out at 130°C and does not eliminate all free radicals, although the number is substantially reduced by the elevated temperature.

The choice of thermal treatment has a significant impact on the crystallinity and mechanical properties of highly crosslinked UHMWPE (Kurtz et al., 2002). At a dosage of 100 kGy, the elastic modulus, yield stress, and ultimate stress of a remelted material is significantly lower than the respective properties for an annealed material (Table 6.3).

Table 6.3

Effect of Postirradiation Thermal Treatment on Uniaxial Mechanical Properties. Note that these Irradiation Treatments Were Achieved with a Single Dose. Properties Were Determined from Treated Rods of GUR 1050.

Dose (gamma)	Heat treatment	Yield stress (MPa)	Ultimate stress (MPa)	Elongation to failure (%)
100 kGy	None	23.2 ± 0.2	47.6 ± 2.0	238 ± 13
100 kGy	110°C anneal	23.0 ± 0.3	47.3 ± 1.5	230 ± 12
100 kGy	130°C anneal	22.6 ± 0.2	48.5 ± 1.5	231 ± 13
100 kGy	150°C remelt	19.5 ± 0.3	43.9 ± 3.9	246 ± 12

From Kurtz et al. 2003.

Figure 6.11 compares the uniaxial tensile behavior of unirradiated UHMWPE material with conventionally sterilized (30 kGy, in N2) polyethylene, and with both annealed and remelted highly crosslinked polyethylenes (100 kGy).

For the two highly crosslinked UHMWPEs shown in Figure 6.11, the annealed material has an average degree of crystallinity of 60%, whereas the remelted material has a crystallinity of 43%. Throughout the entire stress–strain curve, the higher crystallinity of the annealed material results in a greater resistance to plastic deformation when compared with remelted material. Therefore, the

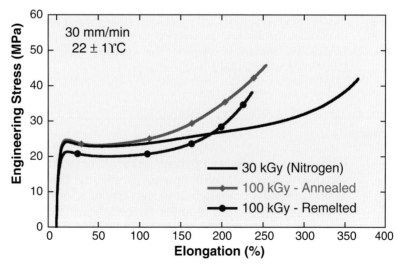

Figure 6.11

Effect of radiation and thermal processing on uniaxial tensile behavior for GUR 1050. For the highly crosslinked UHMWPE, the annealed material has an average degree of crystallinity of 60%, whereas the remelted material has a crystallinity of 43%.

selection of postirradiation thermal treatment is the second most important decision for an implant designer, because it will influence not only the free radical content, but also the crystallinity, yield strength, and ultimate tensile strength of the highly crosslinked polyethylene. These reduced mechanical properties may not influence wear but will certainly influence the resistance of the material to damage caused by impingement or bearing lift off.

Current Clinical Outlook for Highly Crosslinked UHMWPEs

The clinical performance of highly crosslinked UHMWPE is currently being tracked with anticipation by members of the orthopedic community. The short-term clinical results of three types of highly crosslinked UHMWPE materials were reported in 2002 and 2003 (Bragdon et al. 2002, Hopper et al. 2003, Nivbrant et al. 2003). In a U.S. prospective randomized trial, the two-dimensional linear wear rate for one form of highly crosslinked UHMWPE, which was irradiated with a total dosage of 100 kGy and then annealed, was found to be significantly lower (50%) than in patients implanted with conventional, gamma-sterilized UHMWPE (controls) after 2 to 3 years of follow-up (Martell, Verner, and Incavo 2003). An Australian clinical trial employing the same annealed highly crosslinked material reported an 85% reduction in wear rates at 2 years follow-up relative to controls (Nivbrant et al. 2003). Other formulations of highly crosslinked UHMWPE have also showed reportedly excellent performance at 2 years *in vivo* (Bragdon et al. 2002, Hopper et al. 2003). Despite these encouraging short-term clinical results, it is currently not known whether crosslinked UHMWPE improves the long-term clinical wear rate of hip replacements, or whether any differences in wear will, in turn, influence the clinical incidence of osteolysis and reduce the need for revision surgery. At least another decade of clinical results will be needed to evaluate whether the highly crosslinked UHMWPE materials surpass the current gold standard of conventional UHMWPE used in joint replacements.

Summary

Orthopedic surgeons and patients currently have many alternatives to the gold standard CoCr/UHMWPE bearing couple that has historically been used for THR. Currently, one alternative consists of a conventional UHMWPE liner articulating against an alumina femoral head. However, despite the many theoretical advantages demonstrated in laboratory testing, clinical studies have generally not demonstrated a significant reduction of *in vivo* wear rates associated with changing femoral head material alone.

MOM, COC, and highly crosslinked acetabular liners—used in conjunction with either a CoCr or ceramic femoral head—have the potential to significantly reduce the clinical wear rates of THRs relative to the existing gold standard of conventional UHMWPE. All three of these alternative bearings incorporate the successful elements of historical precedents. Among these three choices, highly crosslinked UHMWPE is now by far the most widely used alternative to conventional UHMWPE, especially in the United States.

The use of alternative bearings entails potential risks for the patient. With MOM bearings, the concern is the potential for cancer associated with long-term elevated metal ion exposure. With COC, the concern is the risk of fracture for the femoral head and/or the acetabular liner. With highly crosslinked UHMWPE, following extensive multi-institutional testing, researchers have not yet been able to determine the risks relative to conventional UHMWPE for hip replacements.

The three main alternatives to conventional UHMWPE represent the current state of the art in orthopedics, and are all expected to result in significant reductions in osteolysis. The ultimate goal for all of these alternative bearings is to reduce the incidence of revision for THA. Many years of clinical follow-up are still needed to verify the attainment of this long-term objective.

Acknowledgments

Special thanks are due to Professor Clare Rimnac (Case Western Reserve University) for her editorial assistance with this chapter and for many helpful discussions.

References

Amstutz H.C., P. Campbell, H. McKellop, et al. 1996. Metal on metal total hip replacement workshop consensus document. *Clin Orthop* 329 Suppl:S297–303.

August A.C., C.H. Aldam, and P.B. Pynsent. 1986. The McKee-Farrar hip arthroplasty. A long-term study. *J Bone Joint Surg* 68:520–527.

Bizot P., R. Nizard, S. Lerouge, et al. 2000. Ceramic/ceramic total hip arthroplasty. *J Orthop Sci* 5:622–627.

Boehler M., H. Plenk, Jr., and M. Salzer. 2000. Alumina ceramic bearings for hip endoprostheses: The Austrian experiences. *Clin Orthop* 379:85–93.

Bohler M., W. Schachinger, G. Wolfl, et al. 2000. Comparison of migration in modular sockets with ceramic and polyethylene inlays. *Orthopedics* 23:1261–1266.

Boutin P. 1971. Alumina and its use in surgery of the hip (experimental study). *Presse Med* 79:639–640.

Boutin P. 1972. Total arthroplasty of the hip by fritted aluminum prosthesis. Experimental study and 1st clinical applications. *Rev Chir Orthop Reparatrice Appar Mot* 58:229–246.

Bragdon C.R., G. Digas, J. Karrholm, et al. 2002. RSA evaluation of wear of conventional vs. highly crosslinked polyethylene acetabular component in vivo. *Trans American Association of Hip and Knee Surgeons* 12:23.

Brodner W., P. Bitzan, V. Meisinger, et al. 1997. Elevated serum cobalt with metal-on-metal articulating surfaces. *J Bone Joint Surg* 79:316–321.

Cales B. 2000. Zirconia as a sliding material: Histologic, laboratory, and clinical data. *Clin Orthop* 379:94–112.

Clarke I.C., V. Good, P. Williams, et al. 2000. Ultra-low wear rates for rigid-on-rigid bearings in total hip replacements. *Proc Inst Mech Eng [H]* 214:331–347.

Cuckler J.M., J. Bearcroft, and C.M. Asgian. 1995. Femoral head technologies to reduce polyethylene wear in total hip arthroplasty. *Clin Orthop* 317:57–63.

Dearnley P.A. 1999. A review of metallic, ceramic and surface-treated metals used for bearing surfaces in human joint replacements. *Proc Inst Mech Eng [H]* 213:107–135.

Dobbs H. 2003. Quality improvement resulting from legal and regulatory developments. In *Bioceramics in joint arthroplasty, 8th BIOLOX symposium proceedings*. H. Zippel and M. Dietrich, Eds. Darmstadt, Germany: Steinkopff Verlag.

Doorn P.F., P.A. Campbell, J. Worrall, et al. 1998. Metal wear particle characterization from metal on metal total hip replacements: Transmission electron microscopy study of periprosthetic tissues and isolated particles. *J Biomed Mater Res* 42:103–111.

Doorn P.F., P. Campbell, and H. Amstutz. 1999. Particle disease in metal-on-metal total hip replacements. In *METASUL: A metal-on-metal bearing*. C.B. Rieker, M. Windler, and U. Wyss, Eds. Bern, Switzerland: Hans Huber.

Dorlot J.M., P. Christel, and A. Meunier. 1989. Wear analysis of retrieved alumina heads and sockets of hip prostheses. *J Biomed Mater Res* 23:299–310.

Dumbleton J.H., M.T. Manley, and A.A. Edidin. 2002. A literature review of the association between wear rate and osteolysis in total hip arthroplasty. *J Arthroplasty* 17(5):649–661.

Firkins P.J., J.L. Tipper, M.R. Saadatzadeh, et al. 2001. Quantitative analysis of wear and wear debris from metal-on-metal hip prostheses tested in a physiological hip joint simulator. *Biomed Mater Eng* 11:143–157.

Fisher J., E. Ingham, M. Stone, et al. 1999. Wear particle morphologies in artificial hip joints: Particle size is critical to the response of macrophages. In *METASUL: A etal-on-metal bearing*. C.B. Rieker, M. Windler, and U. Wyss, Eds. Bern, Switzerland: Hans Huber.

Griesmayr G., M. Dietrich, J. Kasprowitsch, and H. Dobbs. 2003. Improvements in processing and manufacturing at Ceramtec. In *Bioceramics in joint arthroplasty, 8th BIOLOX symposium proceedings*. H. Zippel and M. Dietrich, Eds. Darmstadt, Germany: Steinkopff Verlag.

Grobbelaar C.J., T.A. Du Plessis, and F. Marais. 1978. The radiation improvement of polyethylene prostheses: A preliminary study. *J Bone Joint Surg* 60-B:370–374.

Harris W.H. 1991. Aseptic loosening in total hip arthroplasty secondary to osteolysis induced by wear debris from titanium-alloy modular femoral heads. *J Bone Joint Surg* 73:470–472.

Harris W.H. 1994. Osteolysis and particle disease in hip replacement. *Acta Orthop Scand* 65:113–123.

Hendrich C., S. Goebel, C. Roller, et al. 2003. Wear performance of 28 millimeter femoral heads with the Harris-Galante cup: Comparison of alumina and cobalt chrome. In *Bioceramics in joint arthroplasty, 8th BIOLOX symposium proceedings*. H. Zippel and M. Dietrich, Eds. Darmstadt, Germany: Steinkopff Verlag.

Hopper R.H., Jr., A.M. Young, K.F. Orishimo, and J.P. McAuley. 2003. Correlation between early and late wear rates in total hip arthroplasty with application to the performance of highly crosslinked polyethylene liners. *J Arthroplasty* 18(7 Suppl 1): 60–67.

Jacobs J.J., A.K. Skipor, P.F. Doorn, et al. 1996. Cobalt and chromium concentrations in patients with metal on metal total hip replacements. *Clin Orthop* 329 Suppl: S256–263.

Jacobs J.J., A.K. Skipor, L.M. Patterson, et al. 1998. Metal release in patients who have had a primary total hip arthroplasty. A prospective, controlled, longitudinal study. *J Bone Joint Surg Am* 80:1447–1458.

Jacobs J.J., N.J. Hallab, A.K. Skipor, et al. 1999. Metallic wear and corrosion products: Biological implications. In *METASUL: A metal-on-metal bearing.* C.B. Rieker, M. Windler, and U. Wyss, Eds. Bern, Switzerland: Hans Huber.

Jacobsson S.A., K. Djerf, and O. Wahlstrom. 1990. A comparative study between McKee-Farrar and Charnley arthroplasty with long-term follow-up periods. *J Arthroplasty* 5:9–14.

Jacobsson S.A., K. Djerf, and O. Wahlstrom. 1996. Twenty-year results of McKee-Farrar versus Charnley prosthesis. *Clin Orthop* 329:S60–68.

Jazrawi L.M., E. Bogner, C.J. Della Valle, et al. 1999. Wear rates of ceramic-on-ceramic bearing surfaces in total hip implants: A 12-year follow-up study. *J Arthroplasty* 14:781–787.

Kothari M., D.L. Bartel, and J.F. Booker. 1996. Surface geometry of retrieved McKee-Farrar total hip replacements. *Clin Orthop* 329 Suppl:S141–147.

Krikler S., and J. Schatzker. 1995. Ceramic head failure. *J Arthroplasty* 10:860–862.

Kurtz S.M., O.K. Muratoglu, M. Evans, and A.A. Edidin. 1999. Advances in the processing, sterilization, and crosslinking of ultra-high molecular weight polyethylene for total joint arthroplasty. *Biomaterials* 20:1659–1688.

Kurtz S.M., M.L. Villarraga, M.P. Herr, J.S. Bergstrom, C.M. Rimnac, and A.A. Edidin. 2002. Thermomechanical behavior of virgin and highly crosslinked ultra-high molecular weight polyethylene used in total joint replacements. *Biomaterials* 23: 3681–3697.

Kurtz S.M., C. Cooper, R. Siskey, and N. Hubbard. 2003. Effects of dose rate and thermal treatment on the physical and mechanical properties of highly crosslinked UHMWPE used in total joint replacements. *Transactions of the 49th Orthopedic Research Society* 28.

Lancaster J.G., D. Dowson, G.H. Isaac, and J. Fisher. 1997. The wear of ultra-high molecular weight polyethylene sliding on metallic and ceramic counterfaces representative of current femoral surfaces in joint replacement. *Proc Inst Mech Eng [H]* 211:17–24.

Lemons J.E. 1991. Metals and alloys. In *Total joint replacement.* W. Petty, Ed. Philadelphia: W.B. Saunders.

Mahoney O.M., and J.H. Dimon, III. 1990. Unsatisfactory results with a ceramic total hip prosthesis. *J Bone Joint Surg* 72:663–671.

Martell J., J.J. Verner, and S.J. Incavo. 2003. Clinical performance of a highly crosslinked polyethylene at two years in total hip arthroplasty: A randomized prospective trial. *J Arthroplasty* 18 (7 Suppl 1):S55–59.

McKee G.K., and J. Watson-Farrar. 1966. Replacement of arthritic hips by the McKee-Farrar prosthesis. *J Bone Joint Surg* 48:245–259.

McKellop H., S.H. Park, R. Chiesa, et al. 1996. In vivo wear of three types of metal on metal hip prostheses during two decades of use. *Clin Orthop* 329 Suppl: S128–140.

McKellop H., F.W. Shen, B. Lu, et al. 1999. Development of an extremely wear-resistant ultra high molecular weight polyethylene for total hip replacements. *J Orthop Res* 17:157–167.

Merkert P. 2003. Next generation ceramic bearings. In *Bioceramics in joint arthroplasty, 8th BIOLOX symposium proceedings.* H. Zippel and M. Dietrich, Eds. Darmstadt, Germany: Steinkopff Verlag.

Mittelmeier H. 1984. Ceramic prosthetic devices. *Hip* 146–160.

Morlock M., R. Nassutt, M.A. Wimmer, and E. Schneider. 2002. Influence of resting periods on friction in artificial hip joint articulations. In *Bioceramics in joint arthroplasty, proceedings of the 7th international BIOLOX symposium.* J.P. Garino and G. Willmann, Eds. Stuttgart, Germany: Thieme.

Muller M.E. 1995. The benefits of metal-on-metal total hip replacements. *Clin Orthop* 311:54–59.

Muratoglu O.K., C.R. Bragdon, D.O. O'Connor, et al. 2001. A novel method of cross-linking ultra-high-molecular-weight polyethylene to improve wear, reduce oxidation, and retain mechanical properties. Recipient of the 1999 HAP Paul Award. *J Arthroplasty* 16:149–160.

Muratoglu O.K., and S.M. Kurtz. 2002. Alternative bearing surfaces in hip replacement. In *Hip replacement: Current trends and controversies.* R. Sinha, Ed. New York: Marcel Dekker.

Nevelos A.B., P.A. Evans, P. Harrison, and M. Rainforth. 1993. Examination of alumina ceramic components from total hip arthroplasties. *Proc Inst Mech Eng [H]* 207: 155–162.

Nevelos J.E., E. Ingham, C. Doyle, et al. 1999. Analysis of retrieved alumina ceramic components from Mittelmeier total hip prostheses. *Biomaterials* 20:1833–1840.

NIH Consensus Statement. 1994. Total hip replacement. National Institutes of Health Technology Assessment Conference.

Nivbrant B., S. Roerhl, B.J. Hewitt, and M.G. Li. 2003. In vivo wear and migration of high crosslinked poly cups: A RSA study. *Transactions of the 49th Orthopedic Research Society* 28:358.

Oonishi H. 1995. Long term clinical results of THR. Clinical results of THR of an alumina head with a cross-linked UHMWPE cup. *Orthopaedic Surgery and Traumatology* 38:1255–1264.

Oonishi H., Y. Takayama, and E. Tsuji. 1992. Improvement of polyethylene by irradiation in artificial joints. *Radiation Physics and Chemistry* 39:495–504.

Oonishi H., Y. Takayama, and E. Tsuji. 1995. The low wear of cross-linked polyethylene socket in total hip prostheses. In *Encyclopedic handbook of biomaterials and bioengineering. Part A: Materials.* D.L. Wise, D.J. Trantolo, D.E. Altobelli, et al., Eds. New York: Marcel Dekker.

Park J.B. 1995. Metallic biomaterials. In *The biomedical engineering handbook.* J.D. Bronzino, Ed. Boca Raton, FL: CRC Press.

Peters P.C., Jr., G.A. Engh, K.A. Dwyer, and T.N. Vinh. 1992. Osteolysis after total knee arthroplasty without cement. *J Bone Joint Surg* 74:864–866.

Piconi C., and G. Maccauro. 1999. Zirconia as a ceramic biomaterial. *Biomaterials* 20:1–25.

Rieker C.B., H. Weber, R. Schön, et al. 1999. Development of the METASUL articulations. In *METASUL: A metal-on-metal bearing.* C.B. Rieker, M. Windler, and U. Wyss, Eds. Bern, Switzerland: Hans Huber.

Ring P.A. 1968. Complete replacement arthroplasty of the hip by the ring prosthesis. *J Bone Joint Surg* 50:720–731.

Sauer W.L., and M.E. Anthony. 1998. Predicting the clinical wear performance of orthopaedic bearing surfaces. In *Alternative bearing surfaces in total joint replacement.* J.J. Jacobs and T.L. Craig, Eds. West Conshohoken, PA: American Society for Testing and Materials.

Saint-Gobain Céramiques Avancées Desmarquest. 2002a. Information on breakages reported on Prozyr zirconia heads: Key dates. *http://www.prozyr.com/PAGES_UK/Biomedical/historic.htm.* Vincennes Cedex, France (Accessed: April 28, 2003).

Saint-Gobain Céramiques Avancées Desmarquest. 2002b. Information on breakages reported on Prozyr zirconia heads: Key figures. *http://www.prozyr.com/PAGES_UK/Biomedical/figures.htm.* Vincennes Cedex, France (Accessed: April 28, 2003).

Saint-Gobain Céramiques Avancées Desmarquest. 2003. Information on breakages reported on Prozyr zirconia heads: Batches & product configurations concerned by breakages. *http://www.prozyr.com/PAGES_UK/Biomedical/breakages.htm*. Vincennes Cedex, France (Accessed: April 28, 2003).

Schmalzried T.P., D. Guttmann, M. Grecula, and H.C. Amstutz. 1994. The relationship between the design, position, and articular wear of acetabular components inserted without cement and the development of pelvic osteolysis. *J Bone Joint Surg* 76:677–688.

Schmidt M., H. Weber, and R. Schon. 1996. Cobalt chromium molybdenum metal combination for modular hip prostheses. *Clin Orthop* 329 Suppl:S35–47.

Schwämmlein D., R. Schmidt, N. Schikora, et al. 2002. Migration patterns of press-fit cups with polyethylene or alumina liner—a randomized clinical trial using radiostereoanalysis. In *Bioceramics in joint arthroplasty, proceedings of the 7th international BIOLOX symposium*. J.P. Garino and G. Willmann, Eds. Stuttgart, Germany: Thieme.

Sedel L., L. Kerboull, P. Christel, et al. 1990. Alumina-on-alumina hip replacement. Results and survivorship in young patients. *J Bone Joint Surg* 72:658–663.

Shikata T., H. Oonishi, Y. Hashimato, et al. 1977. Wear resistance of irradiated UHMW polyethylenes to Al2O3 ceramics in total hip prostheses. *Transactions of the 3rd Annual Meeting of the Society for Biomaterials* 3:118.

Sieber H.P., C.B. Rieker, and P. Kottig. 1999. Analysis of 118 second-generation metal-on-metal retrieved hip implants. *J Bone Joint Surg Br* 81:46–50.

Smith-Petersen M.N. 1948. Evolution of mould arthroplasty of the hip joint. *J Bone Joint Br* 30-B:59–75.

Sugano N., T. Nishii, K. Nakata, et al. 1995. Polyethylene sockets and alumina ceramic heads in cemented total hip arthroplasty. A ten-year study. *J Bone Joint Surg* 77:548–556.

Sychterz C.J., C.A. Engh, Jr., A.M. Young, et al. 2000. Comparison of in vivo wear between polyethylene liners articulating with ceramic and cobalt-chrome femoral heads. *J Bone Joint Surg* 82:948–951.

Tharani R., F.J. Dorey, and T.P. Schmalzried. 2001. The risk of cancer following total hip or knee arthroplasty. *J Bone Joint Surg* 83-A:774–780.

United States Food and Drug Administration. 2001. Recall of zirconia ceramic femoral heads for hip implants. *http://www.fda.gov/cdrh/recalls/zirconiahip.html* (Accessed: April 28, 2003).

Varano R., S. Yue, J.D. Bobyn, and J. Medley. 1998. Co-Cr-Mo alloys used in metal-metal bearing surfaces. In *Alternative bearing surfaces in total joint replacement*. J.J. Jacobs and T.L. Craig, Eds. West Conshohocken, PA: American Society for Testing and Materials.

Visuri T., E. Pukkala, P. Paavolainen, et al. 1996. Cancer risk after metal on metal and polyethylene on metal total hip arthroplasty. *Clin Orthop* 329 Suppl:S280–289.

Wagner M., and H. Wagner. 1996. Preliminary results of uncemented metal on metal stemmed and resurfacing hip replacement arthroplasty. *Clin Orthop* 329 Suppl: S78–88.

Wagner M., and H. Wagner. 2000. Medium-term results of a modern metal-on-metal system in total hip replacement. *Clin Orthop* 123–133.

Wang A., A. Essner, V.K. Polineni, et al. 1998. Lubrication and wear of ultra-high molecular weight polyethylene in total joint replacements. *Tribology International* 31:17–33.

Weber B.G. 1999. METASUL from 1988 to today. In *METASUL: A metal-on-metal bearing*. C.B. Rieker, M. Windler, and U. Wyss, Eds. Bern, Switzerland: Hans Huber.

Wiles P. 1957. The surgery of the osteo-arthritic hip. *Br J Surg* 45:488–497.

Willert H.G. 1977. Reactions of the articular capsule to wear products of artificial joint prostheses. *J Biomed Mater Res* 11:157–164.

Willert H.G., H. Bertram, and G.H. Buchhorn. 1990. Osteolysis in alloarthroplasty of the hip. The role of ultra-high molecular weight polyethylene wear particles. *Clin Orthop* 258:95–107.

Willert H.G., G.H. Buchhorn, D. Gobel, et al. 1996. Wear behavior and histopathology of classic cemented metal on metal hip endoprostheses. *Clin Orthop* 329 Suppl: S160–186.

Willert H.G., and G.H. Buchhorn. 1999. Retrieval studies on classic cemented metal-on-metal hip endoprostheses. In *METASUL: A metal-on-metal bearing*. C.B. Rieker, M. Windler, and U. Wyss, Eds. Bern, Switzerland: Hans Huber.

Willert H.G., G.H. Buchhorn, and A. Fayyazi. 2003. Hypersensitivity to wear products in metal-on-metal articulation. In *Bioceramics in joint arthroplasty, 8th BIOLOX symposium proceedings*. H. Zippel, and M. Dietrich, Eds. Darmstadt, Germany: Steinkopff Verlag.

Willmann G. 1998. Ceramics for total hip replacement—what a surgeon should know. *Orthopedics* 21:173–177.

Willmann G. 2000. Ceramic femoral head retrieval data. *Clin Orthop* 379:22–28.

Willmann G. 2003. Fiction and facts concerning the reliability of ceramics in THR. In *Bioceramics in joint arthroplasty, 8th BIOLOX symposium proceedings*. H. Zippel and M. Dietrich, Eds. Darmstadt: Steinkopff Verlag.

Winter M., P. Griss, G. Scheller, and T. Moser. 1992. Ten- to 14-year results of a ceramic hip prosthesis. *Clin Orthop* 73–80.

Wroblewski B.M., P.D. Siney, D. Dowson, and S.N. Collins. 1996. Prospective clinical and joint simulator studies of a new total hip arthroplasty using alumina ceramic heads and cross-linked polyethylene cups. *J Bone Joint Surg* 78 B:280–285.

Zahiri C.A., T.P. Schmalzried, E. Ebramzadeh, et al. 1999. Lessons learned from loosening of the McKee-Farrar metal-on-metal total hip replacement. *J Arthroplasty* 14:326–332.

Chapter 6. Reading Comprehension Questions

6.1. Contemporary metal-on-metal hip replacements are fabricated from
a) Wrought stainless steel alloy
b) Forged titanium alloy
c) Commercially pure titanium
d) Wrought cobalt chromium alloy
e) Cast cobalt chromium alloy

6.2. Which of the following issues and concerns contributed to the abandonment of first-generation metal-on-metal components?
a) Biocompatibility
b) Carcigenicity
c) Acetabular loosening
d) Manufacturing tolerances
e) All of the above

6.3. Which of the following bearing surface combinations was first used clinically as an alternative to metal-on-UHMWPE in hip replacements?
a) Metal-on-metal
b) Alumina-on-alumina
c) Alumina-on-UHMWPE
d) Zirconia-on-zirconia
e) Zirconia-on-UHMWPE

6.4. Contemporary ceramic-on-ceramic hip replacements are fabricated from
a) Zirconia
b) Biolox Forte
c) Prozyr
d) ZrO$_2$
e) Aluminum

6.5. The primary motivation to use alternative bearings is
a) To reduce loosening
b) To reduce impingement
c) To reduce wear
d) To reduce infection
e) All of the above

6.6. Today's ceramic-on-ceramic components differ from first-generation implants in which of the following respects?
a) Reduced ceramic grain size
b) Modular metal backing
c) Porous or bioactive coatings at the bone-implant interface
d) Increased ceramic density
e) All of the above

6.7. Highly crosslinked UHMWPE materials were first clinically introduced in the
a) 1960s
b) 1970s
c) 1980s
d) 1990s
e) 2000s

6.8. One of the reasons for thermally treating highly crosslinked UHMWPE is to
a) Reduce fracture
b) Reduce osteolysis
c) Reduce wear
d) Reduce oxidation
e) Reduce crosslinking

Chapter 7

The Origins and Adaptations of UHMWPE for Knee Replacements

Introduction

Knee arthroplasty, referring to surgical reconstruction of the knee joint, has its origins in the late 19th century as the treatment for severe joint degeneration resulting from tuberculosis (Robertsson 2000, Robertsson et al. 2000). In 1890, Gluck—from the Charité hospital in Berlin—described his design of a fixed hinged knee replacement with components fashioned from ivory (1890). These overly constrained hinged knee replacements suffered from short-term failure, and Gluck later retracted his endorsement of this surgical procedure. In the early twentieth century, attempts at knee arthroplasty also involved the implantation of autogenous tissue (such as muscle fascia), as well as chromicized pig's bladder, to serve as articulating surfaces of a reconstructed knee joint (Speed and Smith 1940).

Ultra-high molecular weight polyethylene (UHMWPE) has been used in knee replacements since the late 1960s, when Frank Gunston developed a cemented implant design at Wrightington Hospital (1971, 1973). This early knee replacement resurfaced the individual condyles of the femur and the tibia. Total knee arthroplasty (TKA), which replaces the articulation between the femur and tibia, as well as between the femur and the patella, was developed in the 1970s, primarily at surgical centers in North America. The basic anatomical landmarks and implant features of a typical total knee replacement (TKR) are illustrated in Figure 7.1.

Contemporary knee arthroplasty includes a broad range of surgical procedures, which are tailored by the physician for the specific needs of the patient. For patients with mild arthritis, which is confined to one of the condyles of the

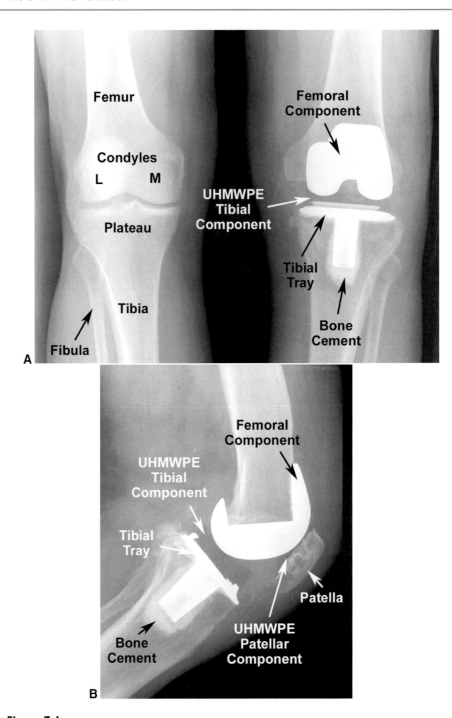

Figure 7.1

(**A**) Anterior–posterior and (**B**) lateral radiographic views of an Insall/Burstein II TKR, with associated anatomical landmarks and implant terminology. In A, the femoral condyles are designated as medial (M) and lateral (L).

knee, the surgeon might decide to perform a unicondylar knee arthroplasty (UKA). If both condyles of the knee are diseased, but the patella remains intact, the surgeon may perform a bicondylar (also referred to as bicompartmental) TKA. When both condyles, as well as the patello–femoral joint are diseased, a tricompartmental total knee replacement (illustrated in Figure 7.1) is performed. If only the patella is diseased, a surgeon might opt to implant a patellar component (this procedure is also referred to as patellar resurfacing). Finally, in the case of extreme circumstances, such as a salvage revision operation or in the event of tumor resection, a semiconstrained hinged knee design might be employed. In all of these procedures, UHMWPE plays a primary role as a polymeric component, articulating either against a metallic component, or in some cases (such as in a patellar resurfacing) the UHMWPE may articulate against cartilage.

The manner in which UHMWPE was fundamentally adapted for TKA evolved most quickly during the 1970s. In 1975, Ewald (1975) wrote, "The problem we are faced with today is to select the best design among the 300 total knee prostheses currently commercially available or in the process of development around the world." Although the main adaptations of UHMWPE for TKA were firmly established by the end of the 1970s, the 1980s, and 1990s were still associated with continuous incremental improvements in the design of the femoral, tibial, and patellar components to address recurring problems with positioning and loosening. During the 1980s and 1990s, there were also major strides in the surgical instrumentation and the techniques used to implant the artificial knee components, which, when coupled with improved implant designs and fixation methods, have contributed to improved survivorship.

Because so many different surgeons and engineers have designed knee replacements since the 1970s, the history of TKA is much more complex than THA (Vince 1994, Walker 1977). In this chapter, we start with tracing the origins of how UHMWPE came to be used in knee arthroplasty when Frank Gunston worked at Wrightington. The remaining sections of this chapter focus specifically on five fundamental adaptations of UHMWPE for knee replacement since the 1970s (Figure 7.2). These five evolutionary stages for UHMWPE in TKA include: 1) Gunston's initial design concept for the polycentric TKR, which replaced both condyles of the femur individually; 2) the adaptation of Gunston's design to UKA for carefully selected groups of patients; 3) the evolution to a bicondylar total knee, in which the tibial and femoral components were joined for ease of insertion and anatomical positioning; 4) the resurfacing of the patello–femoral joint; and 5) the incorporation of metal backing in the design of UHMWPE components. It should be emphasized that all of these major evolutionary steps in the clinical application of UHMWPE for TKA were initiated in the 1970s, even if the final embodiments of these design concepts did not reach fruition until the following decades. Even today, surgeons and biomechanical engineers continue to debate and refine their understanding of these fundamental adaptations of UHMWPE for knee arthroplasty.

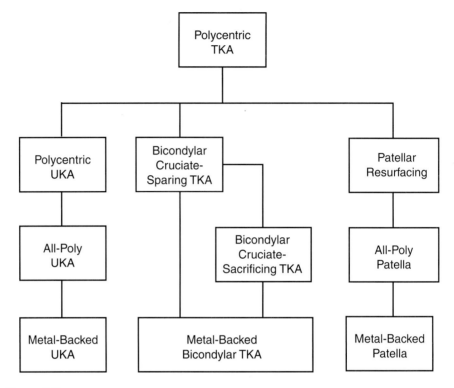

Figure 7.2

Flowchart schematic illustrating the major adaptations of UHMWPE for unicondylar, bicondylar, and patello–femoral replacements.

Frank Gunston and the Wrightington Connection to Total Knee Arthroplasty

In the 1960s, the mainstream treatments for knee arthritis included fusion, replacement with a metallic hinged prosthesis, or implantation of a metallic tibial resurfacing component (Figure 7.3). Hinged prostheses, like the Shiers or the Walldius knee for example, were state-of-the-art at that time (Figure 7.3A, B). These fixed hinge knee prostheses required substantial bone resection and were associated with problems of loosening because of their overconstraint and their inability to accommodate internal–external rotation. Metallic tibial resurfacing components, like the MacIntosh and McKeever prosthesis (Figure 7.3C, D), as well as the Townley resurfacing prosthesis, were also employed during this period (Walker 1977). Unfortunately, the knee implant solutions available in the mid-1960s were fraught with high complication rates and unacceptable long-term functionality.

UHMWPE was introduced for knee arthroplasty at the same place, and at around the same time, as it was introduced for hip replacement. As we have seen in Chapter 4, John Charnley introduced UHMWPE for hip arthroplasty

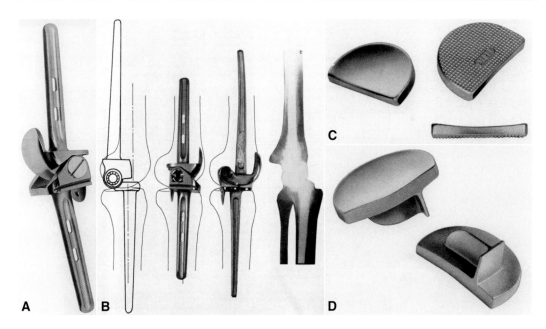

Figure 7.3

Examples of knee arthroplasty during the 1960s. (**A**) Walldius hinged knee replacement; (**B**) the Shiers, Walldius, and Guépar hinged knee replacements, superimposed over the anatomy of the knee (reprinted with permission from Walker P.S. 1977. *Human joints and their artificial replacements.* Springfield, IL: CC Thomas Publisher); (**C**) MacIntosh tibial plateau; and (**D**) McKeever tibial plateau.

in November 1962 at Wrightington Hospital, in Lancashire, England. By the mid-1960s, hip arthroplasty using UHMWPE had become routine at Wrightington. Although Charnley's hip implants were not widely available during the 1960s, Wrightington nevertheless quickly evolved into a training center for orthopedic surgeons, who traveled worldwide to learn the latest techniques in hip arthroplasty.

In 1967, Frank H. Gunston, an orthopedic surgeon from Winnipeg, Canada, was granted a traveling fellowship to study hip arthroplasty at Wrightington. Gunston was initially trained as an engineer, and he was especially attracted to Wrightington because of its machine shop and unique experimental facilities. Like all of the visiting registrars, Gunston learned hip arthroplasty by assisting with the hip surgeries being performed at Wrightington. He also helped with Charnley's ongoing projects related to *in vitro* testing of hip replacements.

During his fellowship at Wrightington, Gunston was struck by the problem of treating rheumatoid arthritis patients, who were afflicted at both the hip and the knee. These patients continued to be debilitated after their hip replacement because of their ongoing knee arthritis. In addition, for these rheumatoid patients, the pain relief associated with their hip arthroplasty made them dissatisfied with the prevailing treatment options for knee arthritis.

In this context, Gunston developed a design for knee arthroplasty reflecting his exposure to UHMWPE and hip arthroplasty at Wrightington (Figure 7.4).

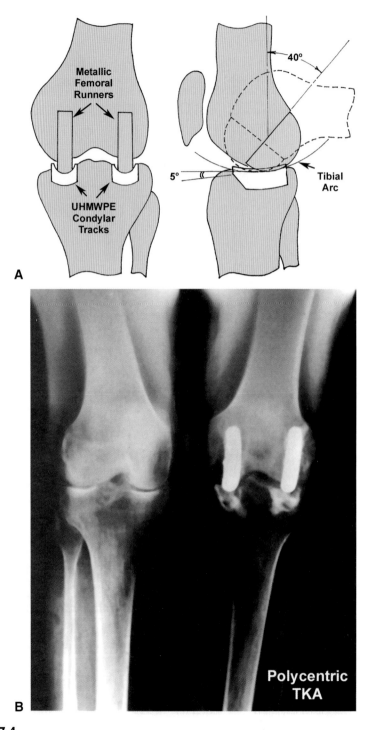

Figure 7.4
(**A**) Schematic of the Polycentric TKA design by Frank Gunston. (Adapted from Gunston 1971.)
(**B**) Radiograph of a Polycentric TKA.

Gunston's design incorporated two separate condylar replacements, each consisting of a convex metallic component (or "runner"), which was implanted on the posterior aspect of the femur, and a concave UHMWPE component (or "track"), which was implanted in the tibia. The implants were cemented into place without disturbing the cruciate ligaments.

In 1971, Gunston wrote, "The biomechanical principles and experience gained from total hip arthroplasty were combined with an analysis of normal knee movement to determine a solution [for knee replacement]." Gunston himself machined the first UHMWPE components for knee arthroplasty out of the RCH 1000 material that was available in the machine shop at Wrightington. In his 1971 paper, Gunston acknowledges Charnley "for his continued encouragement and the use of the facilities of the Center for Hip Surgery." Later, when Gunston returned from his fellowship to begin his orthopedic practice at Winnipeg, he continued to machine all of his own UHMWPE tibial components for his use, as well as for colleagues who requested them.

Although Gunston was clearly influenced by his fellowship experience at Wrightington, Charnley himself did not actively participate in the design of the first artificial knee. Charnley's interests at the time were firmly directed toward improving hip replacement surgery. In 1970, after Gunston had returned to Canada, Charnley developed his own independent TKR design, which was also intended for patients at Wrightington with rheumatoid arthritis. Charnley's knee design, which was distributed by Thackrays in the 1970s as the Load Angle Inlay, had a convex UHMWPE component articulating against a flat metallic tibial plateau. However, Charnley's knee design, although unique, was not successful and never became widely adopted.

Polycentric Knee Arthroplasty

Unlike previous hinged prostheses, Gunston's design attempted to incorporate the complex kinematics of the knee joint. Gunston recognized that during flexion, the femur rolls and slides posteriorly back across the tibial condyles about successive instant centers of rotation (Figure 7.5). In his 1971 paper, Gunston described the motion of the knee during flexion as rotation about a "Polycentric pathway." Consequently, Gunston's knee design was named the "Polycentric."

Gunston decided to publish his experience with the Polycentric rather than patent the design. In 1969, at a Canadian orthopedic meeting, Gunston met Lowell Peterson from the Mayo Clinic (Rochester, MN). Subsequently, Peterson's colleague, Richard Bryan, traveled to Winnipeg in December 1969 to study the procedure and to further evaluate Gunston's clinical results (Bryan and Peterson 1979).

Gunston freely shared his drawings with Bryan and Peterson, and after some design modifications, the first Polycentric TKA was performed at the Mayo Clinic by July 1970. The first 81 Polycentric knees at the Mayo Clinic were fabricated by the hospital (Bryan and Peterson 1979). However, until 1978, the vast majority of the 1938 Polycentric components implanted at the Mayo Clinic were manufactured by orthopedic companies, such as Howmedica (Rutherford, NJ).

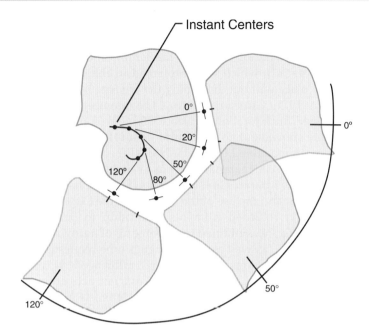

Figure 7.5
Instant centers of rotation during flexion of the natural knee joint. (Adapted from Gunston 1971.)

The Polycentric design was further modified by researchers at Mayo in 1973 to provide wider UHMWPE tracks (Bloom and Bryan 1977). Additional refinements of the Polycentric design, such as the introduction of UHMWPE runners with variable height, were proposed by Cracchiolo (1973). A review by Peterson and associates in 1975 provides an instructive summary of the different types of Polycentric knee components, which were available from implant companies such as Howmedica and DePuy at the time.

The UHMWPE material used for Polycentric components was RCH 1000. Gunston sterilized the UHMWPE tibial components he personally manufactured using an autoclave. However, orthopedic companies like Howmedica, which fabricated most of the Polycentric components implanted during the 1970s, sterilized their UHMWPE components with gamma radiation and used air-permeable packaging.

Proper placement of the four implant components was a major challenge for Polycentric TKA, and even with the specialized instruments that were developed in the 1970s, intraoperative alignment could be difficult (Figure 7.6). Polycentric TKA was initially performed on rheumatoid patients. With experience, the physicians at Mayo started implanting in patients with other forms of arthritis. Because of the shape of the components, Polycentric TKA was contraindicated in patients with severe deformity of the femur or tibia.

Short-term results of the Polycentric were highly encouraging, especially considering the improved mobility of most patients, who had previously

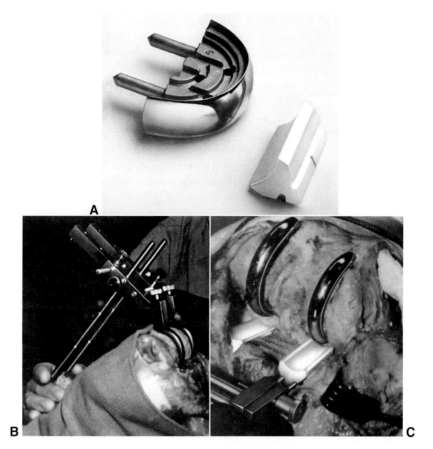

Figure 7.6

Geometry (**A**) and implantation technique (**B, C**) of Polycentric TKA components.

been debilitated by knee pain (Bryan and Peterson 1973, Bryan, Peterson, and Combs, Jr. 1973a, Bryan, Peterson, and Combs, Jr. 1973b, Gunston 1971, Gunston 1973). Unfortunately, loosening, infection, subsidence of the implant components, and instability proved to be important complications for Polycentric TKA. In 1976, Gunston and MacKenzie reported on 89 patients, which were followed from 2 to 7.5 years. They reported nine cases of loosening (10%) and six cases of infection (6%). Although the Polycentric provided pain relief and improved mobility for 81% of their patients, the amount of flexion that could be achieved with this design was generally lower than what would be needed for stair climbing and getting out of a chair. For instance, the average range of motion for Gunston's patients was 91 degrees before the operation, but was 89.9 degrees after Polycentric TKA.

Wear and fatigue damage were observed on UHMWPE tracks of the Polycentric TKA, but the observations of surface damage were typically associated with suboptimal placement of the components. Shoji and colleagues (1976), for example, observed evidence up to 5.3 mm wear, fatigue cracks, and "crater like defects" in retrieved UHMWPE tracks from Polycentric TKA. The four components analyzed by Shoji and associates (1976) were implanted 2 years or fewer. Noting the generally positive short-term outcomes of the Polycentric, the authors advised that proper surgical technique, not the performance of the UHMWPE per se, was critical to avoiding future failures of this design.

In 1984, Lewallen and colleagues reported that the 10-year survivorship for the Polycentric TKA at the Mayo Clinic was 66%, with a 13% incidence of instability due to ligament laxity, 7% incidence of loosening, 3% incidence of infection, and 4% incidence of tibiofemoral joint pain. Although Polycentric TKA has not been clinically used since the 1970s, it can still be viewed in many respects as the great-grandfather of current knee designs. As we shall see, the fundamental applications for UHMWPE in contemporary TKA have their origins in the Polycentric, if only in response to the problems encountered with this pioneering design.

Unicondylar Polycentric Knee Arthroplasty

Surgeons at the Mayo Clinic started using Polycentric UKAs beginning in 1971 (Skolnick, Bryan, and Peterson 1975). This procedure used identical implant components as the TKR version of the Polycentric. Stolnick and colleagues (1975) described the 1-year results for 14 knees in 13 patients in 1975. By 1979, Bryan and Peterson (1979) had reported that 338 Polycentric UKAs had been performed at the Mayo Clinic.

At the Mayo Clinic, Polycentric UKA was performed only on patients with unicondylar disease (osteoarthritis) (Jones et al. 1981, Skolnick, Bryan, and Peterson 1975). An initial assessment was based on radiographic screening. However, the UKA implantation technique involved the same exposure as in TKA, so that both compartments could be directly examined and evidence of unicompartmental disease visually confirmed. Mallory also reported the utility of Polycentric UKA in the treatment of post-traumatic arthritis resulting from fracture-induced deformity of the knee (1973).

The main feature of Polycentric UKA was the relief of pain and restoration of mobility in the majority of patients (Mallory 1973, Skolnick, Bryan, and Peterson 1975). In Stolnick and colleagues' study, the average range of motion was 119 degrees before the procedure and 116 degrees after (1975).

Because the implants and the procedure were identical, Polycentric UKA can be considered a special case of the Polycentric TKA. The primary difference is in the patient selection. Although superceded in the 1970s by other UKA designs, such as the Geomedic, Savastano, and Marmor prostheses (Marmor 1988), the Polycentric UKA is nonetheless the earliest example for the use of UHMWPE to treat unicompartmental disease in the knee.

Examples of first-generation cemented designs from the mid-1970s, intended specifically for UKA, are shown in Figures 7.7, 7.8, 7.10G, 7.12. These UKA designs differ from the Polycentric in several key respects. First, the tibial–femoral contact is less constrained than the rail-in-track geometry of the Polycentric.

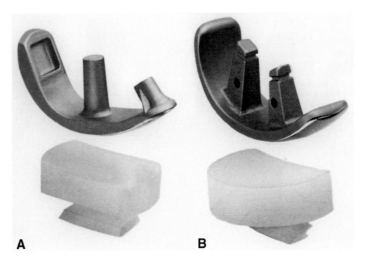

Figure 7.7
First-generation, cemented unicondylar knee designs from the mid-1970s. (**A**) Geomedic;
(**B**) Savastano unicondylar hemiknee prostheses.

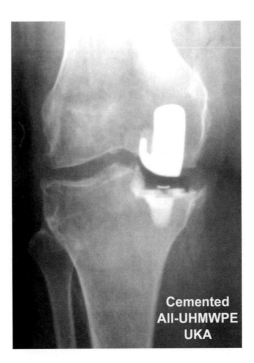

Figure 7.8
Radiograph of a knee implanted with the HSS unicondylar prosthesis, which included an all-UHMWPE tibial component. (Image provided courtesy of Professor Clare Rimnac, Case Western Reserve University, Cleveland, OH.)

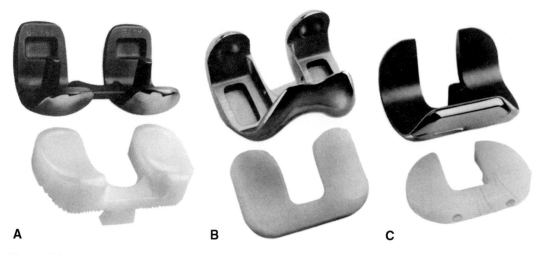

A   B   C

Figure 7.9
Cruciate-sparing, bicondylar knee designs from the mid-1970s. (**A**) Geomedic; (**B**) Townley; and (**C**) Freeman-Swanson total knee prostheses.

In addition, the tibial component is broader, providing greater coverage of the plateau during internal–external rotation. Finally, the femoral component also provides greater coverage of the femoral condyle than the Polycentric, enabling contact throughout a larger range of motion. Further details about the clinical performance of UHMWPE in UKA can be found in Chapter 8.

Bicondylar Total Knee Arthroplasty

Bicondylar knee replacements evolved from difficulties with implanting two sets of condylar prostheses. Problems with surgical positioning of four individual femoral and tibial components prompted implant designers to physically join the separate compartments on the tibial and femoral side, respectively. Bicondylar knee replacements can be classified as "cruciate sparing" or "cruciate sacrificing" depending on whether the posterior cruciate ligament (PCL) was excised during the installation of the UHMWPE tibial component. In this section, we highlight some of the features of early bicondylar knee designs, which were introduced in the 1970s (Figures 7.9–7.13).

Cruciate-Sparing Bicondylar Prostheses

The evolutionary step from the Polycentric knee to bicondylar knee arthroplasty is perhaps best appreciated in the "Geometric" knee design, developed by a group of five surgeons at the Mayo Clinic and clinically introduced in April 1971 (Skolnick et al. 1976). The design was marketed under the Geomedic

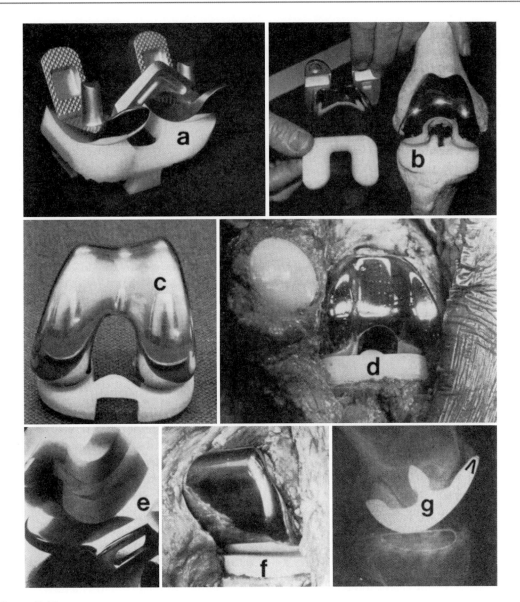

Figure 7.10
Knee arthroplasty designs from the 1970s. (**A**) Modified geometric; (**B**) Townley; (**C**) Leeds; (**D**) Total Condylar; (**E**) Charnley load-angle inlay; (**F**) Freeman-Swanson; (**G**) Marmor modular. (Reprinted with permission from Walker P.S. 1977. *Human joints and their artificial replacements*. Springfield, IL: CC Thomas Publisher.)

trade name (Howmedica, Rutherford, NJ) (Figures 7.9A and 7.11A). Like the Polycentric, the Geomedic design preserved the PCL structures of the knee, with only an anterior UHMWPE bar joining the two condyles.

Although the two components of Geometric knee were somewhat easier to implant than the four components of the Polycentric knee, long-term fixation

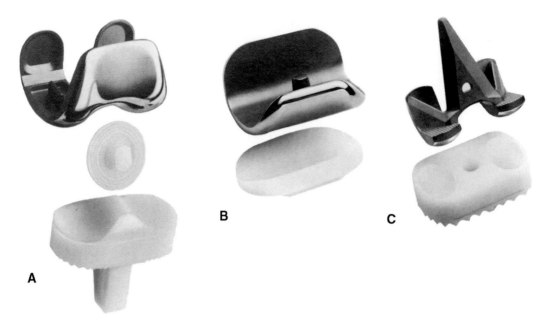

Figure 7.11

Cruciate-sacrificing, bicondylar knee designs from the mid-1970s. (**A**) Total Condylar; (**B**) Freeman-Swanson; and (**C**) Fowler total knee prostheses.

continued to present an important problem for the Geometric design, as well as for other all-UHMWPE tibial components that were developed in this period. Under radiographic examination, radiolucent lines of at least 1 mm were observed in the cement layer underneath the tibial component in 38% of long-term implanted Geometric knees (Rand and Coventry 1988). Using revision or moderate-to-severe pain as the end point, Rand and Coventry (1988) reported that the 10-year survivorship of the Geometric knee was 69%.

During the 1970s, many other types of bicondylar cruciate-retaining TKR designs were developed, as reviewed by Walker (1989). The Townley knee, developed in 1972, is another example of a well-known first-generation, PCL-sparing knee design from this period (Townley 1985). Figure 7.11 provides examples of three cruciate-retaining designs from a 1975 Howmedica catalog; Walker's book provides additional examples of designs from this time period (Figure 7.9). Although the geometry of these designs varied, they provided greater coverage of the condylar surfaces than the original Polycentric dual condylar components conceived by Gunston.

The Total Condylar Knee

Knee implant designers in the 1970s did not universally agree on the need to preserve the cruciate ligament, especially in light of problems encountered with tibial component fixation. In addition, many failures of bicondylar cruciate-sparing designs, like the Polycentric, resulted from instability and ligament laxity during the progression of arthritis. By sacrificing the PCL, implant designers could not only geometrically resurface the ends of the tibia and femur, thereby gaining improved fixation, but they also had the potential to correct anatomical deformities produced by disease and old age. Also, it was not clear that the PCL continued to function competently in elderly patients.

A team of surgeons and engineers, including Peter Walker, John Insall, and Chitranjan Ranawat, from the Hospital for Special Surgery (HSS), developed the Total Condylar knee arthroplasty in 1973 (Walker 1977, Insall, Tria, and Scott 1979, Insall et al. 1976) (Figures 7.10D, 7.11A, 7.12, and 7.13). This tricompartmental

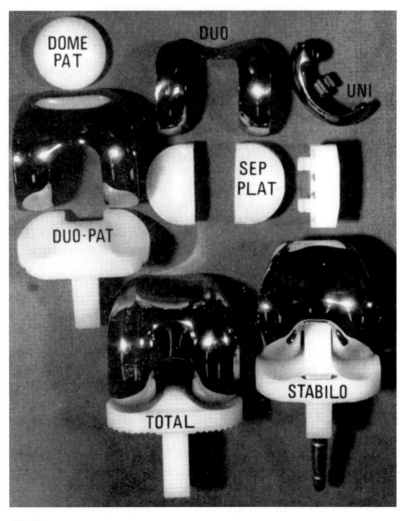

Figure 7.12
Family of unicondylar and bicondylar knee prostheses designed at the HSS by Insall, Ranawat, and Walker during the 1970s. (Reprinted with permission from Walker P.S. 1977. *Human joints and their artificial replacements.* Springfield, IL: CC Thomas Publisher.)

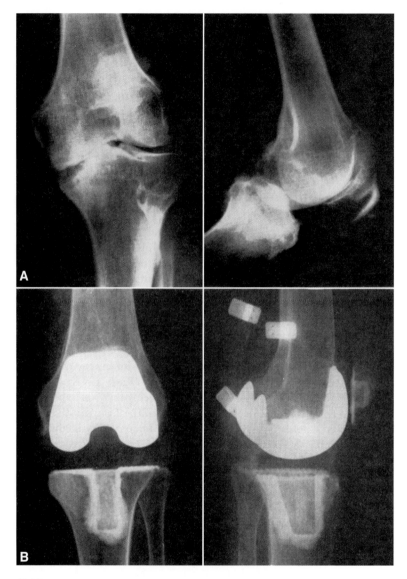

Figure 7.13
Preoperative (**A**) and postoperative (**B**) radiographs of a Total Condylar knee prosthesis. The components are well positioned. (Reprinted with permission from Walker, P.S. 1977. *Human joints and their artificial replacements.* Springfield, IL: CC Thomas Publisher.)

design advanced the state of the art of knee design in several respects, most notably by developing the procedure, instrumentation, and components necessary to simultaneously replace both the tibiofemoral compartments and the patello–femoral compartment. All of the implant components in this design were cemented into place. The geometry of the UHMWPE tibial component included an innovative stem to improve cemented fixation and to reduce

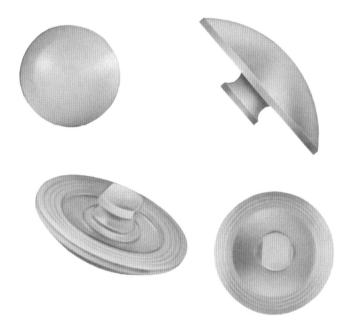

Figure 7.14
UHMWPE patellar component designs for the Total Condylar knee from the mid-1970s.

subsidence into the femur. In addition, the femoral component included a wide anterior flange for articulation with the UHMWPE patellar button (Figures 7.11A, 7.12, and 7.14).

The surgeon–designers have reported impressive clinical results with the Total Condylar Prosthesis (TCP), especially when compared with the outcomes from previous knee designs (Insall and Kelly 1986, Insall, Scott, and Ranawat 1979, Insall, Tria, and Scott 1979, Insall et al. 1976, Ranawat, Rose, and Bryan 1984, Scuderi et al. 1989, Vince, Insall, and Kelly 1989). The first cohort of patients implanted with the Total Condylar knee, between March 1974 and December 1977, have been followed closely for up to 20 years (Ranawat et al. 1993, Rodriguez, Bhende, and Ranawat 2001, Scuderi et al. 1989, Vince, Insall, and Kelly 1989). Depending on whether evidence of radiographic loosening or clinical failure was used for the endpoint, Ranawat and colleagues reported a 91–94% survivorship at 15 years (1993).

The TCP design was modified several times in the 1970s by the surgeon–inventors, giving rise to a family of TCPs (e.g., TCP I, II, III, etc.) (Insall, Tria, and Scott 1979, Walker 1977). According to Insall and colleagues, "The Total Condylar knee prosthesis II was developed because, at times, the articular geometry of the Total Condylar Prosthesis did not provide sufficient anterior–posterior stability, and a few cases of posterior subluxation of the tibia occurred" (1979). For one of the designs (referred to by Insall as the TCP II), a special type of cruciate-sacrificing knee replacement, incorporating a vertical UHMWPE post into the UHMWPE tibial component, was devised to provide greater joint stability, as well as to extend the range of motion during flexion activities (Figure 7.15).

The posterior–stabilized (PS) Total Condylar Prosthesis II (TCP II) was clinically introduced in 1978 (Figure 7.15) (Insall, Lachiewicz, and Burstein 1982). In this design, the vertical UHMWPE post of the tibial component makes contact with a horizontal cam when the knee is loaded and flexed, such as during stair climbing or rising from a chair. The PS design resulted in marked improvements in functional capabilities of joint replacement patients. The average postoperative range of motion for the PS design was 115 degrees (Insall, Lachiewicz, and Burstein 1982), whereas with the previous, unstabilized Total Condylar knee, an average range of motion of 90 degrees had been reported (Insall, Tria, and Scott 1979). With the PS knee, 76% of the patients could now climb stairs normally or walk an unlimited distance.

As discussed by Walker (1989), many other types of bicondylar cruciate-sacrificing knee designs were conceived in the 1970s (Figure 7.12). The Freeman-Swanson knee prosthesis (Freeman, Swanson, and Todd 1973), developed at

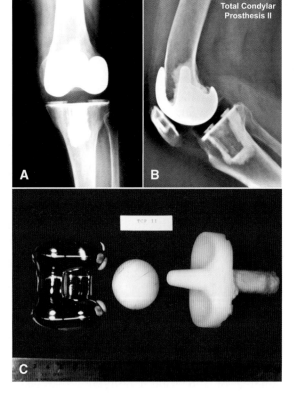

Figure 7.15

Radiographs and photographs of a retrieved Total Condylar Prosthesis II (TCP II), clinically introduced in 1978. (Images provided courtesy of Professor Clare Rimnac, Case Western Reserve University, Cleveland, OH.)

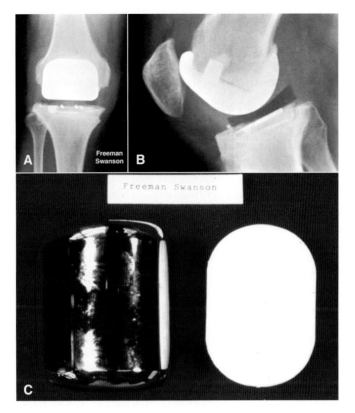

Figure 7.16

Radiographs and photographs of a retrieved Freeman-Swanson TKA, clinically introduced in the 1970s. (Images provided courtesy of Professor Clare Rimnac, Case Western Reserve University, Cleveland, OH.)

London Hospital, is another example of a well-known design from this period (Figure 7.16). However, the Total Condylar knee is considered by many orthopedic researchers and surgeons to be an archetype or "gold standard" among this first-generation of bicondylar knee designs. The excellent clinical results reported by the designing surgeons of the Total Condylar knee have been confirmed by orthopedic centers around the world (Aglietti and Rinonapoli 1984, Borden et al. 1982).

Patello–Femoral Arthroplasty

The Polycentric knee design, as originally conceived by Gunston, did not address the patello–femoral joint. Gunston mentions in his 1971 paper that "the patella may be retained because no impingement of the patella on the prosthesis occurs." Over time, the limitation of this approach became increasingly apparent, because one of the major reasons for revising the Polycentric was due to patella pain (Lewallen, Bryan, and Peterson 1984). Gunston later noted that "marked patello-femoral involvement in osteolysis may require trimming of the patella or patellectomy with usually unsatisfactory clinical result" (1973).

Gunston and MacKenzie's response to severe patello–femoral pain was to develop a separate patello–femoral arthroplasty, which consisted of a metallic patellar button articulating against an UHMWPE track implanted in the femur (1976). Because this solution required the implantation of two additional components, they cautioned against using this solution for patello–femoral replacement "indiscriminately" (1976).

Within this context, the patello–femoral arthroplasty offered by the Total Condylar knee was a far more elegant and simpler solution than previous designs. The general concept of the convex UHMWPE patellar component, originally developed for the Total Condylar knee (see Figure 7.14), remains relevant today. In the hands of the surgeons at the HSS, this patellar implant design initially provided good or excellent results in 95% of patients (Ranawat, Rose, and Bryan 1984).

The design of the patello–femoral arthroplasty has undergone many evolutionary design changes since the days of the Total Condylar knee. On the patellar side, the profile of the domed articulating surface, as well as the size and number of cement fixation pegs, has been tailored to improve conformity of contact, as well as to improve alignment of the patellar component with respect to the femoral condyle. On the femoral side, the design of the anterior flange of the femoral component has evolved to more closely reproduce the anatomic tracking of the patella during deep flexion activities such as stair climbing or squatting. Despite these evolutionary improvements in the design of the articulation, the dome-shaped patellar component included with the Total Condylar knee replacement is considered the foundation of current design concepts incorporating UHMWPE in patello–femoral arthroplasty.

UHMWPE with Metal Backing

The fifth major evolutionary step in the use of UHMWPE for knee arthroplasty was the incorporation of metal backing into tibial component designs during the late 1970s (Figure 7.17). In the Total Condylar knee, and other similar knee designs, the UHMWPE tibial component was mechanically fixed to the metallic tibial tray. Designs of this type are currently referred to as "fixed bearing" designs, because the UHMWPE tibial component remains fixed with respect to the metal backing. In addition to providing integral fixation between the tibial tray and the underlying cement, the introduction of metal backing also made it possible to conceive of a unique family of mobile bearing knee designs, which also have their origins in the late 1970s.

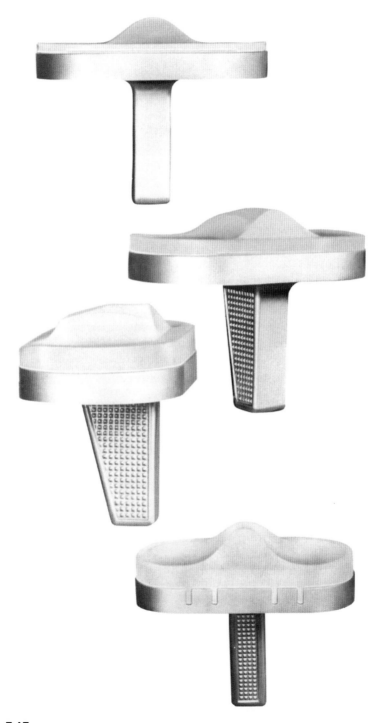

Figure 7.17
Metal backing of the Total Condylar knee replacement from the late 1970s.

Fixed-Bearing Total Knee Arthroplasty

Although the cemented all-polyethylene total condylar prosthesis was intended to provide improved fixation over previous designs, troubling radiolucencies were nonetheless observed at the cement interface within the first 10 years of implantation (Ecker et al. 1987). To address the problem of implant fixation, metal tibial trays were initially used to improve the integrity of the cement-prosthesis interface. Finite element analyses later demonstrated that metal backing had the further theoretical benefit of lowering the stresses in the cement and in the subchondral bone (Bartel et al. 1982, Lewis, Askew, and Jaycox 1982). The use of metal backing was later adapted to patellar and unicondylar components for similar reasons.

Notwithstanding the potential advantages afforded by improved fixation and modularity, the use of metal backing with UHMWPE components for knee replacement has been controversial (Rodriguez et al. 2001). Human anatomy imposes geometric space constraints on the overall size of orthopedic components. Consequently, the inclusion of metal backing requires the use of a thinner UHMWPE insert than would otherwise be possible without a metallic tibial tray. Under comparable joint loading scenarios, the contact and subsurface stresses in UHMWPE tibial components increase as the thickness decreases (Bartel, Bicknell, and Wright 1986). Based on elasticity and finite element solutions of the UHMWPE tibial component in the Total Condylar knee, Bartel and colleagues recommended in 1986 that "a thickness of more than eight to ten millimeters should be maintained when possible."

In the 1990s, clinical failures were reported with metal-backed tibial and patellar components, in part because of the reduced thickness of UHMWPE that could be accommodated with such designs (Collier et al. 1990). Consequently, all-UHMWPE patellar components are currently more widely used by orthopedic surgeons than metal-backed designs. Today, the FDA recommends a minimum UHMWPE thickness of 6 mm for metal-backed tibial components. Despite some early setbacks with first-generation thin inserts, fixed-bearing metal-backed UHMWPE tibial components currently represent the standard of care in TKA.

Mobile Bearing Total Knee Arthroplasty

The mobile bearing is conceptually an interesting and unique adaptation of UHMWPE for metal-backed components in TKA, which have their genesis in the late 1970s. In mobile bearing tibial designs, the UHMWPE insert articulates against both a polished femoral component and a polished metal tibial tray (Figure 7.18). Mobile knee bearings comprise two families of designs, including meniscal bearings and rotating platform knees. In meniscal bearings, the UHMWPE portion of the knee implant consists of two individual condylar components, which are constrained to slide in a polished anterior–posterior groove in the metal tray. In rotating platform bearings, the UHMWPE tibial component contains an inferior stabilization peg, which articulates with a

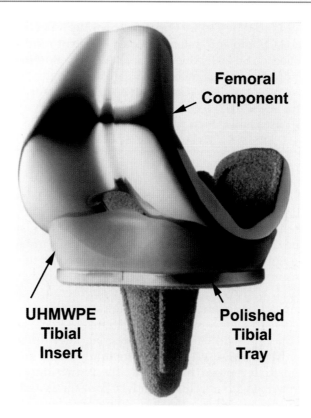

Femoral Component

UHMWPE Tibial Insert

Polished Tibial Tray

Figure 7.18
Photograph of an LCS rotating platform mobile bearing total knee replacement. (Images provided courtesy of J.B. VanMeter, DePuy Orthopedics, Warsaw, IN.)

centralizing hole in the tibial tray. A mobile bearing patellar component, based on the rotating platform design, has also been clinically introduced.

One of the first meniscal bearings, known as the Oxford Knee, was developed as a unicondylar prosthesis in 1977 (Biomet, Warsaw, IN). Starting in 1977, Michael J. Pappas and Frederick F. Buechel from Newark, New Jersey, independently developed several designs of low contact stress (LCS) mobile bearing total joint replacements (DePuy Orthopedics, Warsaw, IN) (2002). The philosophy of these mobile bearing designs was to reduce contact stresses at the articulating surface by making the tibial component and femoral components highly conforming. To provide rotational range of motion for the knee, the designers incorporated a second articulation between the UHMWPE component and the polished metal tray. A unique mobile bearing patellar component, which enabled rotation between the UHMWPE patellar component and the metal backing, was also designed by Buechel and Pappas (2002).

Mobile bearings were sufficiently different from fixed bearing devices that the FDA required a multicenter clinical study to be performed before they could be marketed in the United States. After the successful conclusion of the clinical trials in the early 1980s, the LCS mobile bearing knees were clinically introduced by DePuy Orthopedics in 1985 (Buechel, Jr. 2002). The designers of the LCS implant system reported excellent results after 20 years of follow-up (Buechel, Sr. 2002, Buechel, Sr. et al. 2001, Buechel, Sr. et al. 2002). For further information about mobile bearing knee designs, the reader may wish to consult two recent reviews (Callaghan et al. 2001, Stiehl 2002) and a book (Hamelynck and Stiehl 2002) based on the LCS experience.

Conclusion

Within a decade of incorporating UHMWPE into TKA, this new form of surgery reached the same level of consistency and success as THA. However, the introduction of UHMWPE influenced the historical development of THA and TKA in different ways. With THA, Charnley had already perfected the design of the low friction arthroplasty, but the introduction of UHMWPE ensured the long-term durability of the prosthesis. With TKA, on the other hand, UHMWPE was accepted from the outset as the material of choice by implant designers, and the design concepts, not the material, were forced to evolve to achieve successful performance in the knee.

Starting with Gunston's pioneering knee design, the principal adaptations for UHMWPE used in TKA were firmly established in the 1970s. In 1986, Insall and Kelly opined that "very little future improvement can be expected by tinkering with the [Total Condylar] prosthesis itself, especially for routine cases. What is needed is better surgical training, better instruments, and wider availability of custom designs for special circumstances." Today, UHMWPE continues to serve as the only widely used bearing material for articulation with metallic components in TKA. In Chapter 8, we review aspects related to the clinical performance of UHMWPE in the knee.

Acknowledgments

Many thanks to Professor Clare Rimnac, Case Western Reserve University; Professor Donald Bartel, Cornell University; Frank Gunston, Brandon, Manitoba; and Professor David Lyttle, University of Manitoba, for helpful advice and discussions. Thanks also to Paul Serekian, Howmedica Osteonics, for providing access to the catalog archives for knee replacement during the 1960s, and to J.B. VanMeter and Donald McNulty, DePuy Orthopedics, for assistance with researching the background of the LCS design.

References

Aglietti P., and E. Rinonapoli. 1984. Total condylar knee arthroplasty. A five-year follow-up study of 33 knees. *Clin Orthop* 186:104–111.

Bartel D.L., A.H. Burstein, E.A. Santavicca, and J.N. Insall. 1982. Performance of the tibial component in total knee replacement. *J Bone Joint Surg* 64:1026–1033.

Bartel D.L., V.L. Bicknell, and T.M. Wright. 1986. The effect of conformity, thickness, and material on stresses in ultra-high molecular weight components for total joint replacement. *J Bone Joint Surg* 68:1041–1051.

Bloom J.D., and R.S. Bryan. 1977. Wide-track polycentric total knee arthroplasty: One year follow-up study. *Clin Orthop* 128:210–213.

Borden L.S., T. Heyne, G. Belhobek, et al. 1982. Total condylar prosthesis. *Orthop Clin North Am* 13:123–130.

Bryan R.S., and L.F. Peterson. 1973. Polycentric total knee arthroplasty. *Orthop Clin North Am* 4:575–584.

Bryan R.S., and L.F. Peterson. 1979. Polycentric total knee arthroplasty: A prognostic assessment. *Clin Orthop* 145:23–28.

Bryan R.S., L.F. Peterson, and J.J. Combs, Jr. 1973a. Polycentric knee arthroplasty. A review of 84 patients with more than one year follow-up. *Clin Orthop* 94:136–139.

Bryan R.S., L.F. Peterson, and J.J. Combs, Jr. 1973b. Polycentric knee arthroplasty. A preliminary report of postoperative complications in 450 knees. *Clin Orthop* 94:148–152.

Buechel F.F., Jr. 2002. The LCS story. In *LCS mobile bearing knee arthroplasty: 25 years of worldwide experience.* K.J. Hamelynck and J.B. Stiehl, Eds. Berlin: Springer.

Buechel F.F., Sr. 2002. Long-term followup after mobile-bearing total knee replacement. *Clin Orthop* 404:40–50.

Buechel F.F., Sr., F.F. Buechel, Jr., M.J. Pappas, and J. Dalessio. 2001. Twenty-year evaluation of meniscal bearing and rotating platform knee replacements. *Clin Orthop* 388:41–50.

Buechel F.F., Sr., F.F. Buechel, Jr., M.J. Pappas, and J. Dalessio. 2002. Twenty-year evaluation of the New Jersey LCS rotating platform knee replacement. *J Knee Surg* 15:84–89.

Callaghan J.J., J.N. Insall, A.S. Greenwald, et al. 2001. Mobile-bearing knee replacement: Concepts and results. *Instr Course Lect* 50:431–449.

Collier J.P., M.B. Mayor, V.A. Surprenant, et al. 1990. The biomechanical problems of polyethylene as a bearing surface. *Clin Orthop* 261:107–113.

Cracchiolo A., III. 1973. Polycentric knee arthroplasty using tibial prosthetic units of a variable height. A preliminary report of design characteristics and a concept of clinical use. *Clin Orthop* 94:140–147.

Ecker M.L., P.A. Lotke, R.E. Windsor, and J.P. Cella. 1987. Long-term results after total condylar knee arthroplasty. Significance of radiolucent lines. *Clin Orthop* 216:151–158.

Ewald F.C. 1975. Metal to plastic total knee replacement. *Orthop Clin North Am* 6:811–821.

Freeman M.A., S.A. Swanson, and R.C. Todd. 1973. Total replacement of the knee using the Freeman-Swanson knee prosthesis. *Clin Orthop* 94:153–170.

Gluck T. 1890. Die invaginationsmethods der osteo- und arthroplastik. *Berl Klin Wschr* 19:732.

Gunston F.H. 1971. Polycentric knee arthroplasty. Prosthetic simulation of normal knee movement. *J Bone Joint Surg* 53:272–277.

Gunston F.H. 1973. Polycentric knee arthroplasty. Prosthetic simulation of normal knee movement: Interim report. *Clin Orthop* 94:128–135.

Gunston F.H., and R.I. MacKenzie. 1976. Complications of polycentric knee arthroplasty. *Clin Orthop* 120:11–17.

Hamelynck K.J., and J.B. Stiehl. 2002. *LCS mobile bearing knee arthroplasty: 25 years of worldwide experience.* Berlin: Springer.

Insall J., C.S. Ranawat, W.N. Scott, and P. Walker. 1976. Total condylar knee replacement: Preliminary report. *Clin Orthop* 120:149–154.

Insall J., W.N. Scott, and C.S. Ranawat. 1979. The total condylar knee prosthesis. A report of two hundred and twenty cases. *J Bone Joint Surg* 61:173–180.

Insall J., A.J. Tria, and W.N. Scott. 1979. The total condylar knee prosthesis: The first 5 years. *Clin Orthop* 145:68–77.

Insall J.N., P.F. Lachiewicz, and A.H. Burstein. 1982. The posterior stabilized condylar prosthesis: A modification of the total condylar design. Two- to four-year clinical experience. *J Bone Joint Surg* 64:1317–1323.

Insall J.N., and M. Kelly. 1986. The total condylar prosthesis. *Clin Orthop* 205:43–48.

Jones W.T., R.S. Bryan, L.F. Peterson, and D.M. Ilstrup. 1981. Unicompartmental knee arthroplasty using polycentric and geometric hemicomponents. *J Bone Joint Surg* 63:946–954.

Lewallen D.G., R.S. Bryan, and L.F. Peterson. 1984. Polycentric total knee arthroplasty. A ten-year follow-up study. *J Bone Joint Surg* 66:1211–1218.

Lewis J.L., M.J. Askew, and D.P. Jaycox. 1982. A comparative evaluation of tibial component designs of total knee prostheses. *J Bone Joint Surg* 64:129–135.

Mallory T.H. 1973. The use of polycentric knee arthroplasty in the treatment of fracture deformities of the knee. *Clin Orthop* 97:114–116.

Marmor L. 1988. Unicompartmental knee arthroplasty. Ten- to 13-year follow-up study. *Clin Orthop* 226:14–20.

Peterson L.F., R.S. Bryan, and J.J. Combs, Jr. 1975. Polycentric knee arthroplasty. *Curr Pract Orthop Surg* 6:2–10.

Ranawat C.S., H.A. Rose, and W.J. Bryan. 1984. Replacement of the patello-femoral joint with the total condylar knee arthroplasty. *Int Orthop* 8:61–65.

Ranawat C.S., W.F. Flynn, Jr., S. Saddler, et al. 1993. Long-term results of the total condylar knee arthroplasty. A 15-year survivorship study. *Clin Orthop* 286:94–102.

Rand J.A., and M.B. Coventry. 1988. Ten-year evaluation of geometric total knee arthroplasty. *Clin Orthop* 232:168–173.

Robertsson O. 2000. The Swedish knee arthroplasty register: Validity and outcome. Ph.D. Diss., Lund University.

Robertsson O., S. Lewold, K. Knutson, and L. Lidgren. 2000. The Swedish knee arthroplasty project. *Acta Orthop Scand* 71:7–18.

Rodriguez J.A., N. Baez, V. Rasquinha, and C.S. Ranawat. 2001. Metal-backed and all-polyethylene tibial components in total knee replacement. *Clin Orthop* 392:174–183.

Rodriguez J.A., H. Bhende, and C.S. Ranawat. 2001. Total condylar knee replacement: A 20-year followup study. *Clin Orthop* 388:10–17.

Scuderi G.R., J.N. Insall, R.E. Windsor, and M.C. Moran. 1989. Survivorship of cemented knee replacements. *J Bone Joint Surg* 71:798–803.

Shoji H., R.D. D'Ambrosia, and P.R. Lipscomb. 1976. Failed polycentric total knee prostheses. *J Bone Joint Surg* 58:773–777.

Skolnick M.D., R.S. Bryan, and L.F. Peterson. 1975. Unicompartmental polycentric knee arthroplasty: description and preliminary results. *Clin Orthop* 112:208–214.

Skolnick M.D., R.S. Bryan, L.F. Peterson, et al. 1976. Polycentric total knee arthroplasty. A two-year follow-up study. *J Bone Joint Surg* 58:743–748.

Speed J.S., and H. Smith. 1940. Arthroplasty: A review of the past ten years. *Surg Gynec Obstet* 70:224–230.

Stiehl J.B. 2002. World experience with low contact stress mobile-bearing total knee arthroplasty: A literature review. *Orthopedics* 25:S213–217.

Townley C.O. 1985. The anatomic total knee resurfacing arthroplasty. *Clin Orthop* 192:82–96.

Vince K.G. 1994. Evolution of total knee arthroplasty. In *The knee*. W.N. Scott, Ed. St. Louis: Mosby.

Vince K.G., J.N. Insall, and M.A. Kelly. 1989. The total condylar prosthesis. 10- to 12-year results of a cemented knee replacement. *J Bone Joint Surg* 71:793–797.

Walker P.S. 1977. Historical development of artificial joints. In *Human joints and their artificial replacements*. P.S. Walker, Ed. Springfield, IL: Charles C. Thomas.

Walker P.S. 1989. Requirements for successful total knee replacements. Design considerations. *Orthop Clin North* 20:15–29.

Chapter 7. Reading Comprehension Questions

7.1. Total knee replacement with UHMWPE was clinically introduced in the
   a) 1950s
   b) 1960s
   c) 1970s
   d) 1980s
   e) 1990s

7.2. The first knee arthroplasty incorporating UHMWPE resurfaced
   a) One condyle of the knee
   b) The patella
   c) Two condyles of the knee
   d) The posterior cruciate ligament
   e) Two condyles of the knee and the patella

7.3. Resurfacing of the patella was primarily developed to reduce
   a) Patellar pain
   b) Patellar infection
   c) Patellar osteolysis
   d) Patellar fracture
   e) All of the above

7.4. The vertical post in the Total Condylar Prosthesis is intended to
   a) Improve stability of the knee joint
   b) Replace the posterior cruciate ligament
   c) Contact the femoral component
   d) Increase joint flexion
   e) All of the above

7.5. In which of the following types of patients has UKA been most successful?
   a) All osteoarthritis patients
   b) Patients with osteoarthritis limited to only one compartment of the knee and the patella
   c) Patients with osteoarthritis limited to only one compartment of the knee
   d) Patients with osteoarthritis limited to only the patella
   e) All rheumatoid arthritis patients

7.6. Metal backing was introduced for which of the following reasons?
   a) Decrease infection of the tibial component
   b) Decrease anterior knee pain
   c) Increase the magnitude of the stresses in the cement
   d) Evenly distribute stresses to the bone
   e) All of the above

7.7. The primary design objective of mobile bearing knee replacements is to
   a) Increase knee flexion
   b) Reduce the contact stress between the bearing components
   c) Enhance knee stability
   d) Improve the mobility of the posterior cruciate ligament during load bearing activities
   e) All of the above

7.8. Which of the following components are no longer widely used today in contemporary knee replacements?
   a) Metallic femoral components
   b) Metal-backed tibial trays
   c) Metal-backed patellar components
   d) UHMWPE tibial components
   e) None of the above

Chapter 8

The Clinical Performance of UHMWPE in Knee Replacements

Introduction

Total knee arthroplasties (TKAs) and unicondylar knee arthroplasties (UKAs) are highly successful surgical procedures. As seen in Chapter 7, the clinical track record for TKA was established during 1970s, when it gained the same high level of reliability as hip arthroplasty. Since that time, most TKAs have enjoyed a long-term survivorship of higher than 90% after 10 years of implantation. UKA was also established in the 1970s (see Chapter 7) and has enjoyed a resurgence of clinical interest since the 1990s with the advance of minimally invasive surgical techniques.

Fixed-bearing and mobile-bearing total knee designs currently share the clinical spotlight as safe and effective treatment options for patients requiring total knee replacement (TKR). Figure 8.1 shows examples of contemporary fixed-bearing knee-replacement components that are commercially available today. Chapter 7 contains an example of a contemporary mobile-bearing knee prosthesis (see Figure 7.14). Despite decades of continuous advancement in design, instrumentation, and surgical technique for UKA and TKA, conventional ultra-high molecular weight polyethylene (UHMWPE) remains the gold standard polymeric bearing material for use in both unicondylar and bicondylar knee athroplasty, whether in fixed-bearing or mobile-bearing designs.

The American Academy of Orthopedic Surgeons has estimated that 257,000 primary knee procedures were performed in the United States in 2000, and that number is expected to grow to 457,000 primary procedures by 2030 (Frankowski and Watkins-Castillio 2002). As already discussed in Chapter 4, the incidence of knee arthroplasty is greater than hip arthroplasty (see Figure 5.2). In the United States, more than 60% of primary knee procedures are performed on women (Frankowski and Watkins-Castillio 2002).

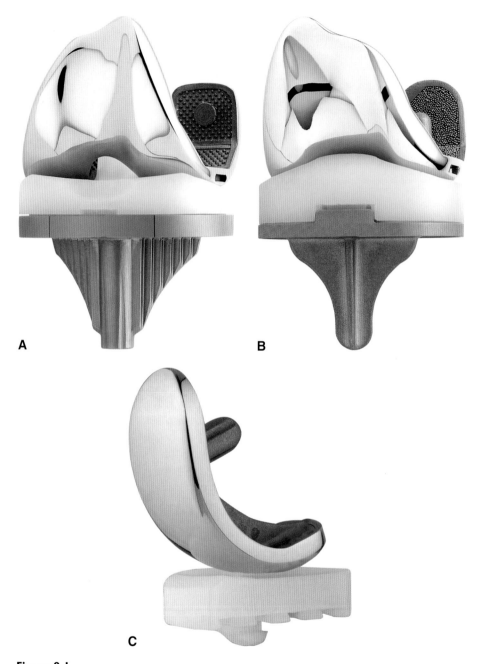

Figure 8.1
Examples of contemporary fixed-bearing knee replacement components. (**A**) Scorpio PS total knee prostheses; (**B**) Duracon total knee prostheses; and (**C**) Eius unicondylar prostheses. Images provided courtesy of Stryker Howmedica Osteonics (Mahwah, NJ).

Although knee arthroplasty enjoys a remarkable clinical track record, problems with wear and fatigue damage of UHMWPE continue to limit the longevity of both unicondylar and bicondylar knee replacement components. Unlike in the hip, where radiographic techniques have been developed to quantify *in vivo* wear rates, there currently exist no standard and widely accepted techniques for tracking the clinical performance of UHMWPE in patients with knee replacement. Thus, today the most effective way to evaluate the *in vivo* performance of UHMWPE continues to be the analysis of retrieved components from revision surgery or from autopsy donations.

This chapter contains four main sections covering TKA and, where applicable, UKA. The first section reviews the biomechanical considerations of knee arthroplasty that distinguish it from hip replacement. The second section describes the survivorship of TKA and UKA, and outlines measures of clinical performance for UHMWPE in knee arthroplasty. The third section is devoted to wear and osteolysis in TKA. In the final section of this chapter, alternatives to metal-on-conventional UHMWPE articulation for knee arthroplasty are described.

Biomechanics of Total Knee Arthroplasty

Anatomical Considerations

The knee is one of the most complex joints in the body. Unlike the hip, in which the joint surfaces are highly conforming, the articulating surfaces in the knee are more nonconforming. The geometry of the knee permits extreme flexion (up to 140 degrees), which may be needed for activities such as getting out of a chair or for squatting (Dahlkvist, Mayo, and Seedhom 1982). On the other hand, when the joint is unloaded, there is more than 5 mm of laxity in the soft tissue structures surrounding the knee (Piziali et al. 1980) to enable complex rotations and relative sliding of the joint surfaces. The principal ligaments of the knee, including the collaterals and the cruciates, which play an important role in the function of knee replacements, are shown in Figure 8.2. The soft tissues surrounding the joint permit a wide range of complex motions of the knee in six degrees of freedom.

The importance and biomechanical role of the soft tissues surrounding the knee joint continues to be debated (Sathasivam and Walker 1999, Trousdale and Pagnano 2002). The posterior cruciate ligament (PCL), in particular, has been an ongoing source of controversy in TKA. This ligament inserts on the posterior aspect of the tibia, and during activities such as stair climbing or rising from a chair, prevents the condyles of the femur from sliding forward off the anterior edge of the tibial plateau. The controversy over the PCL, which started in the 1970s and continues to this day, centers around whether it should be sacrificed during knee arthroplasty (Trousdale and Pagnano 2002). Proponents of PCL sacrifice maintain that the implant should provide the constraint and posterior stabilization of the anatomic knee. Such total knee designs are referred to as posterior-stabilized knees (or more simply, PS knees), and include a central

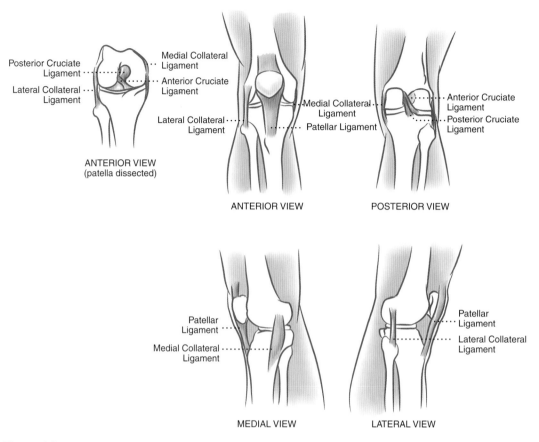

Figure 8.2
Anatomy of cruciate and collateral ligaments of the knee.

post of the tibial component, which engages with a cam in the femoral component to constrain anterior relative motion. Opponents of posterior stabilization maintain that the role of knee arthroplasty is to mimic the anatomic knee as closely as possible, which includes preserving the original PCL. The controversy over conformity and constraint is reflected in the wide range of TKR designs currently available to orthopedic surgeons. As we shall see later, both PS and PCL-sparing designs of knee replacement have successful survivorships that are statistically indistinguishable. Consequently, today the decision to sacrifice the PCL remains a function of the patient (i.e., whether the PCL is too diseased to function) and the personal preference of the surgeon.

Knee Joint Loading

It is difficult to address the clinical performance of UHMWPE in TKA without reference to the demanding functional environment of the knee joint. Starting in the 1960s (Morrison 1969), researchers began to develop models to calculate the load transmission across the contacting surfaces of the knee joint for various activities. Early models to predict the forces across the knee joint are statically indeterminate because the number of independent soft tissue structures acting across the knee joint is greater than what can be solved for using the equations of static equilibrium. However, by making simplifying assumptions in the number of active muscle groups across the joint during a particular activity, it turns out that reasonable estimates of joint reaction forces may be obtained for activities such as walking, stair climbing/descent, squatting, and rising from a chair (Table 8.1). For instance, the early models of Morrison suggested that the tibiofemoral contact force ranged from 2 to 4 times body weight (BW) during regular gait, with an average of 3 times BW (1969, 1970). For certain activities, such as rising from a chair or squatting, the forces across the knee can be up to 7.6 times BW (Table 8.1).

Surprisingly, consistent conclusions are reached using more complex biomechanical models of the knee (Amstutz et al. 1998, Davy and Audu 1987, Olney and Winter 1985, Patriarco et al. 1981, Rohrle et al. 1984), as well as by

Table 8.1
Summary of Average Knee Joint Loading for Activities of Daily Living

Activity	Reference	Patello–femoral joint		Tibiofemoral joint (compression)		Tibiofemoral joint (anterior shear)	
		Knee angle (degree)	Force (x BW)	Knee angle (degree)	Force (x BW)	Knee angle (degree)	Force (x BW)
Walking	Morrison (1970), Reilly and Martens (1972), Harrington (1976)	10	0.5	15	3.0–3.5	5	0.4
Squatting	Dahlkvist, Mayo, and Seedhom (1982)	140	6.0–7.6	140	5.0–5.6	140	2.9-3.5
Rising from a chair	Ellis, Seedhom, and Wright (1984)	120	3.1	120	3–7	120	2.3
Stair climbing/ descent	Reilly and Martens (1972)	60	3.3	45–60	3.8–4.3	5	0.6

Force is Expressed in Units of BW.

measuring *in vivo* forces from telemeterized knee replacements (Lu et al. 1998, Taylor and Walker 2001). For further discussion of the biomechanics of the knee joint, the reader is referred to the works of Morrison (1969), Paul (1976), Walker (1977a), as well as to review articles (Andriacchi and Mikosz 1991, Burstein and Wright 1994).

Stresses in UHMWPE Tibial and Patellar Components for Total Knee Replacement

Since the 1980s, researchers have associated the stress levels acting on UHMWPE components with the extent and severity of surface damage observed in retrieved components. However, the precise relationship among wear, damage, and stress acting on the UHMWPE has remained elusive, because of the numerous factors contributing to clinical performance. The stresses acting on UHMWPE components for total joint replacement are design specific. Today, finite element analysis is the primary method for calculating stress distributions in specific designs of UHMWPE components. Experimentally, pressure-sensitive film may also be used to quantify contact stress, but not other stress components, which may be associated with surface damage.

Contemporary finite element models take into account the complex three-dimensional geometry of tibial, femoral, and patellar components (Figure 8.3). The models must incorporate relevant boundary conditions, including the appropriate joint loading, as was discussed in the previous section.

A finite element simulation of a total joint simulation must also incorporate a realistic material model for UHMWPE. Depending on the scope and question posed by a finite element analyst, a simple linear elastic or isotropic plasticity material model for UHMWPE may be sufficient. These simple material models, however, are not suitable for simulation of cyclic loading, for determining the time-dependent response to loading, or for prediction of large deformation behavior leading up to failure in UHMWPE. Recently, a more advanced material model has been developed and validated for conventional and highly crosslinked UHMWPE (see Chapter 14). Based on the principles of polymer physics, this Hybrid model accurately incorporates rate effects, viscoplasticity, and evolution of anisotropy during large deformations for a wide range of UHMWPE materials.

Beyond the complexities introduced by geometry, boundary conditions, and material properties, the surgical malpositioning of components, or geometric discontinuities for a particular hip or knee design, may also produce regional stress concentrations. However, in this section of the chapter we will focus on describing the macroscopic stress state of UHMWPE within the contact area of the articulating surface of well-aligned components.

The magnitude and distributions of stress in the knee are different from the hip. In the hip, the spherical contacting surfaces are highly conforming, and the effective (von Mises) stress levels are below yield, and, thus, below the onset of irrecoverable plastic deformation. Consequently, for hip components, UHMWPE can reasonably be considered to behave as an elastic material at the

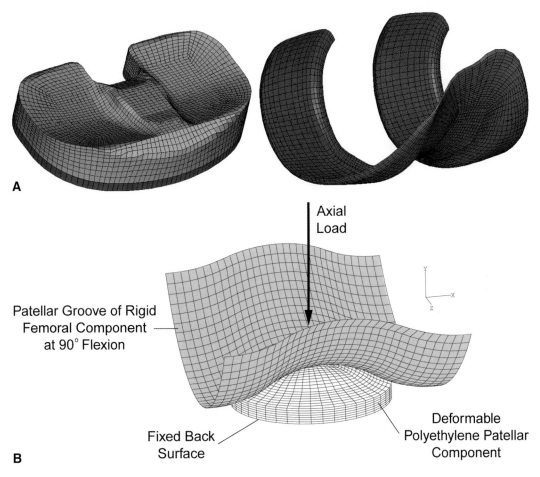

A

Axial
Load

Patellar Groove of Rigid
Femoral Component
at 90° Flexion

Fixed Back
Surface

Deformable
Polyethylene Patellar
Component

B

Figure 8.3
(**A**) Three-dimensional finite element models used to simulate tibio-femoral contact, and (**B**) patello–femoral contact.

continuum level. Elasticity solutions have been developed to calculate the contact stress distributions for metal-backed acetabular components (Bartel et al. 1985). In this regard, a specialized analytical solution is required for UHMWPE acetabular components, because the more widely used Hertzian contact theory does not apply. Finite element analyses are nonetheless helpful to get a complete picture of the stress state in hip replacements and have proved to be necessary when addressing the effects of specific hip component designs. From finite element analysis, researchers have observed that the maximum shear stress occurs at or very near the surface of conforming UHMWPE acetabular bearings. This observation has been used to explain why certain forms of surface damage, such as pitting or delamination, are rarely observed in retrieved UHMWPE acetabular components. As we have seen, the primary damage mode in the hip is adhesive/abrasive wear.

The stress levels for UHMWPE tibial and patellar components are generally higher than in acetabular components. Knee joint forces are comparable to those at the hip. However, because of the nonconformity between the tibial and femoral components surfaces, the joint forces of the knee are distributed over a much smaller area than in the hip. The von Mises stresses in knee replacements are typically greater than the offset yield stress of UHMWPE, and are thus of sufficient magnitude to result in irrecoverable plastic deformation of the component. Figure 8.4 shows the distribution of compressive stress (minimum principal stress) on the surface of a contemporary knee replacement (the corresponding mesh for the analysis is shown in Figure 8.3). If the component is sectioned perpendicular to the surface, one can visualize the stress state in the UHMWPE through the thickness of the tibial component using finite element analysis. Figure 8.5 shows the effective (von Mises) stress in a perpendicular section through the thickness of the tibial component under the center of

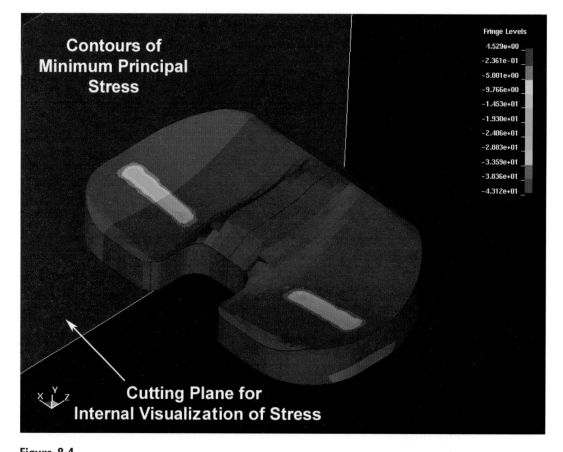

Figure 8.4

Distribution of minimum principal stress on the surface of a tibial component at heel strike during normal gait, using the model shown in Figure 8.3A. Also shown is a cutting plane that will be used to examine the internal stress distribution of the component (see Figure 8.5).

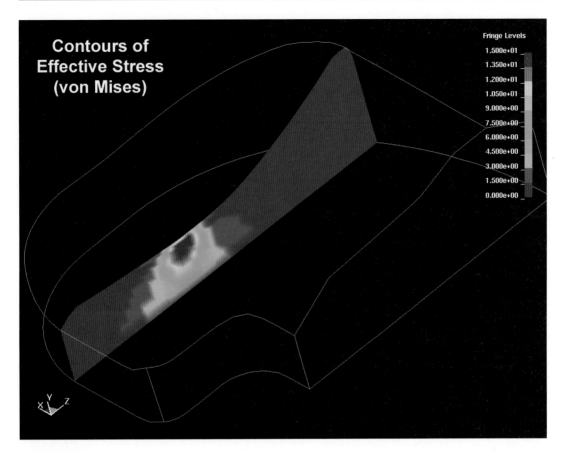

Figure 8.5

Distribution of von Mises (effective) stress through the thickness of a tibial component. Note that the location of the maximum von Mises stress, which also corresponds to the location of maximum shear stress, is located at a depth of 1–2 mm from the articulating surface. The slice is taken perpendicular to the contact surface of a tibial component at heel strike during normal gait, using the model shown in Figures 8.3A and 8.4.

contact. As shown by Bartel and colleagues (1986, 1995), the maximum von Mises stress occurs at a depth of 1 to 2 mm below the articulating surface. The location in maximum von Mises stress also coincides with the location in the maximum shear stress within the UHMWPE component. Thus, not only is the magnitude of stress greater in knee components, but the distribution through the thickness varies substantially from hip replacement components.

Although the computational tools for predicting the stress state within UHMWPE tibial components have advanced considerably since the 1980s, these sophisticated models remain limited to comparative assessment of different design geometries. The ability to extrapolate predicted stress results into the clinical environment remains extremely limited. As stated earlier, the magnitude of stresses predicted in finite element model of knee arthroplasty, while associated with surface damage in a general sense, have not yet been directly linked to long-term survivorship in patients.

It would be logical to suppose that designs that are subjected to a higher stress level would exhibit worse clinical performance. However, this has not necessarily been the case in actual clinical practice. The Insall-Burstein (IB) and the Miller-Gallante (MG) (Figure 8.6) are examples of knee prosthesis designs that have functioned successfully under substantially different stress states. The IB knee is a cruciate-sacrificing design, and the tibial component is designed with concave condylar surfaces for conformity with the femoral component (Thadani et al. 2000). The MG knee, in contrast, is a cruciate-sparing design and includes a relatively flat tibial component to reduce constraint. In the MG knee, the tibial component is nonconforming with the femoral component, especially during activities that involve full flexion.

The contact stresses and von Mises (effective) stresses are up to 20% higher for the nonconforming MG design as compared with the more conforming IB design under identical loading conditions (Kurtz, Bartel, and Rimnac 1998). Depending on the loading, the magnitude of the von Mises stresses for both designs may be sufficiently high to produce localized yielding and permanent deformation of the UHMWPE insert (Kurtz, Bartel, and Rimnac 1998). Despite differences in stress levels, the two designs both exhibit successful clinical performance in terms of long-term survivorship (Berger et al. 2001, Thadani et al. 2000). Interestingly, the designs do exhibit different predominant modes of surface damage. As illustrated in Figure 8.6, IB knees tend to show greater evidence of pitting than MG knees, which tend to delaminate when oxidized.

The magnitude of stresses subjected to an UHMWPE component is not the sole variable governing wear and damage to TKRs. Additional factors, such as the amount of constraint provided by different knee designs (Sathasivam and Walker 1999), the presence of fusion defects in the UHMWPE (Blunn et al. 1997), and the extent of oxidation in the UHMWPE (Bell et al. 1998), have been identified as factors related to the clinical performance of tibial components. Because of the complexity of the multiple factors contributing to *in vivo* wear and surface damage of UHMWPE tibial components, researchers and implant designers in orthopedics have not yet developed the capability to accurately simulate the clinical performance of different knee designs on a computer in advance of human clinical trials.

Clinical Performance of UHMWPE in Knee Arthroplasty

The orthopedic literature contains thousands of articles describing the clinical performance of various knee replacement designs. As in the hip, the clinical performance of knee replacement is most unambiguously defined in terms of survivorship. Clinicians may disagree as to the precise etiology of a TKA failure, but the date of a revision surgery is a precise endpoint for the procedure.

On the other hand, survivorship alone does not fully capture the clinical performance of UHMWPE in the knee. Surface damage and wear of the UHMWPE insert are also important measures of clinical performance of knee arthroplasty. If a knee prosthesis survives the first 10 years of implantation, wear behavior

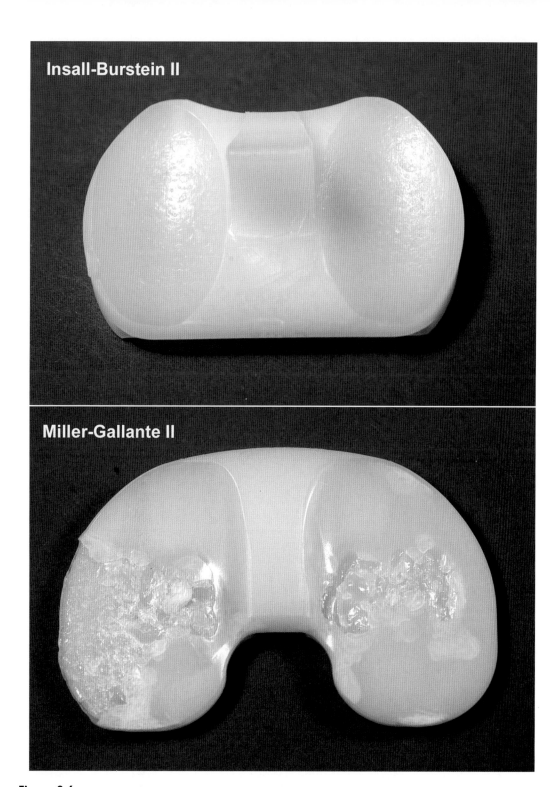

Figure 8.6

Previously implanted Insall-Burstein (IB II) and Miller-Gallante (MG II) tibial inserts with evidence of pitting and delamination, respectively.

of the insert plays an increasing role in the longevity of the joint replacement. In the following discussions, we describe the survivorship of TKA, as well as the assessment of wear and surface damage in knee arthroplasty.

Survivorship of Knee Arthroplasty

TKRs are, in general, highly successful implants. Knee arthroplasty is so successful, in fact, that it is often difficult to discriminate between different knee designs, which generally appear to enjoy comparably high survival rates. The long-term survivorship of TKA, as reflected in the orthopedic literature, has been summarized in a recent meta-analysis performed by Forster (2003). In Forster's study, the knee designs were classified as either PS or nonstabilized. The analysis also evaluated whether metal backing of the tibial component influenced implant survival. The 10- to 11-year survivorship rates from Forster's analysis are summarized in Table 8.2. Although the survival rates for all of the design types considered was higher than 90%, the breadth of the confidence intervals precluded distinguishing among different designs. As discussed by Forster (2003), the orthopedic literature may be limited by publication bias (the tendency to report only successful outcomes), and the vast majority of the published studies had to be excluded from the meta-analysis because of inadequate reporting of survivorship tables in individual studies.

A more comprehensive evaluation of survivorship in contemporary knee arthroplasty (both UKA and TKA) can be derived from the national knee implant

Table 8.2

Summary of Total Knee Replacement Survival at 10 to 11 Years, Covering 5950 Total Knee Arthroplasties[a]

TKA design group	Original cohort size	Success rate at 10 to 11 years (%)	95% Confidence intervals (%)
PS design	1698	92.7	88.0–95.4
Nonstabilized design	2218	92.4	90.3–94.1
Metal-backed tibial component	1561	92.7	89.4–94.6
All-polyethylene tibial component	473	95.9	92.7–97.8
Nonstabilized, metal-backed tibial component	2034	91.4	88.8–93.4
PS, metal-backed tibial component	1272	92.8	83.7–94.2
PS, all-polyethylene tibial component	289	93.3	86.9–96.9
Nonstabilized, all-polyethylene, tibial component	184	98.9	95.7–99.7

[a]Based on a meta-analysis of the orthopedic literature published by Forster (2003).

registries in Sweden and Norway (Robertsson 2000, Robertsson et al. 2001a, Robertsson et al. 2000). For example, as of 1997, the Swedish Knee Arthroplasty Register, started in 1975, had registered 57,533 procedures (Robertsson et al. 2001a). A total of 6865 knee athroplasties were registered in the Swedish Knee Register in 2001 (the year for which the most recent data is available), which corresponds to a 15% increase over 2000. The system used to collect data for the national registry in Sweden has recently undergone extensive reverification and validation (Robertsson 2000).

The Scandinavian national knee registries provide unique, population-based outcome data for TKA and UKA. The Swedish knee registry data, for example, shows that patient age, gender, and disease have a significant effect (osteoarthritis [OA] versus rheumatoid arthritis [RA]) on the survivorship of TKA and UKA (Figures 8.7 and 8.8). The developers of the Scandinavian national registries argue that the early identification of inadequate designs has led to their withdrawal from the marketplace and resulted in national revision rates that have declined steadily over time (Figure 8.9).

Interestingly, the Swedish registry researchers have observed a "learning curve" phenomenon with UKA. For designs with technically demanding implantation procedures, the risk of UKA revision in Sweden was associated with the number of procedures performed at a surgical site (Robertsson et al. 2001b). The negative Swedish experience with UKA in patients with RA also underscores the importance of proper patient selection for this procedure (Figure 8.8).

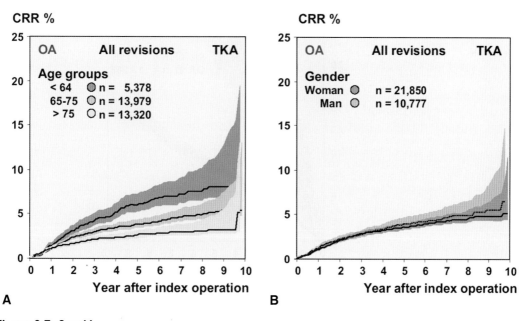

A   B

Figure 8.7 Cont'd

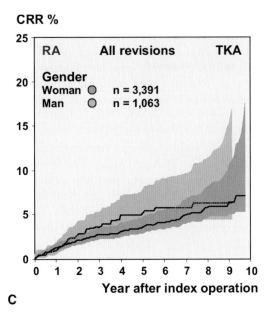

C

Figure 8.7
TKA survivorship, expressed as the cumulative risk of revision, from the Swedish National Knee Arthroplasty Register (Lindgren 2002). Data were compiled for patients as a function of age (**A**) and for those with (**B**) OA and (**C**) RA. (Reprinted with permission.)

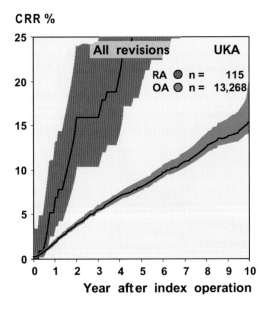

Figure 8.8
Unicompartmental knee arthroplasty survivorship, expressed as the cumulative risk of revision. Data compiled for patients with OA and RA by the Swedish National Knee Arthroplasty Register (Lindgren 2002). (Reprinted with permission.)

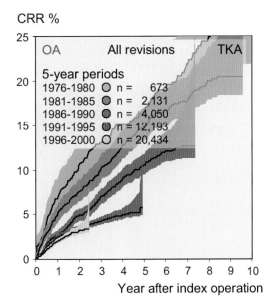

Figure 8.9

Progressive improvement in TKA survivorship in 5-year intervals from 1975 to 2000. Data compiled by the Swedish National Knee Arthroplasty Register (Lindgren 2002). (Reprinted with permission.)

Reasons for Knee Arthroplasty Revision Surgery

A patient's knee replacement may be revised for aseptic loosening, infection, fracture, joint stiffness, tibiofemoral instability due to collateral ligament instability, patellar complications, extensor mechanism rupture, and/or wear or failure of the UHMWPE component (Fehring et al. 2001, Sharkey et al. 2002, Vince 2003). According to Vince, the surgeon revising a knee arthroplasty must first identify and correct the root cause of the previous surgery, not merely treat the symptoms of the failed knee (2003). If the root cause of a revision is malalignment of the femoral component, simply exchanging a worn UHMWPE tibial insert will not address the central problem for the patient and, conversely, will predispose the new tibial component to early failure as well.

Infection, loosening, and patellar complications have been identified as prevalent reasons of TKA revision (Figure 8.10). In a study of 440 revision surgeries performed between 1982 and 1999, Fehring and colleagues reported that infection was the single largest cause of knee arthroplasty revision within the first 5 years of implantation (2001). In a more recent study of 212 knee revisions performed between 1997 and 2000 by Sharkey and associates, infection was responsible for 25% of revisions within the first 2 years of implantation, but only 7.8% of components implanted over 2 years were infected (2002).

Sharkey and colleagues (2002) also found that 25% of knee revisions were associated with wear or surface damage of the tibial or patellar insert. Interestingly, the timing of knee revisions associated with UHMWPE wear has

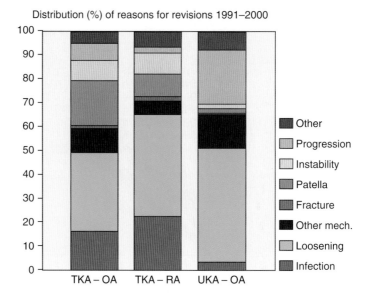

Distribution (%) of reasons for revisions 1991–2000

Figure 8.10

Reasons for knee arthroplasty revision from the Swedish National Knee Arthroplasty Register (Lindgren 2002). Reprinted with permission.

two major peaks (Figure 8.11), the first occurring before 5 years and the second peak in revisions occurring between 6 and 10 years. For components revised for wear within the first 5 years of implantation, patellar problems, loosening, instability, and malalignment were typically also identified as reasons for revision (Sharkey et al. 2002). For knee arthroplasties that survive the first 5 years of implantation, the clinical performance of the UHMWPE plays an increasing role in the longevity of the artificial knee.

Articulating Surface Damage Modes

Wear and damage to UHMWPE components for knee replacement is not a new phenomenon and has been clinically observed since the 1970s. As noted in Chapter 7, in 1976 Shoji and colleagues reported severe wear and fatigue cracks in a small series of Polycentric knee components that were implanted for 2 years or less, but the UHMWPE failures were attributed to surgical misalignment of the implant components. Walker's treatise on natural and artificial joints, published in 1977, also documented examples of pitting, scratching, and burnishing in UHMWPE knee replacement components (Figure 8.12). However, the focus on developing improved designs dominated knee arthroplasty during the 1970s. Consequently, the development of reliable methods to quantify surface damage in artificial knees was delayed until the 1980s.

In 1983, Hood and colleagues published a seminal paper that established a reproducible and semiquantitative method for scoring the modes and prevalence

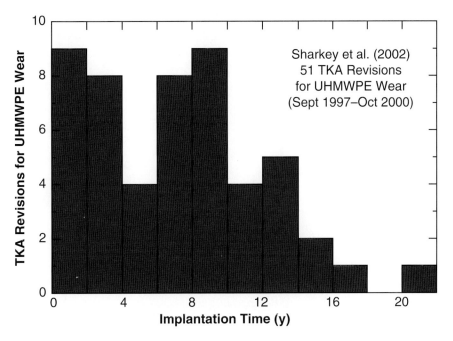

Figure 8.11

Implantation time for TKAs revised at one institution, in which UHMWPE wear was identified as a reason for revision. (Adapted from Sharkey et al. 2002.)

of surface damage in UHMWPE components for knee arthroplasty. Hood and coworkers recognized the immediate need for a system to quantify surface damage and wrote, "Now that metal and polyethylene joint implants are in the second decade of common usage, it would be prudent to develop a method for characterizing the changes in surface and structural conditions of retrieved implants so as to allow the study of the effects of this mechanical degradation on implant performance" (1983). Although originally developed using Total Condylar Prostheses, Hood's general method of classifying and scoring damage modes for UHMWPE components has remained relevant to a surprisingly wide range of different knee replacement designs since the 1980s.

Under a light microscope at 10× magnification, Hood and colleagues (1983) studied 10 regions of the articulating surface of tibial components, and 4 regions of the surface of patellar components, for evidence of wear or surface damage (Figure 8.13). Seven modes of surface damage were identified (Hood, Wright, and Burstein 1983):

1. Pitting. This damage mode is sometimes referred to as "cratering" and is characterized by Hood and associates as surface defects 2–3 mm in diameter and 1–2 mm deep (Figure 8.14). Pitting is classified as a mode of fatigue wear and involves the liberation of millimeter-sized piece of wear debris from the articulating surface. Because the wear debris produced by pitting is considered to be too large to provoke an osteolytic response, from a biological perspective

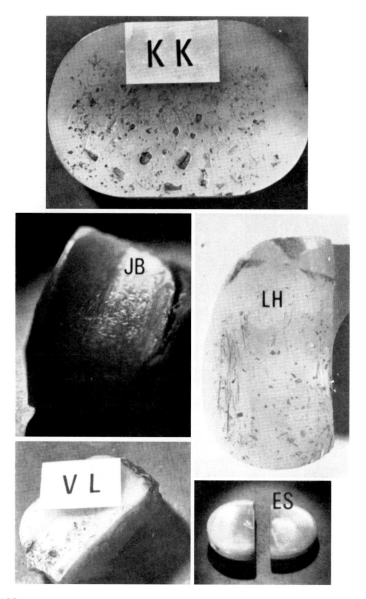

Figure 8.12
Examples of UHMWPE surface damage (e.g., scratching, burnishing, pitting) documented with tibial components during the 1970s. *KK*, Freeman-Swanson; *JB*, Duo-Condylar; *LH*, Geometric; *VL*, Duo-Condylar; and *ES*, Marmor modular knee prosthesis designs. (Reprinted with permission from Walker 1977.)

pitting is a more benign wear mechanism than adhesive/abrasive wear, which produces micrometer-sized debris.

2. Embedded debris. Hood and coworkers initially restricted this damage mode to polymethylmethacrylate (PMMA) debris, but it is possible that bone chips or metallic beads or fragments from the back surface of metallic components could

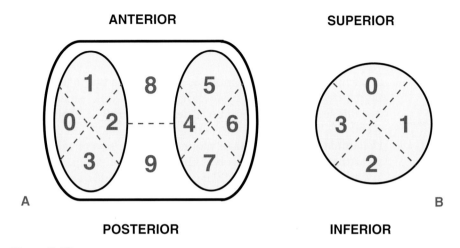

Figure 8.13
The damage scoring method of Hood and associates (1983) involved analyzing wear and surface damage within 10 surface regions of UHMWPE tibial components (**A**) and within 4 surface regions of UHMWPE patellar components (**B**).

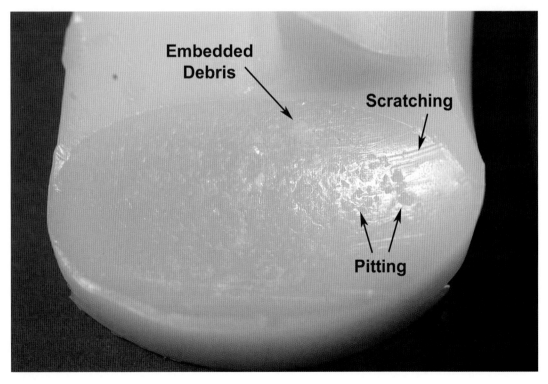

Figure 8.14
Examples of pitting, scratching, and embedded debris surface damage modes.

also become embedded in the UHMWPE (Figure 8.14). Embedded debris can result in third body wear of the UHMWPE and metallic surfaces. In addition, embedded debris can scratch the metallic surface, resulting in further abrasive wear of the UHMWPE.

3. Scratching. This damage mode is identified as linear features on the articulating surface, produced by plowing of microscopic asperities on the opposing metallic surface , or by third body debris (Figure 8.14). This is a mode of abrasive wear.

4. Delamination. This damage mode is a more severe manifestation of fatigue wear than pitting and involves the removal of sheets of UHMWPE from the articulating surface. If the tibial component is sufficiently thick, the UHMWPE underlying the delamination may continue to serve as a functional bearing surface (e.g., the Miller Gallante knee shown in Figure 8.6). In other cases, when the tibial component is too thin and embrittled because of oxidation, delamination can result in catastrophic wear of the UHMWPE, necessitating revision (Figure 8.15).

5. Surface deformation. This damage mode corresponds to a permanent (irrecoverable) change in the surface geometry of the implant (Figure 8.16). It is

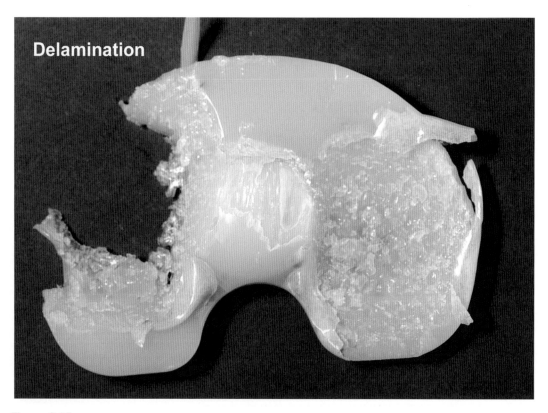

Figure 8.15
Severe delamination of an Ortholoc tibial component (Richards, Memphis, TN).

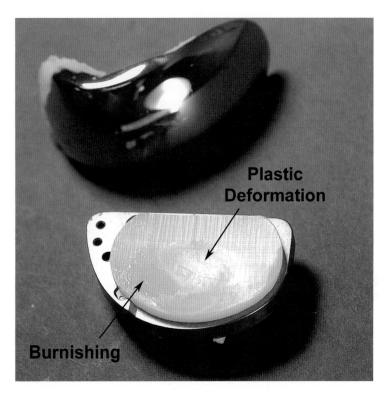

Figure 8.16
Example of burnishing and plastic deformation surface damage modes.

sometimes referred to as "plastic deformation," "cold flow," or "creep." Unlike the other damage modes, plastic deformation does not result in material removal and thus does not strictly correspond to wear.

6. Burnishing. This damage mode is characterized as "wear polishing" and is characteristic of adhesive/abrasive wear (Figure 8.16). From a biological perspective, burnishing produces wear debris that is within the size range that can stimulate an osteolytic response.

7. Abrasion. Hood and coworkers describe abrasion as a shredding or "tufting" of the UHMWPE surface. This is classified as a mode of abrasive wear.

Within each surface region of the tibial or patellar component, Hood and colleagues graded the presence and extent of the seven damage modes on a scale of 0 to 3 (Hood, Wright, and Burstein 1983). A score of 0 corresponds to the absence of the damage mode within the specified region. Scores of 1, 2, and 3 correspond to observation of the damage mode over less than 10%, 10–50%, or more than 50% of the specified region, respectively. An overall assessment of each damage mode was determined by summing the scores across all of the surface regions. For each damage mode, the maximum score was 30 (3 maximum score/region × 10 regions) for the tibial component and 12 (3 maximum score × 4 regions) for the patellar component. A total damage score was computed by adding up the total damage mode scores, corresponding to a maximum score of 210 (3 maximum score/region × 10 regions × 7 damage modes) for the tibial component and 84 (3 maximum score × 4 regions × 7 damage modes) for the patellar component.

The Hood method is semiquantitative and thus allows researchers to compare the location of damage (e.g., medial condyle versus lateral condyle; anterior versus posterior), the prevalence of damage modes within a single design, and also differences in damage between designs. For example, in Hood's assessment of retrieved Total Condylar Prostheses, Hood and coworkers noted that scratching was the most prevalent form of surface damage (observed in 90% of the tibial component retrievals), followed by pitting (81%), burnishing (75%), surface deformation (62%), cement debris (48%), abrasion (41%) and delamination (4%) (1983). Hood and coworkers also noted a significant correlation between patient weight and total damage score, as well as between implantation time and damage score, suggesting that fatigue mechanisms resulting from cyclic loading were likely involved in the generation of wear and damage to knee components.

Although originally developed for the Total Condylar knee, Hood and associates' method has been applied to a wide range of cruciate-sparing as well as cruciate-sacrificing tibial component designs (Kurtz et al. 2000, Won et al. 2000, Wright et al. 1992). Hood's method has also been adapted to quantify damage to vertical stabilizing posts (Furman et al. 2003). After 20 years, Hood's method continues to serve as the fundamental reference for damage modes in UHMWPE components for knee replacement.

Osteolysis and Wear in Total Knee Arthroplasty

Osteolysis, which may be provoked by exposure of bone to particles of UHMWPE, bone cement, or metallic debris, has been a major concern for hip replacement. However, only recently has osteolysis been regarded as an important complication of TKA. An example of severe osteolysis in TKA is shown in Figure 8.17.

Before 1992 (Peters et al. 1992), osteolysis was generally not noted in the orthopedic knee literature except as isolated case reports. Within the past 10 years, however, there has been increased interest in wear and osteolysis as it relates to knee replacement. Several clinical and retrieval studies related to osteolysis in TKA have been summarized in Table 8.3.

Incidence and Significance of Osteolysis in Total Knee Arthroplasty

Whether in the hip or the knee, for osteolysis to expand, wear debris particles need access to the periprosthetic bone. From review of the orthopedic knee literature, it appears that cemented fixation of knee components is an effective barrier to particle access, and explains the lower incidence of osteolysis in studies with cemented components (Table 8.3). In this regard, much of the knee

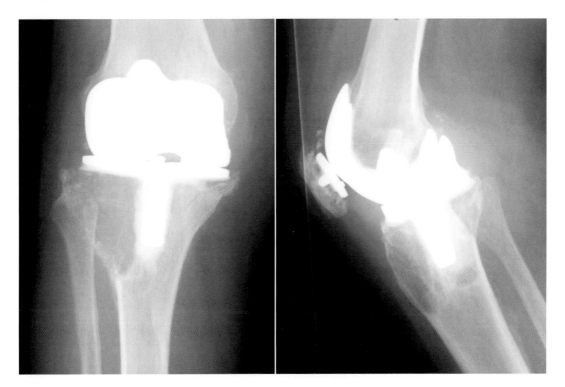

Figure 8.17

Osteolysis in the knee on A–P and lateral radiographs. A large osteolytic lesion is present in the lateral and anterior portions of the tibia.

osteolysis literature, including the 1992 study by Peters and colleagues (1992), relates to this complication in cementless TKA (Table 8.3).

The incidence of osteolysis with cementless fixation has been reported as high as 30% within the first 5 years of implantation (Ezzet, Garcia, and Barrack 1995). In studies of cemented components, in contrast, the incidence of osteolysis is generally lower, and has been reported to range between 0% and 20%. In addition to fixation method, implant factors (e.g., design [Huang et al. 2002, O'Rourke et al. 2002, Whiteside 1995]) and patient factors (e.g., obesity [Spicer et al. 2001]) have been shown to have a significant influence on the incidence of osteolysis in TKA.

Although osteolysis is now recognized as a clinically relevant problem for TKA, a mature understanding of the role of osteolysis in knee implant failure has not yet been reached. Osteolysis may be difficult to visualize using plane radiographs, especially in the femur where the anterior flange of the femoral component obscures the field of view (Huang et al. 2002). Furthermore, osteolysis must be interpreted in the context of normal changes in bone density (bone remodeling) that occurs after TKA.

Not all researchers use the same definition of osteolysis in their radiographic analyses, complicating the comparison of results between studies. Some researchers identify radiolucencies around the margin of an implant as a

Table 8.3
Summary of Osteolysis Studies for Total Knee Arthroplasty

Study	Prosthesis design	Fixation	Mean follow-up (years)	Incidence of osteolysis in study population	Osteolysis diagnosis	Incidence of revisions for osteolysis
Peters et al. (1992)	Synatomic & Arizona (DePuy)	Cementless	3.5	27/174 (16%)	Radiographs	15/27 (56%)
Cadambi et al. (1994)	AMK (DePuy) & PCA (Howmedica)	Cementless	2.6	30/271 (11%)	Radiographs	18/30 (60%)
Ezzet et al. (1995)	AMK (Depuy)	Cemented tibia, Cemented and cementless femur	4.7	17/83 (20%), overall; 0/12 (0%), cemented, no screws; 14/46 (30%), cemented tibia w/screws & uncemented femur	Radiographs	NA
Kim et al. (1995)	PCA (Howmedica)	Cementless	>7	54/60 (90%), tibia; 48/60 (80%), patella; 0/60 (0%) femur	Radiographs	6/48 patellae (13%)
Robinson et al. (1995)	Posterior stabilized (65%), constrained implant (30%)	Cemented and Cementless	4.7	Not studied (revisions only)	Radiographs	17/185 (9%)

Whiteside (1995)	Ortholoc II and Ortholoc Modular, short-stem and long-stem (Wright Medical)	Cementless	3–7 (Ortholoc II) and 2–4 (Ortholoc Modular)	0/675 Ortholoc II; 28/124 (23%) Ortholoc Modular w/long stem; 19/112 (17%) Ortholoc Molar w/short stem	Radiographs	2/47 Ortholoc Modular (4%)
Mikulak et al. (2001)	Posterior-stabilized	Cemented	4.7	Not studied (revisions only)	Radiographs	16/557 (2.9%)
Spicer et al. (2001)	PFC (Johnson & Johnson)	Cemented and cementless	6	29/751 (3.9%)	Radiographs	11/751 (1.5%)
Huang et al. (2002)	LCS (DePuy), PCA (Howmedica), Miller Gallante (Zimmer), Tricon (Smith & Nephew)	Cemented and cementless	8	Not studied (only revisions due to wear and osteolysis studied)	Radiographs & Intraoperative observations	16/34 (47%), mobile bearing group; 6/46 (13%), fixed-bearing group.
O'Rourke et al. (2002)	IB II (Zimmer)	Cemented	6.4	17/105 (16%)	Radiographs	2/17 (12%)
Weber et al. (2002)	AGC (Biomet)	Cemented	6.3 (Monoblock design), 5.5 (Modular design)	40/698 (5.7%), monoblock design; 73/353 (20.7%), modular design	Radiographs	1/40 (2.5%), monoblock design; 6/73 (8.2%), modular design

"linear lytic defect" (Weber et al. 2002), whereas, in other studies, only a focal lesion or cyst is classified as osteolysis (Peters et al. 1992). Of greatest clinical concern are unstable or expansile lesions (so-called balloon lesions) that grow over time and lead to aseptic loosening of a prosthetic component.

Because of difficulty in unambiguously diagnosing osteolysis in the knee, other complications (e.g., infection, instability, malalignment, patellar tracking) may result in implant revision before osteolysis has progressed to the point that it can be identified radiologically. Because of the number of unanswered questions regarding the etiology and significance of osteolysis in TKA, clinical research on this topic is likely to continue for the foreseeable future.

Methods to Assess *In Vivo* Wear in Total Knee Arthroplasty

Clinical interest in osteolysis has led to research on improved clinical methods to quantify UHMWPE wear for TKA. The most widely used clinical measure of wear in the knee is based on analysis of anterior–posterior (A–P) radiographs (Fukuoka, Hoshino, and Ishida 1999) by estimating the minimum distance between the femoral condyles and the tibial base plate (Figure 8.18). The separation between the tibial base plate and the condyles of the femoral component gives an indication of the relative wear between the medial and lateral condyles. If the magnification of the radiograph is known or can be ascertained (based on the known width of the baseplate or tibial stem, for instance), absolute distance measurements can be performed from the radiograph. By comparing changes in apparent component thickness in a series of radiographs over time, the *in vivo* wear rate for a particular TKA patient can be calculated using this "minimum distance" method. However, tilting of the tibial base plate relative to the plane of the radiograph leads to false shortening, as described by Fukuoka (1999), and greatly complicates this radiographic wear assessment technique.

An example of radiographic false shortening produced by tibial tilt is illustrated in Figure 8.18. Without knowing the clinical history of these two radiographs, it is difficult to judge based on direct comparison of the two-dimensional images alone, whether a difference in tibial component height has occurred. In fact, the two radiographs shown in Figure 8.18 are taken of the same patient within 4 months of implantation, and hence the actual amount of wear in Figure 8.18A and 8.18B is actually expected to be negligible, notwithstanding the false shortening captured in Figure 8.18A. Consequently, properly aligned prosthetic components, relative to the plane of the radiograph, are essential for direct quantitative measurement of radiographic wear in TKA.

Despite its limitations, the minimum distance method of estimating wear in TKA may be extremely useful clinically, especially in a case of severe wear, when quantitative radiographic measurements may not be necessary for a surgeon to ascertain that a clinical failure has occurred. For example, severe wear isolated to a single compartment can readily be perceived from a well-aligned radiograph, although it may not always be possible to judge if the component has completely worn through if the tibial component is even slightly tilted.

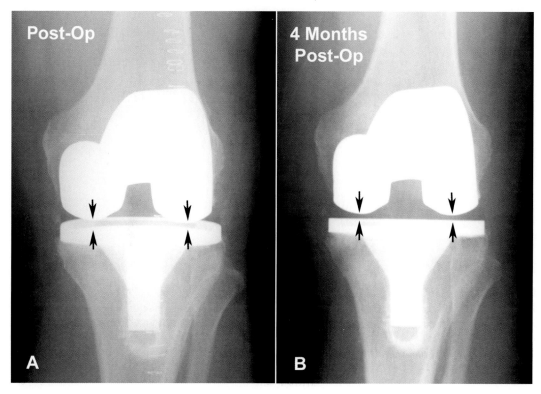

Figure 8.18

Tilting of the femoral and/or tibial components relative to the plane of the x-ray complicates radiographic measurement of wear in TKA. Some studies have measured knee wear as the difference in minimum distance separating the condylar surface of the femur and the tibial tray (denoted with arrows). Note that the initial postoperative radiograph (**A**) shows substantially more tilt of the tibial tray than the radiograph taken 4 months later (**B**).

In the example illustrated in Figure 8.19, revision surgery of the TKA revealed extensive metallosis of the joint space, and examination of the retrieved component confirmed that the component had worn through in the posterior medial compartment. However, except in cases of severe wear, as noted earlier, either careful alignment or computer-aided image analysis techniques are necessary to quantify the minute and gradual changes in tibial component thickness that typically occur over time in TKA based on two-dimensional radiographs.

Sanzén and colleagues (1996) have described a method to measure *in vivo* UHMWPE of tibial components using fluoroscopy-guided A–P radiographs in a series of patients with the same design of knee replacement (PCA, Howmedica, Rutherford NJ). In the first step of Sanzén's procedure, the plane of the A–P radiograph was oriented perpendicular to the plane of the tibial baseplate using fluoroscopy. Image magnification was then corrected by the known width of the baseplate, and the perpendicular distance from the baseplate to the femoral condyles was measured and defined as the "femorotibial distance." In the PCA design, the tibial surface is flat and the component has

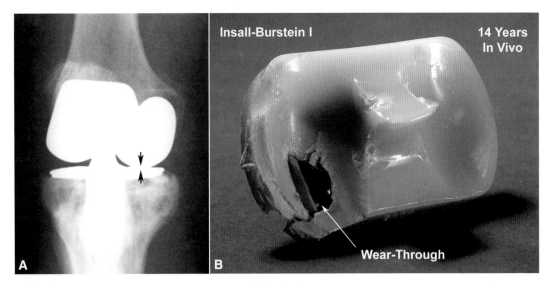

Figure 8.19

Failure of an IB I knee replacement after 14 years *in vivo* due to wear of the UHMWPE component. (**A**) Standing A–P radiograph shows narrowing of the joint space in the medial compartment (indicated by arrows). (**B**) The retrieved UHMWPE component was worn completely through on the medial side. The tibial tray was polished (i.e., burnished) from articulation with the femoral component. Severe metallosis of the surrounding tissues was noted upon revision.

constant thickness in the contact region of the condyles. Consequently, for this particular design, wear was computed by subtracting the two-dimensional tibiofemoral distance from the (known) initial thickness of the component. Using this technique, Sanzén reports absolute *in vivo* wear measurements with a precision of 0.1 mm. Sanzén and colleagues' work does not include a description of the absolute accuracy of their technique, however.

Fukuoka and colleagues (1999) have argued that a two-dimensional radiographic wear measurement technique "seems to include some degree of inaccuracy," even though Sanzén's method accounts for errors introduced by component rotation. In their study, Fukuoka and associates (1999) describes a "simple" method for computing three-dimensional wear from two-dimensional radiographs. Fukuoka's method is referred to in his work as "3-D/2-D matching," in which a three-dimensional computer model of a tibial baseplate and femoral component, along with contours extracted from the A–P radiograph, are used as inputs into optimization software that matches the pose of the implant components necessary to reconstruct the two-dimensional view obtained in the A–P radiograph. Once the "optimal" pose has been found by the software, the three-dimensional wear vectors can be calculated. Fukuoka reports an accuracy (specifically, a root mean square [RMS] error) of 0.04 mm when comparing the tibiofemoral distance calculated using his technique with the physical measurements of actual components in a laboratory setting.

Although fluoroscope-assisted or computer-assisted radiographic techniques are widely employed among the orthopedic community, ultrasound has recently been introduced as a promising alternative for wear assessment in TKA (Sofka, Adler, and Laskin 2003). Using an ultrasound probe, it is possible to visualize not only the femoral and tibial component surfaces, but also to identify the contours of the UHMWPE component (Figure 8.20). Sofka and associates, in a study of 24 TKA patients who were being screened for deep venous thrombosis, found that ultrasonic insert thickness measurements were highly correlated with radiographic measurements ($r^2 = 0.64$) (2003). However, the absolute accuracy of the technique was not reported. Consequently, ultrasound may be regarded as an experimental technique currently under development.

All of the *in vivo* wear measurement techniques described in this section related to TKA have certain drawbacks and unique limitations. Certainly all of these methods are predicated on the presence of a metallic tibial baseplate. In this respect, radiographic wear assessment in the hip and the knee share a common limitation, although the inclusion of a radiographic wire marker in all-UHMWPE cups has somewhat alleviated this concern. In addition, from a practical perspective, both fluoroscope-assisted or computer-assisted radiographic techniques for TKA are more cumbersome and nonstandard than the methods used to obtain radiographs for routine clinical diagnosis. Ultrasound as a wear measurement technique for TKA is in its infancy, insofar as widespread acceptance is concerned. Finally, the absolute accuracy of these *in vivo* wear

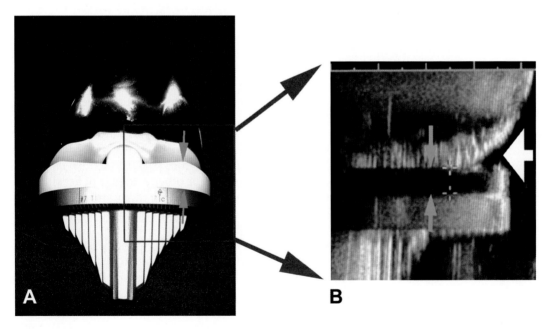

Figure 8.20
Ultrasound as a method to measure UHMWPE wear in TKA. The figure shows a posterior stabilized TKR (**A**) and an ultrasound of the same prosthesis taken in a water bath without soft tissues (**B**). The red inset represents the ultrasound region shown in (**B**). The UHMWPE tibial insert is visible between the green arrows. The large white arrow indicates the direction from which the ultrasound probe was applied. The images were provided courtesy of John Martell, University of Chicago.

measurement techniques, when applied to clinical radiographs, remains as yet unquantified.

For all of these reasons, the most reliable and accurate method of wear assessment in TKA continues to be inspection of UHMWPE components after revision surgery or autopsy removal. The dimensional measurements obtainable from retrieved components using a coordinate measuring machine, for example, have an accuracy at least an order of magnitude better than using radiographic techniques. Further research in knee imaging technology will be needed to close the gap in accuracy between *in vivo* and *ex vivo* wear measurement in TKA.

Backside Wear

We have thus far been concerned primarily with wear at the articulating surface between the femoral condyles and UHMWPE tibial insert. However, recently, researchers have drawn attention to backside articulation (i.e., between the tibial insert and the metallic tray or baseplate) as potentially a clinically relevant source of wear debris (Engh et al. 2001, O'Rourke et al. 2002, Wasielewski et al. 1997). Relative motion between the UHMWPE insert and metallic tray is resisted by the locking mechanism. These mechanisms are proprietary and design specific. Consequently, it is difficult to generalize about the integrity of locking mechanisms and propensity for backside wear.

One well-studied example of a first-generation modular knee replacement is the IB II (Engh et al. 2001, O'Rourke et al. 2002). A retrieved IB II component is illustrated in Figure 8.21 along with the locking mechanism. Backside wear is typically characterized as burnishing or scratching of the UHMWPE component, sometimes with removal of machining marks from the UHMWPE surface. There may also be evidence of extrusion of the UHMWPE into screw holes or recesses on the back surface. All of these features are evident in the retrieved component shown in Figure 8.21. Similar observations have been reported for this design by O'Rourke and colleagues (2002). The tibial baseplate also typically exhibits scratching, burnishing, or other evidence of articulation, especially when the component is fabricated from titanium alloy, as was the case with the IB II. More severe cases of backside wear, not shown in Figure 8.21, include pitting of the UHMWPE surface.

Researchers have used damage-scoring techniques, measured the height of UHMWPE extruded in screw holes, and quantified the relative motion between the insert and the tray (Engh et al. 2001, Wasielewski et al. 1997). Engh and coworkers have recently measured the amount of insert-tray relative motion in 10 different designs of new, retrieved, and autopsy-retrieved tibial inserts (2001). The magnitude of relative motion in the new inserts ranged between 6 and 157 µm. In the retrieved and autopsy-retrieved implants, the magnitude of relative motion ranged between 104 and 718 µm. The researchers concluded that insert-tray relative motion increased after implantation under *in vivo* loading.

The clinical significance of backside wear remains very much open to scientific debate. Engh and coworkers (2001) have postulated that "perhaps the combination of articulating surface wear and backside wear has produced a

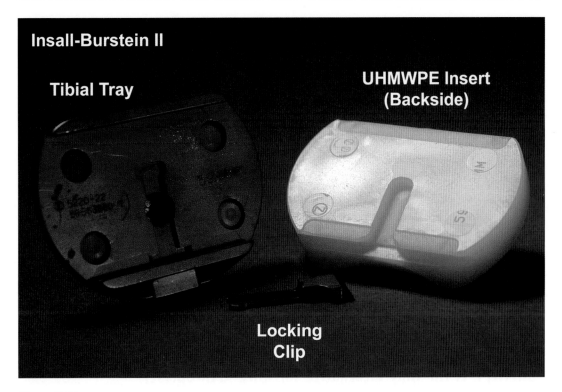

Figure 8.21

Backside wear of an IB II knee replacement. The inferior surface of the tibial tray is burnished from relative motion with respect to the tibial tray. Note the four impressions on the backside of the insert, which correspond to the four screw holes on the tibial tray. Between the screw hole impressions, the machined inscriptions on the backside of the insert have been worn away.

greater volume of debris, which has caused the increased occurrence of osteolysis observed with the use of modular implants." However, the magnitude of relative motion at the back surface is still orders of magnitude less than the sliding that occurs at the articulating surface.

Alternate hypotheses, involving "ease of access" and particle migration, have been suggested to account for differences in osteolysis incidence among different metal-backed tibial designs with otherwise identical locking mechanisms (Whiteside 1995). In other words, modularity may not necessarily increase the total magnitude of wear debris, but may provide easier access of what debris is produced to the bone underlying the tibial baseplate. For these reasons, backside wear in TKA continues to be a controversial topic among members of the orthopedic research community.

Damage to Posts in Posterior-Stabilized Tibial Components

The post in PS condylar components has also been identified as a potential site of impingement and wear in TKA (Callaghan and O'Rourke 2002, Callaghan et al. 2002, Furman et al. 2003). In studies thus far, post damage has been quantified using damage-scoring techniques. Estimates of wear volume generated from the post in PS knees have not yet been reported.

A wide range of damage modes have been observed in posts, the most concerning include fatigue damage, fracture, and adhesive wear (Furman et al. 2003). Although designed to articulate with a cam in the femoral component during flexion, the UHMWPE post may contact the femoral component during hyperflexion, resulting in anterior wear. When axial torque is applied to the knee (from internal/external rotation of the femur with respect to the tibia) the sides and corners of the post may impinge against the sides of the femoral condyles, producing the characteristic "bow-tie" wear scar (Figure 8.22). In more severe cases, fatigue damage or fracture of the post may occur (Furman et al. 2003).

However, it remains to be seen whether PS knees are prone to this type of damage collectively, or if post wear is limited only to certain designs (Callaghan and O'Rourke 2002). Based on a study of 7 retrieved inserts (n = 2 IB/PS II, n = 5 PFC), Callaghan and colleagues suggested that the post served to transmit axial torque across the joint, contributing to backside wear of the inserts (2002). Thus, it is also unclear whether damage to the post is related to wear modes in remote regions of the insert.

UHMWPE Is the Only Alternative for Knee Arthroplasty

UHMWPE plays a crucial role in the long-term success of knee arthroplasty. The articulation of UHMWPE against a stainless steel or cobalt chrome alloy counterface has been the gold standing bearing couple in the knee starting in the 1960s, as was shown in Chapter 7. Without UHMWPE, patients with debilitating knee arthritis would be faced with choosing between a metallic hinge (vintage 1950s technology), with its associated restrictions on activity, or permanent fusion of their joint. It remains a sober fact that there are currently no clinically proven, acceptable alternatives to conventional UHMWPE as a bearing material in the knee.

Despite its successful track record, wear and damage of the UHMWPE insert compromises the longevity of knee arthroplasty. Osteolysis, a rare occurrence with all-UHMWPE tibial components, has been documented with increasing frequency in modular knee replacements. Clearly, improved wear behavior of the UHMWPE insert would be advantageous from the perspective of reducing the risk of osteolysis and aseptic loosening.

Highly crosslinked and thermally treated UHMWPE is a promising wear reduction technology for knee replacements. As covered in Chapter 6, highly crosslinked and thermally treated UHMWPE has been available as a wear-reduction technology for THA since 1998. Because of the reduction in ductility and fracture resistance associated with radiation crosslinking (see Chapter 6), the introduction of this technology to TKA has been somewhat more gradual than what was observed in THA. Recent *in vitro* studies using knee simulators have demonstrated that significant reduction of wear and surface damage can

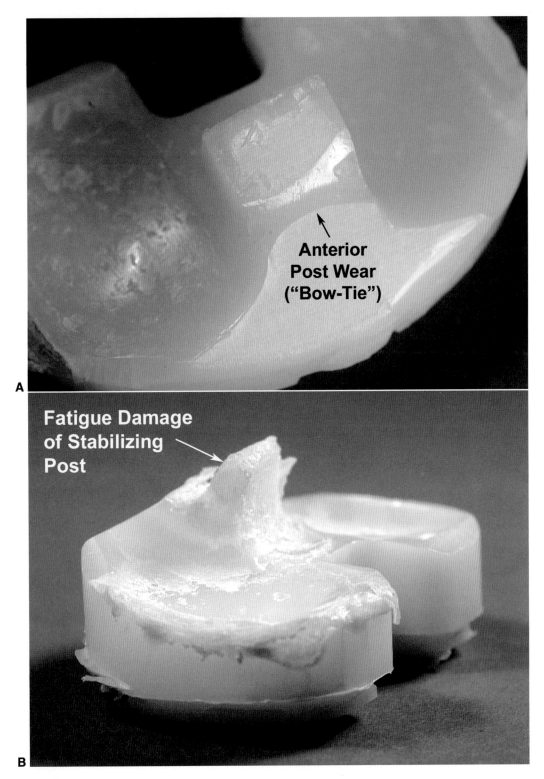

Figure 8.22
Examples of post wear in PFC TKRs (Johnson & Johnson). (**A**) Anterior post wear, producing a "bow tie" wear scar; (**B**) fatigue damage to the stabilizing post.

be achieved using highly crosslinked UHMWPE tibial inserts, even under aggressive loading conditions (Muratoglu et al. 2002, Muratoglu et al. 2003). However, these materials were clinically introduced for knee applications in 2001, and thus it is too early to ascertain whether the benefits observed *in vitro* will be translated to the clinical setting. Additional information about the types of highly crosslinked UHMWPE that are available for use in knee applications can be found in Chapter 15. Because highly crosslinked UHMWPE has not yet been clinically proven in TKA, its use remains controversial and the topic of ongoing orthopedic research.

Acknowledgments

I am grateful to my colleagues at the Rothman Institute, William Hozack, Peter Sharkey, Gina Bissett, James Purtill, and Jay Parvisi. By generous donation of their time and research collaboration, they have deepened my insight into the clinical topics covered in this chapter.

Special thanks to Professor Otto Robertsson, University of Lund, for permission to reproduce the figures related to the Swedish Knee Registry. Thanks also to Clare Rimnac, for editorial assistance with this chapter, and to John Martell, for many helpful discussions and for providing the figure related to ultrasound.

References

Amstutz H.C., P. Grigoris, and F.J. Dorey. 1998. Evolution and future of surface replacement of the hip. *J Orthop Sci* 3:169–186.

Andriacchi T.P., and K.R. Mikosz. 1991. Musculoskeletal dynamics, locomotion, and clinical applications. In *Basic orthopedic biomechanics*. V.C. Mow and W.C. Hayes, Eds. New York: Raven Press.

Bartel D.L., A.H. Burstein, M.D. Toda, and D.L. Edwards. 1985. The effect of conformity and plastic thickness on contact stresses in metal-backed plastic implants. *J Biomech Eng* 107:193–199.

Bartel D.L., V.L. Bicknell, and T.M. Wright. 1986. The effect of conformity, thickness, and material on stresses in ultra-high molecular weight components for total joint replacement. *J Bone Joint Surg* 68:1041–1051.

Bartel D.L., J.J. Rawlinson, A.H. Burstein, et al. 1995. Stresses in polyethylene components of contemporary total knee replacements. *Clin Orthop* 317:76–82.

Bell C.J., P.S. Walker, M.R. Abeysundera, et al. 1998. Effect of oxidation on delamination of ultra-high-molecular-weight polyethylene tibial components. *J Arthroplasty* 13:280–290.

Berger R.A., A.G. Rosenberg, R.M. Barden, et al. 2001. Long-term followup of the Miller-Galante total knee replacement. *Clin Orthop* 58–67.

Blunn G.W., A.B. Joshi, R.J. Minns, et al. 1997. Wear in retrieved condylar knee arthroplasties. A comparison of wear in different designs of 280 retrieved condylar knee prostheses. *J Arthroplasty* 12:281–290.

Burstein A.H., and T.M. Wright. 1994. *Fundamentals of orthopedic biomechanics*. Baltimore.

Cadambi A., G.A. Engh, K.A. Dwyer, and T.N. Vinh. 1994. Osteolysis of the distal femur after total knee arthroplasty. *J Arthroplasty* 9:579–594.

Callaghan J.J., and M.R. O'Rourke. 2002. Picking your implant: All PS knees are not alike! *Orthopedics* 25:977–978.

Callaghan J.J., M.R. O'Rourke, D.D. Goetz, et al. 2002. Tibial post impingement in posterior-stabilized total knee arthroplasty. *Clin Orthop* 404:83–88.

Dahlkvist N.J., P. Mayo, and B.B. Seedhom. 1982. Forces during squatting and rising from a deep squat. *Eng Med* 11:69–76.

Davy D.T., and M.L. Audu. 1987. A dynamic optimization technique for predicting muscle forces in the swing phase of gait. *J Biomech* 20:187–201.

Ellis M.I., B.B. Seedhom, and V. Wright. 1984. Forces in the knee joint whilst rising from a seated position. *J Biomed Eng* 6:113–120.

Engh G.A., S. Lounici, A.R. Rao, and M.B. Collier. 2001. In vivo deterioration of tibial baseplate locking mechanisms in contemporary modular total knee components. *J Bone Joint Surg* 83A:1660–1665.

Ezzet K.A., R. Garcia, and R.L. Barrack. 1995. Effect of component fixation method on osteolysis in total knee arthroplasty. *Clin Orthop* 321:86–91.

Fehring T.K., S. Odum, W.L. Griffin, et al. 2001. Early failures in total knee arthroplasty. *Clin Orthop* 392:315–318.

Forster M.C. 2003. Survival analysis of primary cemented total knee arthroplasty: Which designs last? *J Arthroplasty* 18:265–270.

Frankowski J.J., and S. Watkins-Castillio. 2002. *Primary total knee and total hip arthroplasty projections for the U.S. population to the year 2030*. Rosemont, IL: American Association of Orthopedic Surgeons.

Fukuoka Y., A. Hoshino, and A. Ishida. 1999. A simple radiographic measurement method for polyethylene wear in total knee arthroplasty. *IEEE Trans Rehabil Eng* 7:228–233.

Furman B.D., F. Mahmood, T.M. Wright, and S.B. Haas. 2003. Insall Burstein PS II has more severe anterior wear and fracture of the tibial post than the Insall Burstein I. *Transactions of the Orthopedic Research Society* 28:1404.

Harrington I.J. 1976. A bioengineering analysis of force actions at the knee in normal and pathological gait. *Biomed Eng* 11:167–172.

Hood R.W., T.M. Wright, and A.H. Burstein. 1983. Retrieval analysis of total knee prostheses: A method and its application to 48 total condylar prostheses. *J Biomed Mater Res* 17:829–842.

Huang C.H., H.M. Ma, J.J. Liau, et al. 2002. Osteolysis in failed total knee arthroplasty: A comparison of mobile-bearing and fixed-bearing knees. *J Bone Joint Surg* 84A:2224–2229.

Kim Y.H., J.H. Oh, and S.H. Oh. 1995. Osteolysis around cementless porous-coated anatomic knee prostheses. *J Bone Joint Surg* 77:236–241.

Kurtz S.M., D.L. Bartel, and C.M. Rimnac. 1998. Post-irradiation aging affects the stresses and strains in UHMWPE components for total joint replacement. *Clin Orthop* 350:209–220.

Kurtz S.M., C.M. Rimnac, L. Pruitt, et al. 2000. The relationship between the clinical performance and large deformation mechanical behavior of retrieved UHMWPE tibial inserts. *Biomaterials* 21:283–291.

Lindgren L. 2002. Annual Report 2002—The Swedish Knee Arthroplasty Register—Part II. Lund, Sweden: University of Lund.

Lu T.W., J.J. O'Connor, S.J. Taylor, and P.S. Walker. 1998. Validation of a lower limb model with in vivo femoral forces telemetered from two subjects. *J Biomech* 31(1): 63–69.

Mikulak S.A., O.M. Mahoney, M.A. dela Rosa, and T.P. Schmalzried. 2001. Loosening and osteolysis with the press-fit condylar posterior-cruciate-substituting total knee replacement. *J Bone Joint Surg* 83A:398–403.

Morrison J.B. 1969. Function of the knee joint in various activities. *Biomed Eng* 4:573–580.

Morrison J.B. 1970. The mechanics of the knee joint in relation to normal walking. *J Biomech* 3:51–61.

Muratoglu O.K., C.R. Bragdon, D.O. O'Connor, et al. 2002. Aggressive wear testing of a cross-linked polyethylene in total knee arthroplasty. *Clin Orthop* 404:89–95.

Muratoglu O.K., A. Mark, D.A. Vittetoe, et al. 2003. Polyethylene damage in total knees and use of highly crosslinked polyethylene. *J Bone Joint Surg* 85A Suppl 1:S7–S13.

Olney S.J., and D.A. Winter. 1985. Predictions of knee and ankle moments of force in walking from EMG and kinematic data. *J Biomech* 18:9–20.

O'Rourke M.R., J.J. Callaghan, D.D. Goetz, et al. 2002. Osteolysis associated with a cemented modular posterior-cruciate-substituting total knee design: Five to eight-year follow-up. *J Bone Joint Surg* 84A:1362–1371.

Patriarco A.G., R.W. Mann, S.R. Simon, and J.M. Mansour. 1981. An evaluation of the approaches of optimization models in the prediction of muscle forces during human gait. *J Biomech* 14:513–525.

Paul J.P. 1976. Force actions transmitted by joints in the human body. *Proc R Soc Lond B Biol Sci* 192:163–172.

Peters P.C., Jr., G.A. Engh, K.A. Dwyer, and T.N. Vinh. 1992. Osteolysis after total knee arthroplasty without cement. *J Bone Joint Surg* 74:864–876.

Piziali R.L., W.P. Seering, D.A. Nagel, and D.J. Schurman. 1980. The function of the primary ligaments of the knee in anterior-posterior and medial-lateral motions. *J Biomech* 13:777–784.

Reilly D.T., and M. Martens. 1972. Experimental analysis of the quadriceps muscle force and patello-femoral joint reaction force for various activities. *Acta Orthop Scand* 43:126–137.

Robertsson O. 2000. The Swedish Knee Arthroplasty Register: Validity and outcome. Ph.D. Diss., Lund, Sweden: Lund University.

Robertsson O., S. Lewold, K. Knutson, and L. Lidgren. 2000. The Swedish Knee Arthroplasty Project. *Acta Orthop Scand* 71:7–18.

Robertsson O., K. Knutson, S. Lewold, and L. Lidgren. 2001a. The Swedish Knee Arthroplasty Register 1975–1997: An update with special emphasis on 41,223 knees operated on in 1988-1997. *Acta Orthop Scand* 72:503–513.

Robertsson O., K. Knutson, S. Lewold, and L. Lidgren. 2001b. The routine of surgical management reduces failure after unicompartmental knee arthroplasty. *J Bone Joint Surg* 83:45–49.

Robinson E.J., B.D. Mulliken, R.B. Bourne, et al. 1995. Catastrophic osteolysis in total knee replacement. A report of 17 cases. *Clin Orthop* 321:98–105.

Rohrle H., R. Scholten, C. Sigolotto, et al. 1984. Joint forces in the human pelvis-leg skeleton during walking. *J Biomech* 17:409–424.

Sanzén L., A. Sahlstrom, C.F. Gentz, and I.R. Johnell. 1996. Radiographic wear assessment in a total knee prosthesis. 5- to 9-year follow-up study of 158 knees. *J Arthroplasty* 11:738–742.

Sathasivam S., and P.S. Walker. 1999. The conflicting requirements of laxity and conformity in total knee replacement. *J Biomech* 32:239–247.

Sharkey P.F., W.J. Hozack, R.H. Rothman, et al. 2002. Insall Award paper. Why are total knee arthroplasties failing today? *Clin Orthop* 7–13.

Shoji H., R.D. D'Ambrosia, and P.R. Lipscomb. 1976. Failed polycentric total knee prostheses. *J Bone Joint Surg* 58:773–777.

Sofka C.M., R.S. Adler, and R. Laskin. 2003. Sonography of polyethylene liners used in total knee arthroplasty. *AJR Am J Roentgenol* 180:1437–1441.

Spicer D.D., D.L. Pomeroy, W.E. Badenhausen, et al. 2001. Body mass index as a predictor of outcome in total knee replacement. *Int Orthop* 25:246–249.

Taylor S.J., and P.S. Walker. 2001. Forces and moments telemetered from two distal femoral replacements during various activities. *J Biomech* 34:839–848.

Thadani P.J., K.G. Vince, S.G. Ortaaslan, et al. 2000. Ten- to 12-year followup of the Insall-Burstein I total knee prosthesis. *Clin Orthop* 380:17–29.

Trousdale R.T., and M.W. Pagnano. 2002. Fixed-bearing cruciate-retaining total knee arthroplasty. *Clin Orthop* 404:58–61.

Vince K.G. 2003. Why knees fail. *J Arthroplasty* 18:39–44.

Walker P.S. 1977a. Historical development of artificial joints. In *Human joints and their artificial replacements*. P.S. Walker, Ed. Springfield, IL: Charles C. Thomas.

Walker P.S. 1977b. *Human joints and their artificial replacements*. Springfield, IL: CC Thomas.

Wasielewski R.C., N. Parks, I. Williams, et al. 1997. Tibial insert undersurface as a contributing source of polyethylene wear debris. *Clin Orthop* 345:53–59.

Weber A.B., R.L. Worland, J. Keenan, and J. Van Bowen. 2002. A study of polyethylene and modularity issues in >1000 posterior cruciate-retaining knees at 5 to 11 years. *J Arthroplasty* 17:987–991.

Whiteside L.A. 1995. Effect of porous-coating configuration on tibial osteolysis after total knee arthroplasty. *Clin Orthop* 321:92–97.

Won C.H., S. Rohatgi, M.J. Kraay, et al. 2000. Effect of resin type and manufacturing method on wear of polyethylene tibial components. *Clin Orthop* 376:161–171.

Wright T.M., C.M. Rimnac, S.D. Stulberg, et al. 1992. Wear of polyethylene in total joint replacements: Observations from retrieved PCA knee implants. *Clin Orthop* 276:126–134.

Chapter 8. Reading Comprehension Questions

8.1. Which of the following daily activities can produce the greatest compression forces in the tibiofemoral joint?
a) Walking slowly
b) Walking quickly
c) Rising from a chair
d) Stair climbing
e) Squatting

8.2. Which factors influence the stresses acting on UHMWPE tibial components?
a) Conformity of the joint surfaces
b) Surgical malpositioning
c) Magnitude of the joint reaction force
d) Tibial component thickness
e) All of the above

8.3. What is delamination?
   a) Surface damage in which a sheet of UHMWPE is removed from the surface
   b) Surface damage produced by third bodies
   c) Microscopic damage produced by scratching
   d) Microscopic damage to the crystalline lamellae
   e) None of the above

8.4. Which of the following knee component designs have the best 10-year survivorship?
   a) Posterior stabilized designs
   b) Metal backed tibial component designs
   c) All-UHMWPE tibial component designs
   d) Cruciate sparing designs
   e) None of the above

8.5. Using the Hood method, what is the maximum total damage score for the articulating surface of a bicondylar, cruciate-sparing total knee replacement, not including the patellar component?
   a) 10
   b) 40
   c) 160
   d) 210
   e) 360

8.6. Roughly what proportion of total knee replacement patients are currently considered to develop osteolysis within the first 10 years of implantation?
   a) Greater than 90%
   b) 75–90%
   c) 61–74%
   d) 31–60%
   e) Less than 30%

8.7. Which of the following methods has most recently been developed to measure *in vivo* wear of TKRs?
   a) Direct measurement of calibrated radiographs
   b) Fluoroscopy-assisted measurement of calibrated radiographs
   c) Computer-assisted measurement of calibrated radiographs
   d) Direct measurement of calibrated ultrasound images
   e) None of the above

8.8. Backside wear in TKA
   a) Is not significantly affected by the locking mechanism design
   b) Occurs between the tibial base plate and the tibial component
   c) Occurs between the cam of the femoral component and the posterior aspect of the tibial post
   d) Occurs only after 10 years of implantation
   e) None of the above

Chapter 9

The Clinical Performance of UHMWPE in Shoulder Replacements

Stefan Gabriel
Smith & Nephew Endoscopy
Mansfield, MA

Introduction

Shoulder replacement, although done much less frequently than hip and knee replacement, is the third most prevalent joint replacement procedure worldwide. Current shoulder replacement systems rely on ultra-high molecular weight polyethylene (UHMWPE) components for motion and load bearing. Because of this critical role, the performance of UHMWPE components can determine the overall performance of the replacement system.

When considering the performance of UHMWPE in shoulder replacement components, one should have a basic understanding of the anatomical and biomechanical system into which they are placed, as well as the ways that system can be compromised by disease or trauma. It is also useful to gain a historical perspective on the origins and evolution of design and technique, as well as see a number of systems currently in use. It is especially critical to also examine overall measures of replacement success, as well as specific measures of wear or damage to directly assess the performance of UHMWPE components in particular. This chapter presents information covering these areas of consideration, as well as a discussion of alternatives to standard, contemporary device designs and materials to give an overview of shoulder replacement, and an assessment of UHMWPE performance in clinical use in the shoulder.

The Shoulder Joint

The shoulder is made up of a number of bones, ligaments, and muscles (Figure 9.1). The partial ball defined by the head of the humerus and the partial socket defined by the glenoid of the scapula form the main articulating ball and socket geometry of the joint. Along with the passive and active stability provided by the surrounding joint capsule, ligaments, and the muscles of the rotator cuff, the geometry allows the normal shoulder joint to achieve the largest range of motion of any joint in the human body. Rotation of the humerus relative to the glenoid of the scapula allows positioning of the elbow at any of a number of points on an essentially spherical surface covering nearly a full hemisphere.

Things that can upset the structural balance of the shoulder include arthritis, tendon and ligament abrasions and ruptures, and deterioration and fracture of the bones. Rheumatoid arthritis (RA) affects the soft tissues and cartilage around the joint causing a loss of stiffness, strength, and integrity of the affected structures leading to instability of the shoulder. In the absence of other disease processes, RA commonly results in erosion of the humeral head and central erosion of the glenoid face with a corresponding loss of stabilization (Hill and Norris 2001, Hayes and Flatow 2001).

Osteoarthritis (OA) affects the articular cartilage and underlying bones at the joint. Primary OA is associated with progressive wearing of the humeral head and the posterior aspect of the glenoid (Figure 9.2) (Hill and Norris 2001, Hayes

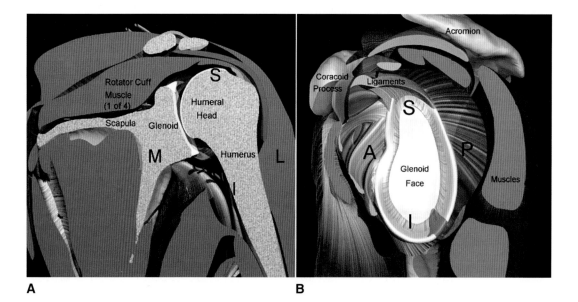

A   **B**

Figure 9.1

Normal anatomy of the glenohumeral joint showing the bones, muscles, tendons, ligaments, and capsule at the shoulder in frontal section (**A**) and glenoid face (**B**) views. The superior (S), inferior (I), medial (M), lateral (L), anterior (A), and posterior (P) directions are also denoted on each view (images used with permission of Primal Pictures, Ltd., London).

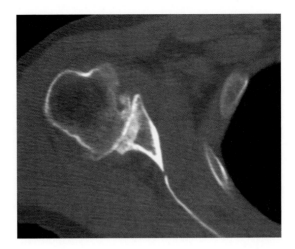

Figure 9.2
Computed tomography image of a shoulder showing an axillary view (looking up under the arm). The lateral direction is to the left, and the anterior direction is up in this view. Note the damage to the humeral head and the severe erosion on the posterior aspect of the glenoid (image courtesy of Jon JP Warner, Massachusetts General Hospital, Boston).

and Flatow 2001). Dislocation caused by laxity or loss of stabilization in one direction or another due to a specific ligament or muscle's weakness or tearing can cause wear of the humeral head and glenoid in various locations, depending on the direction of instability. Fractures due to trauma or the presence of tumors can result in multiple bone fragments. This most often affects the proximal humerus, requiring repair of the bone and reconstruction of the muscle and ligament attachment sites.

Shoulder Replacement

Procedures

When disease, pathology, or trauma lead to debilitating pain or unacceptable loss of function, shoulder arthroplasty, the replacement of one or both of the articular surfaces of the glenohumeral joint, is often performed. In a hemiarthroplasty, only the humeral articular surface is replaced and in a total shoulder arthroplasty (TSA), both humeral and glenoid articular surfaces are replaced (Figure 9.3). To replace the humeral articular surface, the humeral head is resected at its base and the medullary cavity of the proximal humeral bone is prepared to accept the metal stem of a humeral prosthesis. In a fracture repair procedure, the bone fragments are reapproximated and held in place around the stem and under the head of a metal humeral component by wires. Many humeral components include holes in the stem to facilitate this reattachment. In a resurfacing procedure, the humeral head is shaped to accept a metal cap. Biological fixation (bone ongrowth and/or ingrowth), fixation with

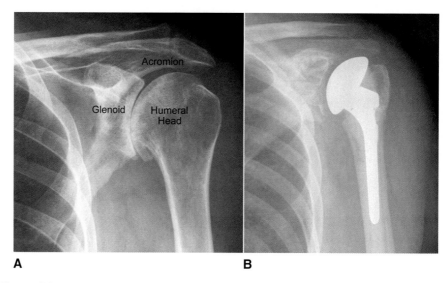

Figure 9.3

Radiographic images of frontal views of shoulders before and after total shoulder arthroplasty. (**A**) The preoperative anatomy and (**B**) the placement of the humeral component can be clearly seen. The UHMWPE glenoid component in (**B**) is radiolucent and is evident only as a space between the humeral component and bone of the glenoid (images courtesy of Jon JP Warner, Massachusetts General Hospital).

polymethylmethacrylate (PMMA) bone cement, and/or fixation via press-fit are relied on to hold the humeral component within or on the bone.

In a TSA, the glenoid articular surface is also removed and the underlying glenoid bone is shaped to receive a glenoid component. In some cases of glenoid erosion, fracture, or malformation, bone graft material is placed under one side or the other of a glenoid component to correct glenoid position. Glenoid components are held in place with biological fixation, screw fixation through holes in the metal backing of a component and into the bone, and/or with PMMA bone cement.

Patient Population

The number of shoulder arthroplasty procedures done yearly has grown from an estimated 5000 per year in the United States between 1990 and 1992 (Wirth and Rockwood, Jr. 1996) to around 15,000 per year between 2002 and 2003 (Jackson 2003). Of these, a substantial portion are hemiarthroplasties. The prevalence of hemiarthroplasties as a percentage of the total number of replacements ranges from 14% (Edwards et al. 2002) in a multicenter study of OA, to 65% in a study of fracture repair (Boileau et al. 2001), to 87% in the Swedish shoulder arthroplasty register started in 1999 (Rahme, Jacobsen, and Salomonsson 2001). Dr. Charles Neer, II, the pioneer of the procedure, estimates that overall, around 20% of shoulder replacements are hemiarthroplasties (2003).

The reported average patient age for a series of shoulder replacements is in the range of 55 to 73 years old with most studies reporting averages from 55 to 65 years (Arredondo and Worland 1999, Boileau et al. 2001, Boileau et al. 2002, Edwards et al. 2002, Godeneche et al. 2002, Goldberg et al. 2001, Hasan et al. 2002, Hill and Norris 2001, Iannotti and Norris 2003, Neer, II, 1955, Neer, II, 1974, Neer, II, Watson, and Stanton 1982, Norris and Iannotti 2002, Sanchez-Sotelo et al. 2001, Sojbjerg et al. 1999, Torchia, Cofield, and Settergren 1997, Worland et al. 1997). Remarkably, this has remained relatively constant since the first published series of hemiarthroplasties (Neer, II, 1955, Neer, II, 1974). Sperling, Cofield, and Rowland (1998) along with Hayes and Flatow (2001) define "young" patients as being younger than 50 years old.

Arthritis is noted as the reason for more than half of all shoulder replacements. A sample of published studies shows a range of 50 to 74% of all indications being arthritis (Neer, II, Watson, and Stanton 1982, Rahme, Jacobsen, and Salomonsson 2001, Snyder 1996, Sojbjerg et al. 1999). Within the arthritis group, the largest number of patients present with RA and the next largest number with OA. Sojbjerg and colleagues report this ratio of RA to OA at 2 to 1 (1999), both Rahme and colleagues (2001) and Snyder (1996) at about 3 to 2, Sanchez-Sotelo and coworkers at 4 to 3 (2001), and Neer and colleagues (1982) and Torchia and associates (1997) at approximately 1 to 1.

Fracture/trauma is the next most common problem addressed by shoulder replacement. Snyder reports the rate as 9% (1996), Torchia and associates as 12% (1997), Neer and colleagues as 23% (1982), Sojbjerg and colleagues as 30% (1999), and Rahme and associates as 35% (2001). In fact, repair of the humerus after fracture was the driving reason for the development of the first modern shoulder replacement component (Neer, II 2003) and the first reported series of modern shoulder replacements were done to correct problems caused by humeral fractures (Neer, II 1974).

All other indications leading to total shoulder replacement (TSR) are generally below approximately 5% each over a range of reported studies (Snyder 1996). It is interesting to note, however, that replacement required for revision of prior surgery has been reported to be from 0% to about 4% (Snyder 1996) to 6% (Rahme, Jacobsen, and Salomonsson 2001) up to 11% (Neer, II, Watson, and Stanton 1982) of all cases.

History

The first recorded TSR was performed by a French surgeon named Péan in 1893 using a constrained design (connected glenoid and humeral components) to treat tuberculous arthritis of the shoulder (Lugli 1978). The patient was a 37-year-old man and the prosthesis components reportedly functioned relatively well until they were removed 2 years later because of infection (Lugli 1978).

The modern era of shoulder replacement was ushered in by Dr. Charles Neer, II. In 1953, in response to the relatively poor results of humeral head resection for patients with proximal humeral fractures, Dr. Neer implanted a vitallium humeral component of his own design in a hemiarthroplasty procedure (Figure 9.4). In 1955, he reported on his first series of 12 patients treated in this way.

Figure 9.4

Humeral component designed and used by Dr. Neer for his first shoulder hemiarthroplasties (image courtesy of Howmedica Osteonics, Allendale, NJ).

Constrained, fixed-fulcrum (captured ball in socket) devices for TSA were also considered by Dr. Neer and others. In 1972, Neer and Robert Averill tried three designs of fixed-fulcrum prostheses, but function was inadequate (Neer, II, Watson, and Stanton 1982).

Successful TSRs were also performed in the early 1970s by Stillbrink (four patients), Kenmore (three patients), and Zipple using polyethylene glenoid components with Neer's original humeral component (Neer, II 1974, Rockwood, Jr. 2000). This is the first recorded use of polyethylene in the shoulder and the date identifies the polyethylene as RCH 1000, the trade name for UHMWPE produced by Hoechst in Germany.

In 1973, Dr. Neer also performed a TSR for the first time (Neer, II 1974). He used a cemented polyethylene glenoid component with his Neer humeral prosthesis that had been modified to more beneficially articulate with a glenoid component (Neer, II 1974). The polyethylene glenoid, an RCH 1000 device, was designed by Robert G. Averill, manufactured by Howmedica, and sterilized by ethylene oxide gas (Neer, II 2003). Between 1973 and 1982, Dr. Neer designed a number of glenoid components, both with and without metal backings (Neer, II, Watson, and Stanton 1982).

The foundation that Neer laid for shoulder replacement can be seen today in many ways. The continued production and use of the 1973 version of his humeral component as the Neer II and the many other humeral and glenoid components based on this original design is testament to its stature in the surgical community. The basic concepts included in the Neer prosthesis such as all-polyethylene and metal-backed keeled glenoid components and wire fixation holes and stabilization flanges on press-fit humeral stems persist in current designs.

Biomechanics of Total Shoulder Replacement

During the course of normal activities of arm use, the loads at the shoulder, and more specifically, the loads between the humeral head and the glenoid can vary widely in both magnitude and direction. Anglin and colleagues give a review of the reported resultant glenohumeral forces from different studies and for different activities as a multiple of body weight (BW) (2000). These range in magnitude from 0.9 × BW during arm abduction to more than 7 × BW during push-ups (Anglin, Wyss, and Pichora 2000). For a realistic BW of 160 pounds for a shoulder patient, these would result in glenohumeral forces ranging from 144 to 1120 pounds. These are relatively large loads that could be a factor in damage to the UHMWPE of a glenoid component.

The direction of this force also varies with different activities. Anglin and associates report a variation of superiorly directed force on the glenoid from 10.9 to 23.6 degrees and a variation of anterior–posteriorly directed force from 2.2 degrees posterior to 17.5 degrees anterior for a variety of common activities (2000). Poppen and Walker report a variation of superior–inferiorly directed force on the glenoid in the plane of the scapula from around 60 degrees inferior to approximately 60 degrees superior during arm abduction (1978). The direction and variation also are shown to depend on the neutral, internal rotation, or external rotation position of the humerus during abduction with internal rotation resulting in more superiorly directed forces and external rotation resulting in more inferiorly and horizontally directed forces (Poppen and Walker 1978). The changing direction of the load, both superior–inferiorly and anterior–posteriorly is noteworthy. That is because of the clinically identified "rocking horse" glenoid, which describes a loosened glenoid component being alternately angled one way and then another by the loads applied to it.

Finite element studies have been conducted using the load magnitudes and directions mentioned earlier in order to examine predicted stresses within all-polyethylene and metal-backed glenoid components, within the cement mantle used for component fixation, and in the surrounding glenoid bone (Lacroix, Murphy, and Prendergast 2000, Lacroix and Prendergast 1997). Of particular interest are stresses in the cement mantle because they can influence the long-term fixation of the polyethylene in the bone. For all-UHMWPE glenoid component designs, increased cement stresses were predicted for keeled versus pegged designs (Lacroix, Murphy, and Prendergast 2000). Decreased cement stresses were predicted for metal-backed versus all-UHMWPE glenoid component designs (Lacroix and Prendergast 1997).

Loads of nontrivial magnitude, such as those described previously, moving across the surface of the glenoid component have the potential to cause stress-related damage and wear to the polyethylene similar to that seen in UHMWPE knee and hip joint replacement components. Stresses in the UHMWPE of the glenoid component articular surface have been studied by Swieszkowski and colleagues (2003) following the methods of Bartel and associates (1986). As shown by Bartel and coworkers and again by Swieszkowski and colleagues, stresses predicted in an UHMWPE component vary with thickness of the polyethylene, conformity between the metal and polyethylene surfaces, and the presence or absence of metal backing (Bartel, Bicknell, and Wright 1986, Swieszkowski, Bednarz, and Prendergast 2003). For loads and geometries corresponding to glenoid and shoulder humeral components, the stresses in the UHMWPE were predicted to be in the range of 5 to 30 MPa (Swieszkowski, Bednarz, and Prendergast 2003). Stresses were substantially higher for less conforming humeral–glenoid component pairs than for more conforming ones, and somewhat higher for glenoid components with metal backings and thinner UHMWPE layers (Swieszkowski, Bednarz, and Prendergast 2003).

The effects of load and conformity between the humeral and glenoid components on joint stability has also been studied. A study by Karduna and coworkers investigated the importance of prosthetic glenohumeral conformity in reproducing the force-displacement relationships of the natural joint as well as its implications for glenoid component stresses (1998). Relatively nonconforming components were shown to develop lower strains and better reproduce natural glenohumeral force-displacement relationships than more conforming components (Karduna et al. 1998). Studies by Warner and associates examined the relative importance of the structures at the shoulder in maintaining that force-displacement relationship for inferior humeral translation (Warner and Warren 2001, Warner et al. 1999). It was shown that increased compressive force across the joint provided by the rotator cuff muscles was the most important factor in maintaining stability (Warner and Warren 2001, Warner et al. 1999).

It should be noted that the loads, motions, and stresses considered and reported in the studies assumed or implied normal kinematics and geometry of the shoulder joint and anatomically correct placement of humeral and glenoid components relative to the other structures at the shoulder. This is noteworthy because, in reality, the disease or trauma state that leads to a TSA most often affects one or more of the surrounding bony, ligamentous, tendinous and muscular structures at the shoulder, leaving it in a deteriorated state. In fact, in some cases, it is a deterioration of one or more of the structures at the shoulder that is the underlying cause of the disease progression to the point of requiring shoulder replacement.

The humeral and glenoid components are not placed into an otherwise well-functioning joint environment, but rather into a surgically disrupted (to implant the components) and surgically repaired joint environment. There is also a good possibility that the humeral and glenoid components do not adequately replicate the geometry, properties, and/or location of the natural articular surfaces relative to the surrounding shoulder structures. The effects of these situations are that, in addition to normal loads and motions, replacement components can be faced with potentially damaging exceptions to the normal shoulder loads and motions.

Contemporary Total Shoulder Replacements

It has been estimated that more than 100 shoulder prosthesis designs are currently in use worldwide (Rockwood, Jr. 2000). Where there is articulation between components, UHMWPE is used on one of the counterfaces, the other being polished metal, usually cobalt chromium alloy. Methods of differentiation between designs include the mechanical interaction between components (e.g., fixed-fulcrum, semiconstrained, unconstrained, etc.), the intended indications to be addressed (e.g., primary, total, revision, fracture, etc.), and/or the overall nature of the design (e.g., monoblock, modular, bipolar, metal-backed glenoid, etc.). These categories can aid in the understanding of the classes of designs that exist, but they are not definitive because of the overlap between groups.

Some currently used prosthesis systems that include UHMWPE components are (alphabetically by manufacturer):

Biomet (Warsaw, IN) (Figure 9.5)

> Bio-Modular®. A modular cobalt chromium humeral component with keeled and three-pegged UHMWPE glenoid components developed with Drs. Russell Warren and David Dines and introduced in 1986. The glenoid components are sterilized by gamma in argon. It is intended for primary, hemiarthroplasty, total, and fracture use with cemented fixation.

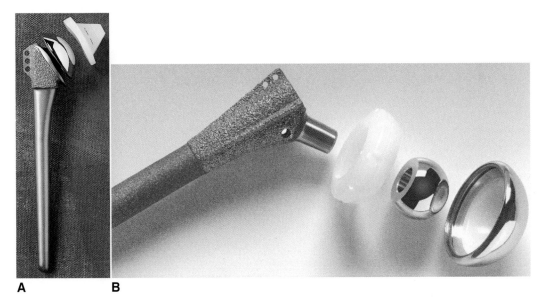

A   **B**

Figure 9.5

BioModular and BiAngular BiPolar contemporary shoulder prosthesis system components. The BioModular components are shown in (**A**) and an exploded view of the BiPolar assembly is shown in (**B**) (images courtesy of Biomet, Warsaw, IN).

Bi-Angular®. A modular cobalt chromium or titanium stem humeral component with UHMWPE glenoid components with an angled keel or metal backed with an angled keel and three pegs sterilized by gamma in argon. The system was developed with Dr. Richard Worland and introduced in the mid-1990s. It is intended for primary, hemiarthroplasty, total, and fracture use with cemented fixation.

BiPolar. A modular head with a cobalt chromium shell and an inner UHMWPE bearing surface and retaining ring capturing an inner cobalt chromium head. The concept was originally developed in the 1970s and was intended for use in a hemiarthroplasty salvage procedure.

Integrated®. A collection of monoblock and modular cobalt chromium humeral prostheses within a single system. Included are the Kirschner II-C monoblock, and the Atlas® modular design. Glenoid components include all-poly UHMWPE keeled and pegged components as well as a metal-backed, screw-fixed design. Components of the system are intended for primary, hemiarthroplasty, total, and fracture use with cemented and press-fit fixation.

Centerpulse (Austin, TX) (Figure 9.6)

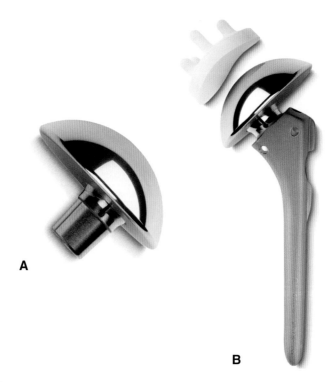

A

B

Figure 9.6

Anatomical Shoulder contemporary shoulder prosthesis system components. An offset humeral head is shown in (**A**) and the noncemented humeral stem assembly is shown in (**B**). (Images courtesy of Centerpulse, Austin, TX).

Select®. Modular titanium stem humeral prostheses with keeled and pegged UHMWPE glenoid components sterilized by gamma in an oxygen-free environment. Developed with Dr. Wayne Burkhead, it was introduced in 1987. It is intended for primary, hemiarthroplasty, total, and fracture use with cemented fixation.

Anatomical Shoulder™. Modular titanium stem and cobalt chromium humeral head prostheses with UHMWPE four-pegged glenoid components sterilized by gamma in an oxygen-free environment. Developed with Drs. Christian Gerber and Jon Warner, the first implantation was performed in 1995. It is intended for primary, hemiarthroplasty, total, and fracture use with cemented and press-fit fixation.

DePuy Orthopaedics (Warsaw, IN) (Figure 9.7)

Global™ and Global™ Advantage® and Global™ FX. Modular cobalt chromium humeral prostheses developed with Drs. Charles Rockwood and Fredrick Matsen, III, and first introduced in 1990. They include UHMWPE keeled and five-pegged glenoid components sterilized by gas plasma. They are intended for primary, hemiarthroplasty, total, and fracture use (FX specifically) with press-fit fixation.

Encore (Austin, TX)

Foundation® and Foundation® fracture. Modular titanium stem humeral prostheses with keeled and pegged UHMWPE glenoid components sterilized

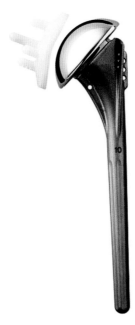

Figure 9.7

Global contemporary shoulder prosthesis system components (image courtesy of DePuy Orthopaedics, Warsaw, IN).

by gamma in nitrogen. Developed with Drs. Richard Friedman and Mark Frankle, it is intended for primary, hemiarthroplasty, total, and fracture use with cemented and press-fit fixation.

Howmedica Osteonics (Allendale, NJ) (Figure 9.8)

Solar®. Modular titanium stem humeral prostheses with UHMWPE angled-peg glenoid components sterilized by gamma in nitrogen and a vacuum. They are intended for primary, hemiarthroplasty, total, and fracture use with cemented and press-fit fixation. The system also includes a bipolar head that can be used with the stems.

Smith & Nephew (Memphis, TN) (Figure 9.9)

Neer II™. The Neer monoblock cobalt chromium prosthesis design as developed in 1973 by Dr. Charles Neer, II (Neer, II 1974, Neer, II, Watson, and Stanton 1982). It includes a keeled UHMWPE glenoid component sterilized by ethylene oxide gas (EtO). The Neer II design is also marketed by Biomet. Dr. Neer originally presented five different UHMWPE glenoid components, two all-poly and three metal-backed that were designed for use with the humeral component and allowed the choice of a range of different constraints (Neer, II, Watson, and Stanton 1982). It is intended

Figure 9.8

Solar contemporary shoulder prosthesis system components (image courtesy of Howmedica Osteonics, Allendale, NJ).

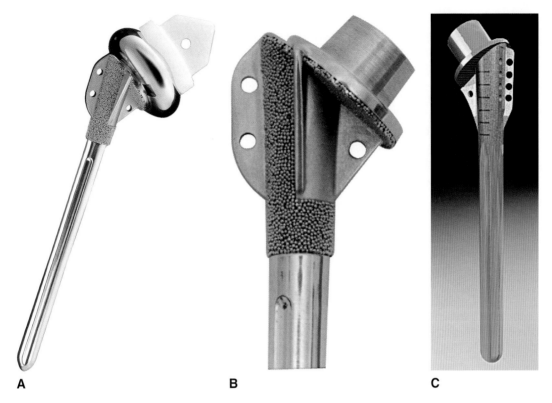

Figure 9.9
Cofield[2] and Neer III contemporary shoulder prosthesis system components. The Cofield[2] is shown in (**A**) and (**B**). A close-up view of the ingrowth surface of the proximal humeral stem is shown in (**B**). The humeral stem of the modular Neer III is shown in (**C**) (images courtesy of Smith & Nephew, Memphis).

for primary, hemiarthroplasty, total, and fracture use with cemented or press-fit fixation.

Neer III™. The update to the Neer humeral component with modified fins and head position relative to the stem and added stem markings. It is intended for primary, hemiarthroplasty, total, and fracture use for cemented or press-fit fixation.

Cofield™ and Cofield[2]™. Monoblock and modular cobalt chromium humeral prostheses developed with Dr. Robert Cofield. First introduced in 1983, the Cofield humeral component provides a surface designed for bone ingrowth. They include a keeled UHMWPE and a metal-backed, screw-fixed glenoid sterilized by EtO. They are intended for primary, hemiarthroplasty, total, and fracture use with cemented fixation.

Modular Shoulder System. A modular titanium stem humeral component with a keeled UHMWPE glenoid component sterilized by EtO. Cobalt chromium, titanium nitrite, and ceramic humeral heads are available as part of the system. It is intended for primary, hemiarthroplasty, total, and fracture use with press-fit fixation.

Tornier (Stafford, TX) (Figure 9.10)

Aequalis™ and Aequalis™ Fracture. Modular titanium or cobalt chromium stem, proximal stem angle adaptor, and humeral head prostheses with UHMWPE keeled and metal-backed, screw-fixed glenoid components. Developed with Drs. Gilles Walch and Pascal Boileau, the system was introduced in the early 1990s. It was the first system to have angular stem and offset head modularity and its introduction marked the beginning of the third-generation, or adaptable prosthesis designs, so-called because it was more possible than in previous designs to adapt the prosthesis to the anatomy of the patient. They are intended for primary, hemiarthroplasty, total, and fracture use with cemented and press-fit fixation.

Zimmer (Warsaw, IN)

Bigliani/Flatow®. Modular cobalt chromium stem humeral prostheses with keeled and pegged UHMWPE glenoid components sterilized by gamma in nitrogen. Developed with Drs. Louis Bigliani and Evan Flatow, it was introduced in 1999. It is intended for primary, hemiarthroplasty, total, and fracture use with cemented and press-fit fixation.

As can be seen from this partial listing of currently used devices, the number of components and designs from which a surgeon can choose is relatively large. There is a full range of choices depending on indication and surgeon preference. First-generation, or monoblock humeral component designs such as the Neer II, Cofield, and Kirschner-II-C, continue to be used along with second-generation, or modular, designs and third-generation, or adaptable, designs

A   **B**   **C**

Figure 9.10

Aequalis and Aequalis Fracture contemporary shoulder prosthesis system components. The noncemented humeral component is shown in (**A**). An underside view of an indexing offset head, and the proximal stem angle adaptors are shown in (**B**). An Aequalis fracture prosthesis in a fracture jig to aid in component positioning and reconstruction of the proximal humerus is shown in (**C**) (images courtesy of Tornier, Stafford, TX).

such as the Aequalis and Anatomical prostheses. Humeral heads are polished cobalt chromium, and humeral stems are cobalt chromium or titanium.

The glenoid components in all the systems are made from UHMWPE. They are available in all-polyethylene keeled and pegged designs, metal-backed keeled and pegged designs, and metal-backed, screw-fixed designs. A range of sterilization methods from gas plasma to EtO to gamma in inert and vacuum environments are used by the manufacturers (Orthopedics Today 1999) with a goal of decreasing the wear of the components as compared with UHMWPE sterilized by gamma in air. The UHMWPE used in bipolar humeral heads as intermediate head bearing and retaining surfaces is manufactured and sterilized in the same way as glenoid components from those companies that offer them.

Clinical Performance of Total Shoulder Arthroplasty

Overall Clinical Success Rates

The goal of TSA is to reduce or remove pain, reconstruct the articular surfaces of the glenohumeral joint, and to restore function to the shoulder by restoring strength and movement. In general, contemporary TSR achieves these goals very well. This has been determined by objective functional measurement as well as subjective measures of patient satisfaction for postoperative follow-up times ranging from 2.8 years to more than 15 years. The success of TSA in restoring function has been reported to be from 42 to 95% (Edwards et al. 2002, Iannotti and Norris 2003, Neer, II 1974, Neer, II, Watson, and Stanton 1982, Norris and Iannotti 2002, Snyder 1996, Sojbjerg et al. 1999, Worland et al. 1997). Patient satisfaction with TSA has been reported to range from 42 to 94% (Antuna et al. 2001, Boileau et al. 2001, Boileau et al. 2002, Godeneche et al. 2002, Hasan et al. 2002, Hayes and Flatow 2001, Hill and Norris 2001, Noble and Bell 1995, Sanchez-Sotelo et al. 2001).

Taking a closer look at these results shows differences with respect to a number of variables. The indications for TSA are varied and therefore present unique challenges to the repair. In general, TSAs performed to address problems caused by primary arthritis (OA or RA) are the most successful (Godeneche et al. 2002, Iannotti and Norris 2003, Neer, II 1974, Norris and Iannotti 2002, Sojbjerg et al. 1999, Trail and Nuttall 2002) and the success rates for TSA performed to address problems due to fracture are substantially lower (Boileau et al. 2001, Hayes and Flatow 2001). In younger patients (those younger than 50 years), the results for TSA are also less successful than those performed in older patients (Hasan et al. 2002, Hayes and Flatow 2001).

As with any mechanical system, time in service also affects the continued effectiveness of the procedure. For one group of studies, a range of 90 to 95% success rate was reported for short-term results (up to 5 years) and a much lower 55 to 88% success rate was reported for results up to and longer than 10 years (Mackay, Hudson, and Williams 2001). Survivorship analyses for TSAs also show this expected drop-off with time implanted (Torchia, Cofield, and

Settergren 1997, Trail and Nuttall 2002). One study showed a 100% survivorship from 0 to 2 years followed by 95% from 4 to 6 years and 92% up to 9 years (Trail and Nuttall 2002), while another shows 93% survivorship up to 10 years and 87% up to 15 years (Torchia, Cofield, and Settergren 1997). In spite of this natural decline with increased postoperative time, the functional benefit from TSA is quantifiable, demonstrable, and long lasting (Goldberg et al. 2001). The fact that the success of TSR is so good overall and that its effectiveness persists to such a degree for so long is testament to the inherent longevity of the replacement components and the excellent surgical techniques used for implantation.

In spite of these good results, there are complications that adversely affect the performance of TSRs. Table 9.1 lists complications seen in TSA in the order of their prevalence (Noble and Bell 1995, Wirth and Rockwood, Jr. 1996). One or more of these complications can cause failure of TSA, leading to revision. Revision rates range from 0 to 11% in reported TSA series (Edwards et al. 2002, Neer, II, Watson, and Stanton 1982, Sanchez-Sotelo et al. 2001, Snyder 1996, Sojbjerg et al. 1999, Trail and Nuttall 2002).

Loosening

Many other clinicians also note glenoid loosening as the primary complication and the major reason for failure of TSA (Boileau et al. 2002, Edwards et al. 2002, Gagey, Pourjamasb, and Court 2001, Hasan et al. 2002, Hayes and Flatow 2001, Lacroix and Prendergast 1997, Nagels et al. 2002, Norris and Iannotti 2002, Sanchez-Sotelo et al. 2001, Skirving 1999, Snyder 1996, Sojbjerg et al. 1999, Sperling, Cofield, and Rowland 1998, Torchia, Cofield, and Settergren 1997, Trail and Nuttall 2002). Among unsatisfactory arthroplasties, Hasan and colleagues report that 59% of them included loose glenoid components (2002).

Table 9.1
Complications with Total Shoulder Arthroplasty Ranked According to Prevalence

Prevalence ranking (approx. % of all cases)[a]	Complication
1 (10% or more)	Glenoid loosening
2 (up to 4%)	Instability
3 (up to 4%)	Postoperative rotator cuff tear/retear
4 (around 3%)	Periprosthetic bone fracture
5 (less than 1%)	Infection
6 (less than 1%)	Neural injuries
7	Prosthesis dissociation/fracture/wear

[a]Wirth and Rockwood, Jr. 1996, Noble and Bell 1995.

Evidence of impending or eventual loosening is also reported frequently in the form of radiolucent lines at prosthesis component fixation locations. Reports of clinical experience note a relatively high incidence of glenoid radiolucencies (Hayes and Flatow 2001, Lacroix and Prendergast 1997, Torchia, Cofield, and Settergren 1997, Trail and Nuttall 2002, Wirth and Rockwood, Jr. 1996), although there is not always a correlation seen between radiolucencies and negative clinical results (Boileau et al. 2002, Godeneche et al. 2002, Sanchez-Sotelo et al. 2001). On the other hand, a roentgen stereophotogrammetric study of glenoid migration indicated that radiolucency may be an underestimating indicator of potential glenoid problems (Nagels et al. 2002).

The complications listed in Table 9.1 are interrelated. Loosening of a glenoid component can cause instability, as can a preexisting or postoperative tear of the rotator cuff muscles. Instability due to lax or torn ligamentous or muscular structures can, on the other hand, lead to glenoid loosening. This instability is identified as the main culprit in TSA failure by some authors (Gerber, Ghalambor, and Warner 2001, Sojbjerg et al. 1999, Warren, Coleman, and Dines 2002). Superior instability of the glenohumeral joint allows higher than normal superior motion of the humeral head, which leads to intermittent superior edge loading of the glenoid component (Antuna et al. 2001, Gerber, Ghalambor, and Warner 2001, Hayes and Flatow 2001, Sojbjerg et al. 1999, Warren, Coleman, and Dines 2002, Wirth and Rockwood, Jr. 1996). Loss of active stabilization of the joint alone can increase the amount of translation of the humeral head on the glenoid from around 3 mm to around 11 mm (Iannotti and Williams 1998). This intermittent loading can lead to rocking of the glenoid component.

Fixation of the glenoid component within the bone of the glenoid is generally good, but as noted earlier, it seems to be the weak link of TSA. Fixation is achieved either with polymethylmethacrylate (PMMA) bone cement or via screws. Cemented all-polyethylene components are widely used, but metal-backed components with bone ingrowth surfaces are also used with cement or supplemental screw fixation. In spite of the radiolucencies seen, cemented all-polyethylene glenoids are seen as the gold standard of fixation (Ibarra, Dines, and McLaughlin 1998).

Wear

Another possible interrelationship between TSA complications is that between component loosening and wear or damage to the UHMWPE component. The association among wear, wear debris, and osteolysis leading to aseptic component loosening has been widely reported for total hip arthroplasty (THA) and total knee arthroplasty (TKA). In comparison, the number of reports of osteolysis or the possibility of osteolysis in TSA is relatively small (Boileau et al. 2002, Gunther et al. 2002, Mabrey et al. 2002, Scarlat and Matsen, III 2001, Wirth et al. 1999). Although it is a relatively small problem overall in TSA, damage to UHMWPE components can be drastic and affect the longevity of the procedure.

Wear of UHMWPE in TSA leading to osteolysis or the possibility of osteolysis has been reported as part of a clinical series (Boileau et al. 2002, Skirving 1999), as part of a study of negative outcomes (Hasan et al. 2002), as part of studies examining retrievals (Gunther et al. 2002, Scarlat and Matsen, III 2001, Weldon et al. 2001), and as part of studies examining wear debris (Mabrey et al. 2002, Wirth et al. 1999). In a report of a series of implants, Boileau and colleagues noted three metal-backed glenoid components that showed complete UHMWPE wear-through and osteolysis at revision (2002). Weldon and associates noted changes in the UHMWPE surfaces of glenoid components that they attributed to wear and deformation (2001). In 60% of the components, these changes substantially altered the glenohumeral stability provided by the glenoid (Weldon et al. 2001). Hasan and coworkers studied unsuccessful arthroplasties after an average of 3.6 years follow-up (2002). They reported a 20% overall incidence of glenoid UHMWPE wear and a 59% overall incidence of loose glenoid components with 77% of the cases with loose glenoids also exhibiting stiffness of the shoulder and/or malpositioning of the implants (Hasan et al. 2002).

Observations of retrieved glenoid components show that wear and damage to the UHMWPE component can vary from being relatively subtle to being catastrophic (Gunther et al. 2002, Rockwood and Wirth 2002, Scarlat and Matsen, III 2001). Relatively severe wear and damage was seen in one group of Global™ (DePuy Orthopaedics, Inc.) glenoid components made of Hylamer UHMWPE (developed by a DePuy Orthopaedics and E. I. DuPont Company joint venture) and sterilized by gamma irradiation in air (Rockwood and Wirth 2002). Hylamer UHMWPE was developed to have a higher yield strength and tensile strength than conventional UHMWPE and it also had a higher tensile modulus. Oxidation damage due to the gamma sterilization in air that becomes progressively worse during shelf storage was more severe in the Hylamer material than in conventional UHMWPE and rendered many of the Hylamer glenoids made and sterilized at the time prone to wear and damage clinically (Rockwood and Wirth 2002). Components affected in this way were Hylamer UHMWPE glenoids made and sterilized from 1990 to 1993. Changes in sterilization methods with Hylamer from 1993 to 1998 and the elimination in 1998 of Hylamer from glenoid components have limited the possibilities of adverse clinical results resulting from this issue (Rockwood and Wirth 2002).

Retrieval studies of glenoid components identify the damage modes on UHMWPE glenoid components. Scarlat and colleagues reported on a group of 43 retrieved glenoid components that had been retrieved after an average of 2.5 years of service, 67% of which were keeled all-polyethylene designs, 21% of which were metal-backed designs, and the remainder of which were nonkeeled all-polyethylene designs (Scarlat and Matsen, III 2001). Of the 21 cases where the reasons for the original arthroplasty and component removal were known, preimplantation instability was noted in 38% and all but three components were removed because of looseness (two were removed at autopsy and one was removed due to infection) (Scarlat and Matsen, III 2001). Of the four damage modes identified, rim wear was the most prevalent (70%), followed by surface irregularity (67%), fracture (30%), and central wear (20%) (Scarlat and Matsen, III 2001).

A damage mode identification and grading scheme used for UHMWPE knee tibial components and hip acetabular components has also been applied to a small group of retrieved UHMWPE glenoid components (10 components) (Gunther et al. 2002). Scratching was noted as the most prevalent damage mode and was seen in 90% of the examined regions (Gunther et al. 2002). Abrasion (68%), pitting (60%), delamination (58%), deformation (40%), embedded debris (28%), and burnishing (8%) were also reported (Gunther et al. 2002). Component fractures of four glenoids, and complete wear-through of one glenoid component were also noted (Gunther et al. 2002). The data shows a combination of abrasive and fatigue wear mechanisms at work in the shoulder. This is similar to the damage modes seen in knee components and in contrast to the predominant wear modes seen in total hip components.

Further analysis of the type of wear in UHMWPE glenoid components has also been conducted by examining the wear debris generated in the shoulder. Isolation and characterization of the UHMWPE wear debris generated in failed shoulder, hip, and knee arthroplasties gives added insight into wear in the shoulder and how it compares with that in the hip and knee. Wirth and coworkers (1999) and Mabrey and associates (2002) have applied American Society for Testing and Materials (ASTM) standard F1877-98 for the quantitative description of wear debris to UHMWPE wear particles from shoulder, hip, and knee components. The presence of substantial numbers of UHMWPE wear particles in the tissues surrounding aseptically loose glenoid components is direct evidence corresponding to clinical observations that wear and loosening are related. UHMWPE wear particles isolated from total shoulders are comparable in size and form to those isolated from total knees and total hips (Mabrey et al. 2002, Wirth et al. 1999). In general, the shoulder particles are more similar to knee particles, being larger and more oblong than hip particles (Mabrey et al. 2002, Wirth et al. 1999) (Figure 9.11). The data is consistent with the different articulation characteristics of hip, knee, and shoulder arthroplasties. In hip arthroplasty, the metal head rotates within the UHMWPE liner. In contrast, in knee and shoulder arthroplasties, both rotation and translation of the metal components occur relative to the polyethylene components.

Controversies in Shoulder Replacement

Although shoulder replacements generally achieve the goal of pain relief, repair of structure, and restoration of function, there are still many ways to go about it and many products and techniques to choose from. The introduction of UHMWPE as a glenoid articular bearing surface allowed surgeons to contemplate and perform TSR for a broader range of indications than was possible with a monoblock humeral component. As noted earlier, however, the addition of a glenoid component opens up the possibility for complications related to its fixation and durability. For some surgeons, this leaves open the question of whether to use a glenoid component.

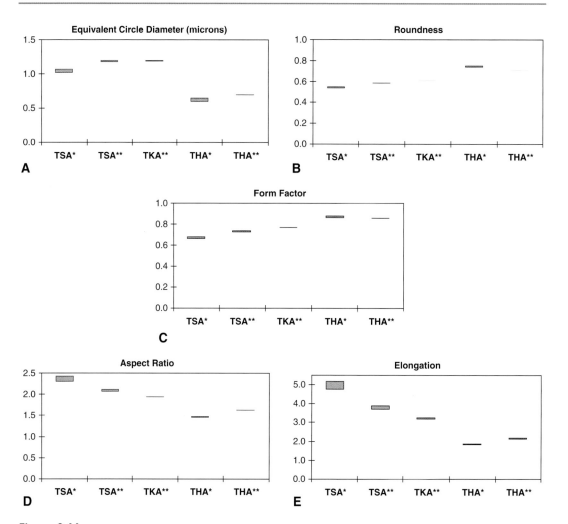

Figure 9.11

Plots of data showing ASTM F1877-98 quantitative description characteristics of retrieved UHMWPE particles from TSA, TKA, and THA. Data from Mabrey et al. (2002) is noted with a double asterisk (**) and data from Wirth et al. (1999) is noted with a single asterisk (*). The boxes in each plot enclose the data values of the mean ± the standard error in the mean reported for each implant type.

The equivalent circle diameter (the diameter, in microns, of the circle with the same area as the projected area of a particle) is plotted in (**A**).

The roundness (how closely the projected shape of a particle resembles a circle on the basis of diameter [perfect match with a circle = 1.0]) is plotted in (**B**).

The form factor (how closely the projected shape of a particle resembles a circle on the basis of perimeter [perfect match with a circle = 1.0]) is plotted in (**C**).

The aspect ratio (maximum straight-line particle length divided by maximum straight-line particle width) is plotted in (**D**).

The elongation (actual particle length divided by mean particle width) is plotted in (**E**).

It is interesting to note that where differences are apparent, the TSA and TKA particles are similar to one another and generally larger (as shown by the equivalent circle diameter data) and more elongated (as shown by the aspect ratio and elongation data) than THA particles.

Results and rates of TSA and hemiarthroplasty are varied, but both definitely seem to have a place in treatment of the shoulder. The Swedish Register of shoulder arthroplasty reports that 87% of shoulder replacements since 1999 have been hemiarthroplasties because of the high rates of radiolucency seen in other studies and because of the relative difficulty of implanting glenoid components (Rahme, Jacobsen, and Salomonsson 2001). On the other hand, in a large, multicenter study (571 cases of a series of 1701 arthroplasties), 86% of the cases were TSAs performed on the basis of surgeon preference (Edwards et al. 2002). A better outcome has been shown for TSA compared with hemiarthroplasty for OA (Edwards et al. 2003) and in cases of glenoid erosion (Iannotti and Norris 2003), although glenoid bone loss has also been noted as a contraindication for TSA (Hill and Norris 2001). For OA with a functioning rotator cuff, the results of total and hemiarthroplasty have been shown to be comparable (Arredondo and Worland 1999). In cases of rotator cuff tear or arthropathy, however, hemiarthroplasty has been noted as a preference because of the association of glenoid loosening with rotator cuff problems (Sojbjerg et al. 1999, Warren, Coleman, and Dines 2002, Worland et al. 1997). It is also interesting to see reports showing half of the revised TSAs being converted to hemiarthroplasties (Antuna et al. 2001, Sojbjerg et al. 1999) although in one study, a larger percentage of patients were satisfied after revision to another TSA than after revision to a hemiarthroplasty (Antuna et al. 2001).

UHMWPE can still play a vital role in hemiarthroplasties, even when a glenoid component is not used. The bipolar prosthesis has an UHMWPE internal liner and retaining ring that holds the inner head within the outer metal shell. Studies report 95% success and 92% satisfaction with a bipolar prosthesis used without a glenoid component, even when used in cases with massive rotator cuff tears (Arredondo and Worland 1999, Worland et al. 1997). These results are still relatively short term (around 3 years follow-up) and the number of cases is relatively small, but the proponents of the bipolar design concept note that the dual articulation of the device is likely to result in less damage to an unresurfaced glenoid than a standard prosthetic humeral head (Arredondo and Worland 1999, Worland et al. 1997).

Even after the choice to use a glenoid component has been made, there are still a number of alternatives to consider. All-polyethylene components are available in both keeled and pegged variants. The keeled design goes back to the time of the first TSA in the 1970s and the pegged design is a more recent development. Fixing the glenoid component in the bone is challenging in part because of the relatively small amount of bone available, especially in cases of erosion or fracture. Keels and pegs are different answers to this challenge. For comparable keeled and pegged designs, clinical results comparing radiolucency rates indicate that the pegged design may be better than the keeled design (Trail and Nuttall 2002). On the other hand, one study notes three incidences of peg breakage (Norris and Iannotti 2002). Theoretical results predicted lower stresses beneath the pegged design compared with the keeled design in normal glenoid bone (Lacroix, Murphy, and Prendergast 2000). In rheumatoid bone, however, theoretical results predicted lower stresses beneath the keeled design (Lacroix, Murphy, and Prendergast 2000).

The important glenoid component questions of all-polyethylene or metal-backed and cemented or noncemented are also not exclusive of one another. Fixation of all-polyethylene components is performed only with bone cement. The addition of a metal backing opens up the possibilities for glenoid fixation. The surfaces of some metal-backed glenoid components are textured to allow for bony ingrowth to achieve biological fixation, but regulatory approval of these devices is currently for use only with PMMA bone cement. The metal backings of some glenoid components also allow for screw fixation. For the most challenging cases of glenoid repair requiring bone grafting to aid in component fixation and to reconstruct the proper location and orientation of the glenoid face, both all-polyethylene as well as metal-backed components have been cemented with some success out to an average of 5 years follow-up (Hill and Norris 2001).

Arguments in favor of metal-backed glenoid components include theoretical tensile stresses that are predicted to be lower beneath metal-backed components than beneath corresponding all-polyethylene glenoid components (Lacroix and Prendergast 1997). Early results for contemporary modular metal-backed designs are also promising (Skirving 1999), and there is the potential for strong fixation via osseointegration offered by metal-backed glenoid components (Antuna et al. 2001, Ibarra, Dines, and McLaughlin 1998). Godeneche and colleagues note an 83% rate of radiolucency for all-polyethylene glenoids compared with a 20% rate of radiolucency for metal-backed glenoids (2002). Boileau and associates also note this same discrepancy in radiolucency rates (85% for all-polyethylene, 25% metal-backed), but go on to report increased wear and loosening in the metal-backed glenoids compared with the all-polyethylene ones (2002).

The preponderance of the evidence is that better clinical results are seen for cemented, all-polyethylene components, especially with third-generation cementing technique, leading Ibarra and colleagues to call them the gold standard for fixation (1998). Clinical reports note incidences of screw breakage with metal-backed components (Antuna et al. 2001, Snyder 1996), incidences of dissociation of the UHMWPE liner from the metal backing (Antuna et al. 2001, Godeneche et al. 2002, Levy and Copeland 2001), and wear of the UHMWPE liner through to the metal backing (Boileau et al. 2002, Gunther et al. 2002). Theoretical results also show larger UHMWPE stresses in metal-backed glenoids than in all-polyethylene glenoids (Swieszkowski, Bednarz, and Prendergast 2003). Also, a metal backing requires a thinner UHMWPE component than a corresponding all-polyethylene glenoid with the same overall component thickness.

Overall, good results are obtained by surgeons on both sides of each of these questions. Glenoid component use and choice of design, material, and fixation method of the glenoid component can be important. Of utmost importance, however, is the technical skill of the surgeon in assessing the joint, positioning the components correctly within the joint, and reconstructing and repairing the bony and soft-tissue structures.

Future Directions in Total Shoulder Arthroplasty

Design

Starting from three sizes of a monoblock humeral component without a glenoid component, TSA design has already evolved a great deal. Third-generation components with modularity, which includes head offsets and head-stem angulation, already have been manufactured and widely used to great success. One area of TSA design, however, has not been so quick to be developed and may have a good future for shoulder replacement.

The reverse total shoulder prosthesis design concept is one in which the humeral head is replaced with a concave polyethylene bearing surface and the glenoid face is augmented by a convex articular metal component. As the name implies, this reverses the normal anatomic geometries of the humeral head and glenoid face. An example of this prosthesis concept is shown in Figure 9.12.

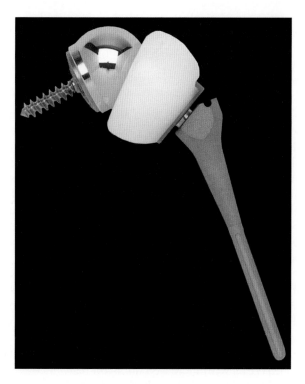

Figure 9.12

Reverse shoulder prosthesis system components. The metal screw-fixed ball is implanted in the scapula to replace the glenoid, and the concave polyethylene component mounted on the stem is implanted into the proximal humerus to replace the humeral head (image courtesy of Encore Medical, Austin, TX).

Constrained versions of this concept have been used earlier and were largely unsuccessful because of loosening (Boulahia et al. 2002). In 1987, Grammont and Baulot designed a semiconstrained reverse prosthesis and recommended it be used prudently and in patients older than 70 years of age (Boulahia et al. 2002). Since then, reports from Europe have noted its success in a limited number of rotator cuff deficient patients after relatively short (2 to 3 years) follow-up periods (Boulahia et al. 2002, De Wilde et al. 2001, Rittmeister and Kerschbaumer 2001).

The concept is, however, very much in its infancy and its place in the future of shoulder repair has yet to be determined. Encore Medical Corp. (Austin, TX) is currently conducting an investigational device exemption (IDE) study of its new Reverse™ Shoulder Prosthesis system (Figure 9.12). The first implantation in North America of another reverse prosthesis design (the Delta® prosthesis [DePuy, Warsaw, IN]) was reported to have been conducted on February 27, 2003, by Dr. Anthony Miniaci in Toronto, Canada (Canada Newswire 2003). The potential benefits of the concept are that it can allow better function than standard prostheses without relying on rotator cuff function and that it can decrease bending moment forces at the prosthesis–glenoid interface. If these benefits are clinically realized, then it could become an important addition to standard prostheses, especially in difficult arthroplasty cases.

Materials

Changes to the humeral head and to the glenoid liner are beginning to be made to address the potential for UHMWPE damage that exists in current TSA prosthesis systems. The standard humeral head material is a highly polished cobalt chromium alloy. Modular heads are also offered with titanium nitrite–bearing surfaces as well as ceramic surfaces, but they are as yet only known to be available from one manufacturer and are not yet widely used. The goal of these alternative surfaces to the standard cobalt chromium is to decrease the friction between the humeral and glenoid articular surfaces and thereby decrease the potential for damage to the UHMWPE. That addresses one side of the current wear couple.

On the other side of the wear couple is the UHMWPE-bearing surface. With the exception of Hylamer, standard UHMWPE has been used and sterilization methods have been updated so that currently, all of the manufacturers process the polyethylene in such a way as to reduce any oxidative degradation. These methods include gamma sterilization in nitrogen or other inert atmospheres, gas plasma and EtO sterilization, and storage in vacuum. Highly crosslinked UHMWPEs with increased abrasion resistance have been used in THA acetabular liners. Decreased mechanical properties of these compared with nonhighly crosslinked UHMWPEs, however, make them less attractive for use in shoulders where, as discussed earlier, the characteristics of the glenohumeral articulation are such that fatigue wear and damage causing pitting, delamination, and fracture are more prevalent. Enhancement or even elimination of UHMWPE, such as has been done for metal-on-metal and ceramic-on-ceramic hips, does not seem to be imminent for TSR components.

Conclusion

In general, the success rate of TSR exceeds 90% for a wide range of common shoulder problems. As in THA and TKA, the aim of TSA is to reduce pain, restore damaged anatomy, and restore function. As for hip and knee arthroplasty, there are continuing concerns about component fixation and persisting instability of the joint.

The introduction of UHMWPE to shoulder arthroplasty via the addition of glenoid components has increased the applicability of the procedure and has been a great success, even though there remains a substantial subset of cases that are successfully addressed via hemiarthroplasty. The addition of UHMWPE has, however, brought along its associated issues. Glenoid fixation is the biggest of these and the continuing persistence of a variety of glenoid component designs and metal backings and fixation methods attests to the fact that a clearly superior answer has not yet been found. Wear and damage is the other issue specific to the UHMWPE glenoid, and, although it remains a relatively small percentage of all cases, the ramifications can be severe.

In spite of these issues, TSA with UHMWPE remains a successful procedure bringing lasting pain relief and restoration of function to tens of thousands of patients worldwide every year.

Acknowledgments

Many thanks to Jon JP Warner of Massachusetts General Hospital, and industry representatives Terry Armstrong, Jodelle Brosig, Masood Durkhshan, Monika Gibson, Elaine Mattheus, Marly Moate, Brian Sauls, and Kate Smith for their assistance with figures of radiographs, anatomy, and current prostheses. Many thanks also to Steven Kurtz of Drexel University for his assistance with copies of reference materials and to Avram Edidin of Drexel University for his discussions of historical information.

References

Anglin C., U.P. Wyss, and D.R. Pichora. 2000. Glenohumeral contact forces. *Proc Inst Mech Eng [H]* 214:637–644.

Antuna S.A., J.W. Sperling, R.H. Cofield, and C.M. Rowland. 2001. Glenoid revision surgery after total shoulder arthroplasty. *J Shoulder Elbow Surg* 10:217–224.

Arredondo J., and R.L. Worland. 1999. Bipolar shoulder arthroplasty in patients with osteoarthritis: Short-term clinical results and evaluation of birotational head motion. *J Shoulder Elbow Surg* 8:425–429.

Bartel D.L., V.L. Bicknell, and T.M. Wright. 1986. The effect of conformity, thickness and material on stresses in ultra-high molecular weight components for total joint replacement. *J Bone Joint Surg* 68:1041–1051.

Boileau P., C. Trojani, G. Walch, et al. 2001. Shoulder arthroplasty for the treatment of the sequelae of fractures of the proximal humerus. *J Shoulder Elbow Surg* 10:299–308.

Boileau P., C. Avidor, S.G. Krishnan, et al. 2002. Cemented polyethylene versus uncemented metal-backed glenoid components in total shoulder arthroplasty: A prospective, double-blind, randomized study. *J Shoulder Elbow Surg* 11:351–359.

Boulahia A., T.B. Edwards, G. Walch, and R.V. Baratta. 2002. Early results of a reverse design prosthesis in the treatment of arthritis of the shoulder in elderly patients with a large rotator cuff tear. *Orthopedics* 25:129–133.

Canada Newswire. 2003. "First North American 'reverse' shoulder replacement surgery performed at Toronto Western Hospital." March 25, 2003.

De Wilde L., M. Mombert, P. Van Petegem, and R. Verdonk. 2001. Revision of shoulder replacement with a reversed shoulder prosthesis (Delta III): Report of five cases. *Acta Orthop Belg* 67:348–353.

Edwards T.B., A. Boulahia, J.F. Kempf, et al. 2002. The influence of rotator cuff disease on the results of shoulder arthroplasty for primary osteoarthritis: Results of a multicenter study. *J Bone Joint Surg* 84:2240–2248.

Edwards T.B., N.R. Kadakia, A. Boulahia, et al. 2003. A comparison of hemiarthroplasty and total shoulder arthroplasty in the treatment of primary glenohumeral osteoarthritis: Results of a multicenter study. *J Shoulder Elbow Surg* 12:207–213.

Gagey O., B. Pourjamasb, and C. Court. 2001. Revision arthroplasty of the shoulder for painful glenoid loosening: A series of 14 cases with acromial prostheses reviewed at four year follow up. *Rev Chir Orthop* 87:221–228.

Gerber A., N. Ghalambor, and J.J. Warner. 2001. Instability of shoulder arthroplasty: Balancing mobility and stability. *Orthop Clin North Am* 32:661–670.

Godeneche A., P. Boileau, L. Favard, et al. 2002. Prosthetic replacement in the treatment of osteoarthritis of the shoulder: Early results of 268 cases. *J Shoulder Elbow Surg* 11:11–18.

Goldberg B.A., K. Smith, S. Jackins, et al. 2001. The magnitude and durability of functional improvement after total shoulder arthroplasty for degenerative joint disease. *J Shoulder Elbow Surg* 10:464–469.

Gunther S.B., J. Graham, T.R. Norris, et al. 2002. Retrieved glenoid components: A classification system for surface damage analysis. *J Arthroplasty* 17:95–100.

Hasan S.S., J.M. Leith, B. Campbell, et al. 2002. Characteristics of unsatisfactory shoulder arthroplasties. *J Shoulder Elbow Surg* 11:431–441.

Hayes P.R., and E.L. Flatow. 2001. Total shoulder arthroplasty in the young patient. *Instr Course Lect* 50:73–88.

Hill J.M., and T.R. Norris. 2001. Long-term results of total shoulder arthroplasty following bone-grafting of the glenoid. *J Bone Joint Surg* 83:877–883.

Iannotti J.P., and G.R. Williams. 1998. Total shoulder arthroplasty—factors influencing prosthetic design. *Orthop Clin North Am* 29:377–391.

Iannotti J.P., and T.R. Norris. 2003. Influence of preoperative factors on outcome of shoulder arthroplasty for glenohumeral osteoarthritis. *J Bone Joint Surg* 85:251–258.

Ibarra C., D.M. Dines, and J.A. McLaughlin. 1998. Glenoid replacement in total shoulder arthroplasty. *Orthop Clin North Am* 29:403–413.

Jackson D., Moderator. 2003. Shoulder surgery round-table–panel talks about arthroplasty problems, solutions. F. Matsen, III, A. Miniaci, C. Neer, II, and G. Williams, Participants. *Orthopedics Today, Slack, Inc.,* May.

Karduna A.R., G.R. Williams, J.P. Iannotti, and J.L. Williams. 1998. Total shoulder arthroplasty biomechanics: A study of the forces and strains at the glenoid component. *J Biomech Eng* 120:92–99.

Lacroix D., L.A. Murphy, and P.J. Prendergast. 2000. Three-dimensional finite element analysis of glenoid replacement prostheses: A comparison of keeled and pegged anchorage systems. *J Biomech Eng* 122:430–436.

Lacroix D., and P.J. Prendergast. 1997. Stress analysis of glenoid component designs for shoulder arthroplasty. *Proc Inst Mech Eng [H]* 211:467–474.

Levy O., and S.A. Copeland. 2001. Rotational dissociation of glenoid components in a total shoulder prosthesis: An indication that sagittal torque forces may be important in glenoid component design. *J Shoulder Elbow Surg* 10:197.

Lugli T. 1978. Artificial shoulder joint by Pean (1893). The facts of an exceptional intervention and the prosthetic method. *Clin Orthop*133:215–218.

Mabrey J.D., A. Afsar-Keshmiri, G.A. Engh, et al. 2002. Standardized analysis of UHMWPE wear particles from failed total joint arthroplasties. *J Biomed Mater Res* 63:475–483.

Mackay D.C., B. Hudson, and J.R. Williams. 2001. Which primary shoulder and elbow replacement? A review of the results of prostheses available in the UK. *Ann R Coll Surg Engl* 83:258–265.

Nagels J., E.R. Valstar, M. Stokdijk, and P.M. Rozing. 2002. Patterns of loosening of the glenoid component. *J Bone Joint Surg* 84:83–87.

Neer C.S., II. 1955. Articular replacement of the humeral head. *J Bone Joint Surg* 37:215–228.

Neer C.S., II. 1974. Replacement arthroplasty for glenohumeral osteoarthritis. *J Bone Joint Surg* 56:1–13.

Neer, C.S., II. 2003. Telephone conversation. June 19.

Neer C.S., II, K.C. Watson, and F.J. Stanton. 1982. Recent experience in total shoulder replacement. *J Bone Joint Surg* 64:319–337.

Noble J.S., and R.H. Bell. 1995. Failure of total shoulder arthroplasty: Why does it occur? *Semin Arthroplasty* 6:280–288.

Norris T.R., and J.P. Iannotti. 2002. Functional outcome after shoulder arthroplasty for primary osteoarthritis: A multicenter study. *J Shoulder Elbow Surg* 11:130–135.

Orthopedics Today. 1999. Guide to polyethylene in joint implants. *Orthopedics Today,* Slack, Inc., Oct–Nov.

Poppen N.K., and P.S. Walker. 1978. Forces at the glenohumeral joint in abduction. *Clin Orthop* 135:165–170.

Rahme H., M.B. Jacobsen, and B. Salomonsson. 2001. The Swedish elbow arthroplasty register and the Swedish shoulder arthroplasty register: Two new Swedish arthroplasty registers. *Acta Orthop Scand* 72:107–112.

Rittmeister M., and F. Kerschbaumer. 2001. Grammont reverse total shoulder arthroplasty in patients with rheumatoid arthritis and nonreconstructible rotator cuff lesions. *J Shoulder Elbow Surg* 10:17–22.

Rockwood C.A., Jr. 2000. The century in orthopedics—a century of shoulder arthroplasty innovations and discoveries, *Orthopedics Today,* Slack, Inc., February.

Rockwood C.A., and M.A. Wirth. 2002. Observation on retrieved Hylamer glenoids in shoulder arthroplasty: Problems associated with sterilization by gamma irradiation in air. *J Shoulder Elbow Surg* 11:191–197.

Sanchez-Sotelo J., T.W. Wright, S.W. O'Driscoll, et al. 2001. Radiographic assessment of uncemented humeral components in total shoulder arthroplasty. *J Arthroplasty* 16:180–187.

Scarlat M.M., and F.A. Matsen, III. 2001. Observations on retrieved polyethylene glenoid components. *J Arthroplasty* 16:795–801.

Skirving A.P. 1999. Total shoulder arthroplasty—current problems and possible solutions. *J Orthop Sci* 4:42–53.

Snyder G. 1996. Shoulder implant system. In *Clinical performance of skeletal prostheses.* L. Hench and J. Wilson, Eds. London, Chapman & Hall.

Sojbjerg J.O., L.H. Frich, H.V. Johannsen, and O. Sneppen. 1999. Late results of total shoulder replacement in patients with rheumatoid arthritis. *Clin Orthop* 366:39–45.

Sperling J.W., R.H. Cofield, and C.M. Rowland. 1998. Neer hemiarthroplasty and Neer total shoulder arthroplasty in patients fifty years old or less: Long-term results. *J Bone Joint Surg* 80:464–473.

Swieszkowski W., P. Bednarz, and P. Prendergast. 2003. Contact stresses in the glenoid component in total shoulder arthroplasty. *Proc Inst Mech Eng [H]* 217:49–57.

Torchia M.E., R.H. Cofield, and C.R. Settergren. 1997. Total shoulder arthroplasty with the Neer prosthesis: Long-term results. *J Shoulder Elbow Surg* 6:495–505.

Trail I.A., and D. Nuttall. 2002. The results of shoulder arthroplasty in patients with rheumatoid arthritis. *J Bone Joint Surg* 84:1121–1125.

Warner J.J., M.K. Bowen, X. Deng, et al. 1999. Effect of joint compression on inferior stability of the glenohumeral joint. *J Shoulder Elbow Surg* 8:31–36.

Warner J.J., and R.F. Warren. 2001. Quantitative determination of articular pressure in the human shoulder joint. *J Shoulder Elbow Surg* 5:496–497.

Warren R.F., S.H. Coleman, and J.S. Dines. 2002. Instability after arthroplasty: The shoulder. *J Arthroplasty* 17 Suppl 1:28–31.

Weldon E.J., III, M.M. Scarlat, S.B. Lee, and F.A. Matsen, III. 2001. Intrinsic stability of unused and retrieved polyethylene glenoid components. *J Shoulder Elbow Surg* 10:474–481.

Wirth M.A., C.M. Agrawal, J.D. Mabrey, et al. 1999. Isolation and characterization of polyethylene wear debris associated with osteolysis following total shoulder arthroplasty. *J Bone Joint Surg* 81:29–37.

Wirth M.A., and Rockwood C.A., Jr. 1996. Complications of total shoulder-replacement arthroplasty. *J Bone Joint Surg* 78:603–616.

Worland R.L., D.E. Jessup, J. Arredondo, and K.J. Warburton. 1997. Bipolar shoulder arthroplasty for rotator cuff arthropathy. *J Shoulder Elbow Surg* 6:512–515.

Chapter 9. Reading Comprehension Questions

9.1. TSA replaces the joint separating which bony anatomical structures?
  a) The glenoid and the scapula
  b) The acromion and the head of the humerus
  c) The coracoid process and the head of the humerus
  d) The glenoid and the head of the humerus
  e) All of the above

9.2. Which method is used to attach components to the underlying bone in TSA?
  a) Biological fixation
  b) Bone screws
  c) Bone cement
  d) Porous coating
  e) All of the above

9.3. The surgeon credited with pioneering TSA was
  a) Sir John Charnley, M.D.
  b) Frank Gunston, M.D.
  c) Charles Neer II, M.D.
  d) Robert Averill, M.D.
  e) All of the above

9.4. Which of the following factors contribute to glenoid component failure?
    a) Inadequate bone stock in the glenoid
    b) Large magnitude of shoulder joint reaction force
    c) Wide variation in the direction of the shoulder joint reaction force
    d) Limited available area over which to distribute the shoulder joint reaction force
    e) All of the above

9.5. What outcome typically results from the "rocking horse" phenomenon in TSA?
    a) Loosening of the glenoid component
    b) Loosening of the humeral component
    c) Wear of the humeral component
    d) Loosening of the scapular component
    e) All of the above

9.6. Which of the following factors is not considered to play a role in magnitude and distribution of UHMWPE stresses acting on the glenoid component?
    a) Component geometry
    b) Patient weight
    c) Patient height
    d) UHMWPE thickness
    e) Incorporation of metal backing

9.7. Which of the following statements about shoulder instability is true?
    a) It is the single most prevalent failure mechanism for TSA
    b) It is caused by imbalance of the soft tissues acting across the joint
    c) Is occurs after a severe joint infection
    d) It occurs when the rotator cuff is intact and not damaged
    e) All of the above statements are true

9.8. What is the current success rate for TSA?
    a) Greater than 90% at 10 years.
    b) 85–90% at 10 years.
    c) 80–85% at 10 years.
    d) 75–80% at 10 years.
    e) 70–75% at 10 years.

Chapter 10

The Clinical Performance of UHMWPE in the Spine

Marta L. Villarraga
Exponent, Inc. and Drexel University
Philadelphia, PA

Peter A. Cripton
University of British Columbia
Vancouver, Canada

Introduction

Ultra-high molecular weight polyethylene (UHMWPE) is the polymeric bearing material of choice in artificial disc replacements. Two total disc replacements (TDRs) that incorporate UHMWPE, the SB Charité™ III (Link® Spine Group, Branford, CT) and the PRODISC® (Spinal Solutions, New York, NY), consist of three components: two endplates made of cobalt chrome molybdenum (CoCr) and an UHMWPE insert or inlay between them. These implants have been used for more than a decade in Europe and are currently undergoing Food and Drug Administration (FDA)-regulated clinical trials in the United States. Figure 10.1 shows radiographs with examples of these two TDRs in place. Not surprisingly, this combination of materials is what has been used extensively in other total joint replacements (TJRs) already reviewed in this book (hip and knee).

Artificial discs for the surgical replacement of pathological intervertebral discs have received increasing attention as a promising alternative to contemporary spinal fusion and spinal discectomy procedures (Eijkelkamp et al. 2001, Kostuik 1998, Lemaire et al. 1997, Pointillart 2001, Zeegers et al. 1999, Whitecloud, III 1999). There is not yet a consensus among clinicians regarding

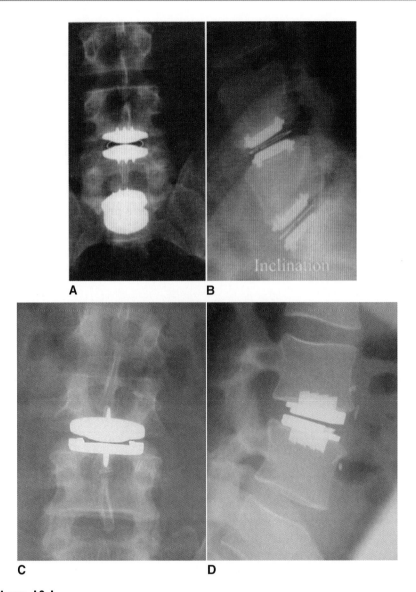

Figure 10.1

(**A**) A–P radiograph and (**B**) sagittal radiograph showing SB Charité in place; (**C**) A–P radiograph and (**D**) sagittal radiograph showing PRODISC in place. (Reprinted with permission Büttner-Janz, Hochschuler, and McAfee 2003, Zigler et al. 2003.)

the exact indications and requirements for these devices. However, there is a growing consensus in the clinical spine community that, for many indications, spinal fusion actually represents the premature advancement of a pathological spinal joint to its ultimate pathological endpoint, which is namely a fused joint with bone spanning the disc space. Spinal fusion is known to affect the biomechanics of spinal levels adjacent to the operated segment, which is suspected to result in deleterious clinical sequelae (Eck, Humphreys, and Hodges 1999).

An artificial disc restoring the natural kinematics of the spine is thought to have the potential to eliminate the disadvantages of fusion, namely preventing the loss of spinal motion and of dynamic spinal balance. In the case of a primary discectomy without fusion, the hope is that replacement of the pathological disc with an artificial one will reduce the relatively high prevalence of recurrent back pain or disc pathologies associated with these procedures (Suk et al. 2001, Yorimitsu et al. 2001).

The concept for artificial discs first appeared in the 1960s (Fernstrom 1966) with the use of steel and vitallium balls implanted in intervertebral spaces. Unfortunately, this design was not successful because these balls penetrated through the endplates and the vertebrae eventually fused. Despite this undesirable consequence, research and development in the area of TDRs has continued since then.

The goals of artificial disc designs are to restore the intervertebral motion of the natural disc (Bao and Yuan 2000, Dooris et al. 2001, Hedman et al. 1991, Langrana et al. 1994). At the same time, many investigators think that the nonlinear stiffness provided by the natural disc should also be incorporated into artificial disc designs (Hedman et al. 1991). Many methods have been proposed to accomplish this. The most prevalent designs borrow materials, geometry, and concepts from the already well-established low-friction articulating artificial joints used for total hip replacement (THR) and total knee replacement (TKR). These designs differ from one another in their choice of articulating biomaterial pairs and in the amount of constraint provided at the articulating interface.

Currently, the precise indications for TDR (also known as total disc arthroplasty [TDA]) for single or multilevel disease are still evolving. The investigational device exemption (IDE) clinical studies approved by the FDA for the first two artificial discs being evaluated in the United States call for similar indications. In general, the main indications include degenerative disc disease at one or two adjacent segments in the lumbar spine, with at least 6 months of failed conservative therapy and long-term chronic back pain (Link 2001, Spine Solutions 2002a). The typical age range for patients considered for this is 18 to 60 years (Spine Solutions 2002a). There are also specific contraindications and exclusion criteria that are being recommended in these studies. The most common include prior lumbar surgery, spondylolisthesis, osteopathies, and infection or tumors. The first SB Charité™ disc was implanted in 1984 and the first PRODISC was implanted in 1990 (Zigler 2002). By the end of 2000, more than 5000 SB Charité discs had been implanted in patients throughout the world (Büttner-Janz, Hochschuler, and McAfee 2003, David 2003), and 1000 PRODISCs had been implanted in patients in Europe.

This chapter provides an overview of the use of UHMWPE in TDRs. First, the biomechanical considerations for the use of UHMWPE in the spine are reviewed, followed by a description of the TDR designs currently using an UHMWPE component, including what is known of the properties of the UHMWPE component in each design. This is followed by a review of the clinical performance of the UHMWPE in TDA, which is mainly based on the long-term experiences in Europe and reports of early experiences in the United States. The chapter concludes with a brief discussion of alternatives to UHMWPE in TDA in the spine.

Biomechanical Considerations for UHMWPE in the Spine

The design of any device to be implanted in the intervertebral space must incorporate considerations of the biomechanics of the particular spinal level to be implanted. Among other factors, the primary biomechanical factors to be considered can be characterized as the kinematics (motion), kinetics (applied forces), and load sharing (distribution of stress between anatomic components). The device should allow the expected kinematics, it should be able to withstand millions of cycles of the expected loads, and it should attempt not to disrupt the distribution of the tissue level stresses and strains experienced in a healthy intervertebral joint. The kinematics, kinetics, and load sharing of the spine vary significantly as one moves from the cervical to the thoracic to the lumbar spine.

Once having determined the device configuration and the likely kinematics, kinetics, and load sharing expected for the device itself, it is then necessary to consider the number of interfaces to be incorporated in the device. Each interface must then be designed to withstand the kinematic, kinetic, and load-sharing conditions identified. For example, many devices use the same material pairs as in the already established THAs or TKAs. In these designs, an UHMWPE component is placed between two metallic endplates (Figure 10.1). There are therefore four critical interfaces to consider: 1) upper metallic endplate with vertebra, 2) upper metallic endplate with UHMWPE component, 3) UHMWPE component with lower metallic endplate, and 4) lower metallic endplate with lower vertebra. Each of these interfaces must be able to withstand the expected kinematic and kinetic load cycles. Good engineering practice dictates that these interfaces should also be designed with an appropriate factor of safety and with some measures taken to ease the removal of the device should it fail in service.

Kinematic Considerations

Each spinal joint allows a specific combination of three-dimensional motion. The motion allowed changes in a continuum from the cervical to the lumbar spine. The extent of motion allowed by each spinal level has been extensively characterized using both *in vivo* and *in vitro* approaches and is commonly termed the range of motion (ROM). For the most part, the *in vitro* data agrees well with the *in vivo* data. Rotational spinal motion is often characterized according to anatomic axes as flexion/extension, lateral bending, and axial torsion (Figure 10.2). A compendium of available data was presented by White and Panjabi (1990) and is reproduced in Figure 10.3.

In addition to ROM data, other characteristics of the natural joint that should be reflected in the reconstructed joint are the center of rotation (if the motion is two-dimensional), or the helical axis of motion (Kinzel, Hall, and Hillberry 1972) (if the motion is three-dimensional), and the extent and nature of the coupled motions that will accompany a specific motion. For example, because of the shape of the facet joints in the lower cervical spine (C3–C7), axial torsion

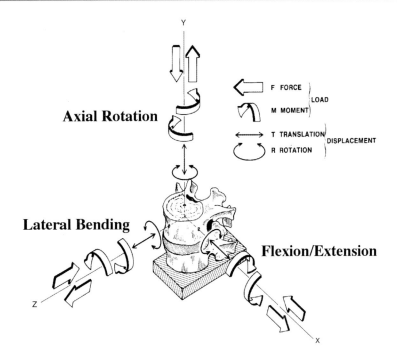

Figure 10.2

Axes used to characterize spinal motion: flexion–extension, lateral bending, and axial torsion. (Reprinted with permission White and Panjabi 1990.)

is always accompanied by significant amounts of coupled lateral bending and vice versa in this region (Wen et al. 1993). Centers of rotation, helical axes of motion, and coupled motions vary as a specific intervertebral joint moves through its ROM. Various authors have studied these parameters for the cervical (Cripton et al. 2001), thoracic, and lumbar spines (Oxland, Panjabi, and Lin 1994). The center of rotation (instantaneous axis of rotation [IAR]) data for the lumbar spine as presented by White and Panjabi in 1990 are presented in Figure 10.4.

Kinetic Considerations

The mechanical environment of the lumbar spine has been studied extensively using biomechanical models (McGill 1990, Calisse, Rohlmann, and Bergmann 1999, McGill and Norman 1986), *in vivo* disc pressure measurements (Nachemson 1981, Wilke et al. 1999), and instrumented orthopedic devices (Rohlmann, Bergmann, Graichen 1997, Rohlmann et al. 1998, Rohlmann et al. 1999). The same approaches have been used in the thoracic and cervical spines, but there is a significantly smaller volume of information available for these regions. In the lumbar spine, compressive loads in excess of 8000 N and

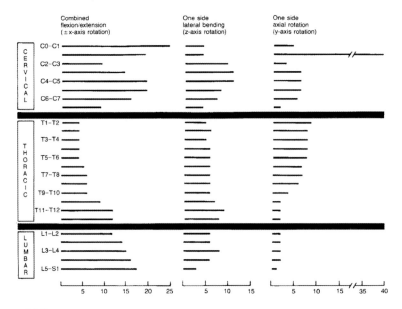

Figure 10.3
Representative values for ranges of rotational motion at the different levels of the spine. (Reprinted with permission White and Panjabi 1990.)

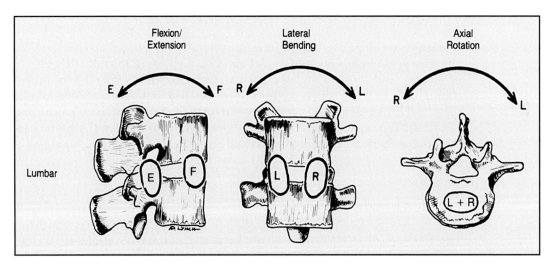

Figure 10.4
Approximate locations of the center of rotation in the lumbar spine. E, location of the axes going from a neutral to an extended position, F, going from a neutral to a flexed position, L, center of rotation in left lateral bending and torsion, and R, right lateral bending and torsion. (Reprinted with permission White and Panjabi 1990.)

anterior–posterior (A–P) shear loads in excess of 1000 N have been predicted, using EMG-driven biomechanical modeling, for healthy volunteers during lifting (McGill and Norman 1986). Lower loads of approximately 2500 N have been measured using *in vivo* disc pressure measurements for volunteers during lifting (Nachemson 1981, Wilke et al. 1999) (Figure 10.5). It is important to also consider the moments present at the intervertebral level of interest during the expected postsurgical activities of the patient. Many of the authors referenced earlier have also reported the expected *in vivo* moments.

Load-Sharing Considerations

At many levels of the spine, the facet joints will support some of a particular applied load, thereby reducing the total load transferred to the device in the intervertebral joint. It is important that the implanted device maintain the overall load sharing of the segment and the loading of the facet joints in both cases, because significant changes in the stresses or strains experienced by the anatomic structures could lead to deleterious changes such as osteophyte formation, degeneration, fusion, or fracture (Kumaresan et al. 2001). Cripton and colleagues (2000) have investigated the load-sharing properties of the lumbar spine stabilized with a posterior fixation system (Figure 10.6). Although not directly applicable to the nonstabilized spines of most patients who will receive

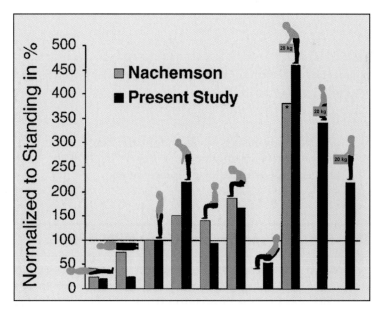

Figure 10.5

Loads in the lumbar spine as measured by Nachemson (1981) and Wilke and colleagues (1999); 450% on the graph corresponds to approximately 2200 N. (Reprinted with permission Wilke et al. 1999.)

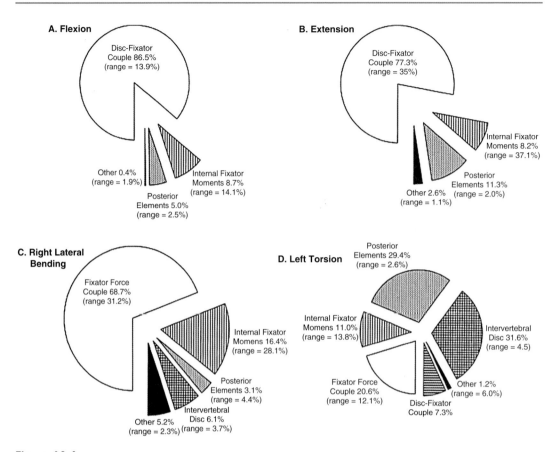

Figure 10.6

Load-sharing properties of the lumbar spine stabilized with a posterior fixation system. (Reprinted with permission Cripton et al. 2000.)

a TDR, these results do highlight the central role that the disc plays under flexion–extension motion, and this is likely to be magnified in unstabilized recipients of artificial disc replacements.

Total Disc Replacement Designs Using UHMWPE

There are currently two designs of total artificial discs that include an UHMWPE component as a bearing surface for their implant. Both the SB Charité™ III (Link Spine Group) and PRODISC (Spinal Solutions) designs have been used clinically since the 1990s in Europe and are currently undergoing clinical trials under approved IDEs in the United States. In 2003, after the clinical trials were already under way, the original companies owning the rights to these two implants were acquired by other orthopedic manufacturers.

The PRODISC was acquired by Synthes USA (Paoli, PA), with the final completion of the negotiation contingent upon successful completion of the clinical trials. The SB Charité III was acquired by Depuy Acromed (Johnson & Johnson), with a similar condition on further contingent payments due upon achievement of regulatory and other milestones.

SB Charité III

Historical Development

In the early 1980s, Schellnack and Büttner-Janz initiated the development of the artificial disc that later became the SB Charité. The naming of the implant is after the initials of the inventors and the hospital where it was designed and first implanted (S, Kurt Schellnack, B, Karin Büttner-Janz, and Charité, site of development and first implantation) (Büttner-Janz, Hochschuler, and McAfee 2003, Link 2002). The first implantation took place on September of 1984 in the Charité Hospital in Berlin. To date, three generations of designs have evolved and the current design SB Charité III was first produced by Waldemar LINK® in 1987 (Büttner-Janz, Schellnack, and Zippel 1987, Link 2002).

Design Concept

The main concept in this design was based on the already existing low-friction principle in the TJR field, using a metal alloy articulating with UHMWPE. The SB Charité artificial disc is made of three components, two of which are metal endplates that attach to the upper and lower vertebral body endplates by means of anchor teeth, and an UHMWPE sliding core that moves in between these two metal endplates. The endplates of the first two models of the SB Charité (I and II) were manufactured from 1-mm thick stainless steel. Specifically, the ones in model Type I were round and later evolved to an oval shape in model Type II (Büttner-Janz, Hochschuler, and McAfee 2003).

Biomaterials

The current SB Charité III version has endplates made from CoCr alloy (ISO 5832/IV, ASTM F 75-82) that include a convexity to improve the fit with the concave surface of the vertebral endplates (Büttner-Janz, Hochschuler, and McAfee 2003, Link and Keller 2003). Each endplate has three anchoring teeth ventrally and dorsally, and different levels of lordotic angulation. In addition, the biconvex UHMWPE sliding core (ISO 5834/II and ASTM F 648-83) is made to fit the metal endplates such that it does not dislocate from its position between the plates and it also features a radiopaque x-ray marker wire. The current set available for implantation (Figure 10.7) includes four sizes of cores, with five different heights, in combination with four sizes of endplates and four lordotic angles to choose from (0, 5, 7.5 or 10 degrees).

The properties for the UHMWPE are listed in Table 10.1. It is identified under the trade name Chirulen®, which (as described in Chapter 2) denotes a

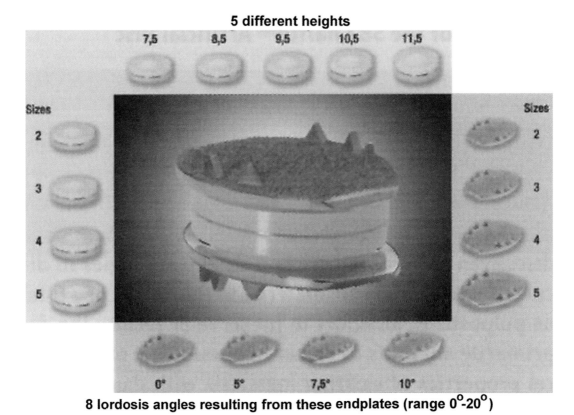

5 different heights

8 lordosis angles resulting from these endplates (range 0°-20°)

Figure 10.7

Variety of sizes of the SB Charité III artificial disc. (Reproduced with permission Büttner-Janz, Hochschuler, and McAfee 2003.)

Table 10.1

Material Properties for UHMWPE (Chirulen) Used in the SB Charité[a]

Property	Value	Test method
Average molecular weight	5×10^6 g/mol	—
Density (compression molded sheet)	930 kg/m^3	ISO 1183 method A
Yield stress	21 MPa	ISO 527
Tensile modulus	720 MPa	ISO 527
Elongation at break	>50%	ISO 527
Notched impact strength	No break	DIN 53453

[a]From Büttner-Janz, Hochschuler, and McAfee 2003.

material produced in Germany by a dedicated press using established parameters. The material used for the insert is compression-molded GUR 1020 from Poly HI Solidur (Link 2003). This material is sterilized using an N-Vac® process proprietary to LINK (LINK N-VAC) in which there is a reduction of oxygen during the sterilization, followed by an infiltration of nitrogen, and then followed by a phase of welding (sealing) of the packaging (Figure 10.8). The goal of this process is to retain the properties of the UHMWPE and reduce the volume of deformation and wear and tear by approximately 20% as compared with conventional methods (Link and Keller 2003). The circumference of the sliding core is surrounded by a rim that aims to prevent the edges of the superior and inferior endplates from touching each other and, as such, avoid metal-on-metal (MOM) articulation and wear (Link and Keller 2003).

Biomechanical Evaluation of SB Charité

Biomechanical testing on the SB Charité concentrated on the UHMWPE sliding core and was performed as part of the evaluation of the device. Initial evaluations of the SB Charité I and II were conducted by Büttner-Janz and colleagues (1989). For these tests the superior and inferior endplates were embedded in polypropylene and the entire implant was subjected to increasing cyclic loads (load–unload) up to 19.5 kN. Hysteresis was observed in the UHMWPE component up to loads of 4.2 kN. With increasing loads between 6 kN and 8 kN, there was irreversible deformation of the core resulting from cold flow. With loads in the 10.5 kN range there was a reduction of the height

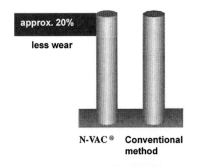

**By reduction of oxygen
the material is optimally
protected against degradation**

(The Volume of deformation and wear and tear
is approx. 20% less than in conventional methods)

Phase of vacuum   Infiltration of nitrogen   Phase of welding (sealing)

Figure 10.8
N-VAC packaging/sterilization process used for the SB Charité III. (Reproduced with permission Büttner-Janz, Hochschuler, and McAfee 2003.)

of the core up to 10%, and with loads up to 19.5 kN there was visible bulging of the core. Testing simulating long-term use was also performed at a rate of 5 to 10 cycles per second (5–10 Hz). Testing was carried out up to 2×10^7 cycles (to be equivalent to 20 years of use) with a preload of 0.7 kN and a maximum weekly load of 8 kN, with several increments throughout. This testing revealed that the sliding cores deformed at most by 10% during the initial simulated two years of use. After testing, track marks were observed on the plates and the core, which were attributed to the repetitive movement imposed during testing (Büttner-Janz, Schellnack, and Zippel 1989).

Further independent dynamic tests were performed on the SB Charité III at the University of Kiel and at Mt. Sinai Medical Center in Cleveland, OH to evaluate the cold flow properties of the sliding core for collection of additional data required in the regulatory submission with the FDA (Link 2002). The objective of these tests was to provide an estimate of permanent deformation and failure location on the UHMWPE core under pure compression. The second smallest size (#2) in two heights (7.5 and 9.5 mm) was evaluated at two compressive loads (2.5 and 4.5 kN) in tests that lasted 24 hours at 37°C in Ringer's solution at three different frequencies. The tests indicated that the maximum expected deformation of the sliding core after 10 years was 0.403 mm and 0.738 mm, for the 7.5 mm and 9.5 mm cores, respectively, which is meant to represent a decrease in height of less than 8%. Based on these tests it was concluded that cold flow was not a major factor in this prosthesis design. It has also been stated by Link and Keller that the cold flow limits of the UHMWPE used in the SB Charité is 22 N/mm². Based on this limit they have estimated that the SB Charité has a safety of margin from 3.9 mm for size #2, to 5.71 mm for size #5 (2003).

Additional functional testing in a simulator has been conducted on the SB Charité up to 10×10^6 cycles alternating from 0.5 to 2.5 kN. Tests were performed simulating axial rotation, lateral bending, and flexion–extension. These tests were conducted on the smallest size implants with medium size thickness (9.5 mm) in a medium consisting of calf serum at 37°C. Results showed that there was no abrasive wear on the contact zones of the UHMWPE sliding core. Figures 10.9 and 10.10 show scanning electron micrographs of the tribologic surface of the superior and inferior aspect of the UHMWPE, respectively. They indicated that these surfaces showed only fine abrasive grooves and "flags" of limited size (<10 microns). They concluded that there was no abrasive wear over 80–90% of the contact area (inferior and superior).

Biomechanics of Implant Performance

The SB Charité is an unconstrained implant. As such, it is capable of reproducing pure A–P and lateral translation of the cephalad vertebral body even without rotation (Huang et al. 2003). The IAR of this disc replacement lies within the disc space, and because of its unconstrained nature it is mobile and as such can compensate for deviations from adequate placement (Huang et al. 2003). In flexion–extension due to its three-component set up including the sliding core,

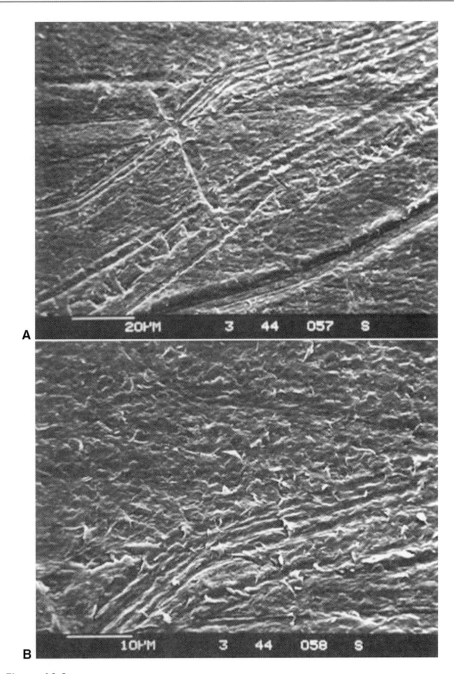

Figure 10.9
Superior polar area of the UHMWPE sliding core at (**A**) 1000 magnification and (**B**) 2000 magnification. (Reproduced with permission Büttner-Janz, Hochschuler, and McAfee 2003.)

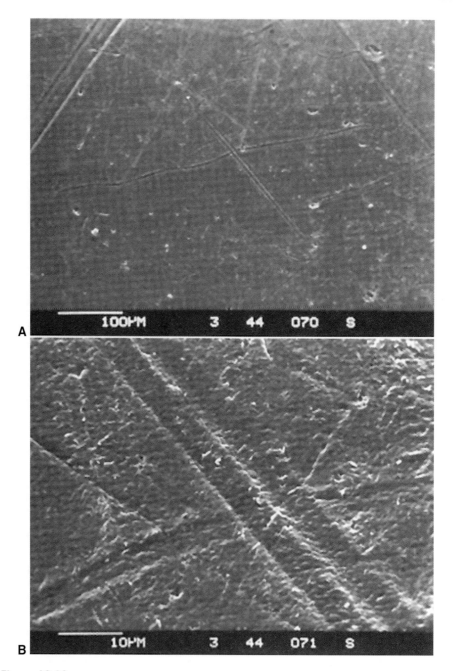

Figure 10.10
Tribologic contact area of the inferior aspect of the UHMWPE sliding core at (**A**) 200 magnification and (**B**) 2000 magnification. (Reproduced with permission Büttner-Janz, Hochschuler, and McAfee 2003.)

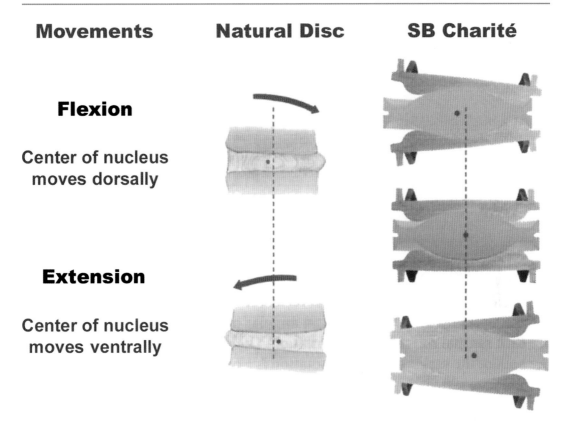

| **Movements** | **Natural Disc** | **SB Charité** |

Flexion

**Center of nucleus
moves dorsally**

Extension

**Center of nucleus
moves ventrally**

Figure 10.11

Flexion and extension produce a small displacement of the nucleus. (Reproduced with permission Büttner-Janz, Hochschuler, and McAfee 2003.)

the SB Charité allows the IAR to change on the joint position permitting a small displacement of the nucleus (Figure 10.11). This movement allows the SB Charité to replicate the sagittal rotation and translation seen during flexion–extension (Link 2002, Link and Keller 2003). This device will likely subject the facet joints and posterior ligaments to increased shear loads due to anterior-posterior mobility (which allows A–P translation) (Huang et al. 2003). The long-term clinical and biomechanical effects of these potential increased shear loads are still unknown. In lateral bending of the lumbar spine, the SB Charité replicates the coronal rotation combined with translation due to its sliding core (Link 2002, Link and Keller 2003). Table 10.2 summarizes data collected on mean ROM in degrees at the L4–5 level comparing fresh frozen cadaveric specimens (with most ligamentous structures intact) with specimens with an implanted SB Charité (Ahrens, Shelokov, and Carver 1998). This data shows similar ROM in all modes of loading with the exception of torsion, which was attributed to the altered anterior ligament (Link 2002).

Table 10.2

Mean Range of Motion in Degrees at the L4–5 Level Comparing Motion in Cadaveric Specimens with Those with an SB Charité

Motion	Max. load (Nm)	Normal disc	SB Charité
Extension	12	3.49 (0.82)	3.27 (0.83)
Flexion	12	7.72 (1.74)	9.78 (1.48)
Left flexion	8	2.78 (1.78)	2.37 (0.57)
Right flexion	8	5.24 (2.54)	7.41 (2.65)
Torsion	7	1.66 (0.74)	3.01 (0.73)

From Ahrens, Shelokov, and Carver 1998, Link 2002.

PRODISC

Historical Development

The first generation PRODISC was designed in France by Dr. Thierry Marnay in the late 1980s and was first implanted in early 1990. In conjunction with Dr. Villette, they implanted this disc replacement in a total of 64 patients from 1990 until 1993 and decided to follow them long term (Marnay 2003, Spine Solutions 2002b). The PRODISC II was described as an improvement over the first design and was launched in the European market in December 1999 (Tropiano et al. 2003).

Design Concept

This TDR has a modular design consisting of three implant components and follows a ball-and-socket joint principle in its design (Figure 10.12). There are two metal endplates and an UHMWPE inlay that is locked into the lower endplate. To date there have been two generations of the design with the updates including changes in materials, anchorage, modularity, and instrumentation. The current PRODISC is modular, allowing the surgeon to customize it to the patient's anatomy. It comes in two sizes (M and L), with each size having two available angulations of the superior plate component (6 and 11 degrees), and three different heights of the polyethylene inlay (10, 12, and 14 mm) (Spine Solutions 2002a).

Biomaterials

The endplates of the original PRODISC were made of titanium and each had two keels for anchorage with the vertebral body. In the PRODISC II, these endplates are now made of CoCr alloy and have only one keel and two spikes. Fixation is achieved through the wedged keel and the two spikes that are anchored by a press-fit. The surfaces of the endplates have a rough plasma-sprayed titanium coating. The goal is that the increased surface roughness

Figure 10.12
Components of the PRODISC: two endplates and UHMWPE inlay. (Reproduced with permission Zigler et al. 2003.)

provided by the plasma-sprayed coating, in addition to the press-fit, will ensure immediate stability by promoting a surface for bony ingrowth (Spine Solutions 2002a). As of 2003, there was no specific information available on the type of UHMWPE, the source of manufacturing, or how it is sterilized.

Biomechanical Evaluation of PRODISC

The shock absorption capacity of the PRODISC has been evaluated to determine its vibration and shock transmissibility in comparison to a TDR with a metal on metal interface (MAVERICK, Medtronic Sofamor Danek, Memphis, TN). One sample for each kind of implant was evaluated. The shock absorption capacity was defined by the shock transmission ratio (output spectrum/input spectrum) and quantified by the percentage of the transmitted load in comparison to the input load. In addition, the phase angle deviation between the input and the output signal (damper coefficient) is indicative of the delay in signal transmission and signal distortion of the device and characterizes the shock absorber effects. The devices were tested under shock impact loads and sinusoidal vibrations with a static compressive preload of 350 N. The testing of the PRODISC, in comparison with the MOM device (cobalt chrome alloy) showed that the difference between the two devices was less than 0.8%, with both devices having practically identical vibration and shock transmissibility. Under the oscillating and shock loading, both devices had a transmission ratio greater than 99.8% and 98%, respectively. Both devices had a phase angle (shock absorption effect) of less than 10 degrees, indicating that neither of the two devices had an effective shock absorption capacity (LeHuec et al. 2003). There is no data that compares the shock absorption abilities of the PRODISC with the human intervertebral disc.

Biomechanics of Implant Performance

Due to its ball-and-socket design, the PRODISC is a constrained TDR. The ball (in the UHMWPE component) and the socket (in the upper endplate) are highly conforming and have the same radius of curvature. As such, pure A–P and lateral translation of the upper vertebral body are not possible (Huang et al. 2003). This implant has an IAR that is slightly below the caudal endplate, similar to that of most lumbar motion segments, with the exception of L5–S1. Rotation of this implant around this IAR provides apparent translation of the upper vertebral body. However, this IAR is fixed in location, making the kinematics of this TDR very dependent on proper surgical placement. The constraint inherent with the PRODISC will lead to absorption of all the shear loads because it prevents pure A–P or lateral translations. It has been suggested that this could potentially result in long-term facet preservation (Huang et al. 2003). It should also be noted, however, that if this design results in facet loads significantly below those experienced before implantation of the artificial disc then this could also result in deleterious effects.

In terms of kinematics, the PRODISC is stated by its manufacturer to have ± 10 degrees in the sagittal and frontal planes (Figure 10.13) (Tropiano et al. 2003). Specifically, it is stated to have a ROM of 13 degrees in flexion and 7 degrees in extension. Translation occurs as a result of flexion–extension and lateral bending. In axial rotation, the implant does not limit the ROM (PRODISC 2002).

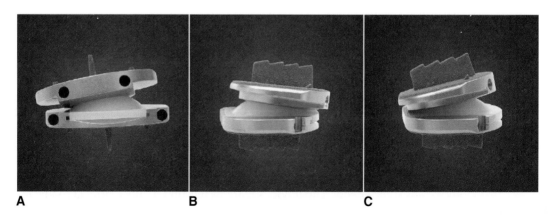

A   **B**   **C**

Figure 10.13
PRODISC shown in (**A**) lateral bending, (**B**) flexion, and (**C**) extension. (Reproduced with permission Tropiano et al. 2003.)

Table 10.3

Range of Motion in Degrees for Intact and Implanted L5–S1 Segments with a PRODISC[a]

Compressive load	State of FSU[b]	Flexion–extension	Lateral bending	Axial torsion
600 N	Intact	4.2 ± 3.6	3.1 ± 2.2	0.9 ± 0.5
	Implanted	2.5 ± 2.0	6.8 ± 2.9	1.8 ± 1.4
1200 N	Intact	3.0 ± 3.3	2.3 ± 1.8	0.7 ± 0.3
	Implanted	1.4 ± 0.5	4.9 ± 4.6	1.0 ± 0.8

[a]From Lipman et al. 2003.
[b]Functional spinal unit.

A cadaveric study comparing the range of motion in the L5–S1 motion segment in five human spines, before and after implantation with the PRODISC, has indicated that implantation of this device did not significantly affect the ROM (Lipman et al. 2003). Specimens were tested in an apparatus that applied pure bending moments. The specimens were cycled in torque to a maximum of ± 10 Nm in flexion–extension, lateral bending, and torsion with 600 N and 1200 N compressive loads. The ROM at 8 Nm after the fifth cycle of loading was documented and used for comparison between intact and implanted specimens. Summary data is shown in Table 10.3.

Clinical Performance of UHMWPE in the Spine

SB Charité III

Artificial discs had been implanted in patients in Europe long before they were implanted in patients in the United States under the IDE studies. There are a number of reports in the literature on the short-term (Griffith et al. 1994, Lemaire 2003, Lemaire et al. 1997, Zeegers et al. 1999) (<5 years) and long-term (Lemaire 2003, Van Ooij, Oner, and Verbout 2003) (>10 years) clinical experience of the SB Charité in Europe. Only recently have we seen reports of the initial clinical experience in the United States (Hochschuler et al. 2002, McAfee et al. 2003).

The long-term experience in Europe with follow-up of 77 patients with 121 artificial discs, reveals that after 10 years of follow-up, results remain good in more than 90% of all cases. Radiographically after 10 years, there was no evidence of loosening or change in the position of the implant, except for one case that subsided because of osteoporosis, no dislocation or fracture of the UHMWPE core, and no significant loss of disc height (Lemaire 2003). Reports on the complications of the SB Charité include a review of 27 patients with unsatisfactory results with a mean follow-up of 7.6 years. Among the most common late complications were incapacitating back and leg pain, degenerative disc disease at other levels (already seen at the time of the disc implantation), facet joint arthrosis, subsidence of the endplates, subluxation of the UHMWPE core, migration of the prosthesis, breakage of the metal wire round the UHMWPE core, and radiographic signs of wear in the UHMWPE after 13 years (Van Ooij, Oner, and Verbout 2003).

Because the follow-up of the SB Charité implantations in the United States are still under the auspices of the IDE, currently there are only preliminary reports available in the literature from two different centers that are participating in the study, with follow-up from 1 to 3 years (Hochschuler et al. 2002, McAfee et al. 1999). The functional outcome measures used in both studies showed significant improvement following implantation of the SB Charité (Hochschuler et al. 2002, McAfee et al. 2003). Neither of the two series report any device failures or dislocations.

PRODISC

In the year 2000, an evaluation was conducted by Spine Solutions to establish a 7- to 11-year follow-up of the first 64 European patients implanted with a PRODISC. These patients had a total of 93 implants, consisting of 39 single-level, 21 two-level, and 4 three-level implantations. Significant differences were found between the contemporary (at the time of follow-up) and preoperative states for the visual analog scores (VASs), and the Oswestry disability (ODQ) scores indicated minimal disability at the time of follow-up. A 92.7% rate of satisfaction was documented among those patients that were followed, including 63.6% being entirely satisfied and 29.1% being satisfied. In general, patients indicated to have less pain and reported being more active. There were no device-related complications, no implant migration or subsidence of any significance, and only two surgical complications of vascular nature. Four cases were subsequently fused posteriorly, and one case had spontaneous fusion, but there were no implant removals (Linovitz 2001).

Another study following 53 patients in Europe implanted with the PRODISC between December 1999 and December 2001 provides a follow-up evaluation after a mean period of 1.4 years (Tropiano et al. 2003). In this group of patients complete satisfaction was reported by 87% with satisfaction in 13%. Complications reported in 9% of the patients included a vertebral body fracture, malposition, and persistent radicular pain. In this group of patients there were statistically significant improvements in the VASs and ODQ scores following implantation. Bertagnoli and Kumar also report on 1-year follow-up on 108 patients who received a total of 134 disc replacements (2002). Of these patients 90.8% showed excellent results with an increased ROM.

Some early results of initial PRODISC implantations in the United States have been reported (Zigler 2002, Zigler et al. 2003). The group of 28 PRODISC patients and 11 360-degree fusion patients showed a significantly greater reduction in ODQ at 3-month follow-up with no difference noted in the VAS. At 6-month follow-up, patient satisfaction was more favorable in the TDR patients, with motion being more significantly improved as well (Zigler et al. 2003). This group of patients will be followed for at least 24 months as part of the IDE.

Alternatives to UHMWPE for Total Disc Arthroplasty in the Spine

Artificial disc replacements have also been developed that do not incorporate UHMWPE components. These have a MOM bearing system, which presumably is to reduce the amount of wear produced. This once again shows the trend of borrowing concepts from the TJR world and applying them to TDRs. In addition, for TDR designs, the wear performance of CoCr has been evaluated and was shown to have less wear rate in comparison with titanium alloy designs (Hellier, Hedman, and Kostuik 1992).

The other TDRs that have also reached clinical trials under an IDE with the FDA in the United States have MOM bearing surfaces. These two TDRs are the MAVERICK (Medtronic Sofamor Danek, Memphis, TN) and FLEXICORE™ (SpineCore, Summit, NJ). The MAVERICK was used in Europe before the IDE in the United States began.

The MAVERICK TDR is made of CoCr alloy, which is also the gold standard in MOM bearings in the hip. It consists of two components that are separate and attach to each endplate (Figure 10.14). The interaction of these two components is in a ball-and-socket design with the socket being on the top component and the ball in the lower component. The only data on this TDR is the study that compared its shock absorption with that of the PRODISC (LeHuec et al. 2003). The FLEXICORE TDR also has MOM bearing surfaces (CoCr) and is made of seven preassembled pieces that include a conical internal spring (Guyer and Ohnmeiss 2003). This disc has a ball-and-socket joint type of design, allowing the vertebral bodies to move about a rotational center between them. There are no reports in the literature on this disc replacement on either laboratory studies or clinical performance.

Figure 10.14
MAVERICK TDR. (Reproduced with permission LeHuec et al. 2003.)

Conclusion

Much like in the total joint surgeries, where motion preservation has been the goal in order to restore physiologic movement and improve the quality of life of patients, it is the goal for TDA to have a similar effect in the spine. Until recently, spinal fusion has been the treatment of choice to alleviate low back pain, with success depending on a series of factors including diagnosis, number of levels fused, and prior surgeries. As has been the experience in the area of total joints (hips and knees), the ultimate clinical performance of UHMWPE in the spine in total artificial discs can be comprehensively evaluated only after long-term follow-up of patients is made available. In addition, examination of retrieved implants (when available) will also provide the scientific community with a better understanding of the performance of UHMWPE *in vivo* in the spine anatomical setting. As of 2003, clinical trials through IDE status with the FDA are still ongoing in the United States for TDRs with UHMWPE components to show the safety and efficacy of these devices before allowing them to be marketed to the general public.

Acknowledgments

We thank Lauren Ciccarelli and Cara Miller for their review of the chapter.

References

Ahrens J., A. Shelokov, and J. Carver. 1998. Normal joint mobilities maintained with an artificial disk. Lecture, International Society for the Study of the Lumbar Spine (ISSLS), Toronto.

Bao Q.B., and H.A. Yuan. 2000. Artificial disc technology. *Neurosurg Focus* 9:1–7.

Bertagnoli R., and S. Kumar. 2002. Indications for full prosthetic disc arthroplasty: A correlation of clinical outcome against a variety of indications. *Eur Spine J* 11Suppl 2: S131–136.

Büttner-Janz K., K. Schellnack, and H. Zippel. 1987. [An alternative treatment strategy in lumbar intervertebral disk damage using an SB Charité modular type intervertebral disk endoprosthesis]. *Z Orthop Ihre Grenzgeb* 125:1–6.

Büttner-Janz K., K. Schellnack, and H. Zippel. 1989. Biomechanics of the SB Charité lumbar intervertebral disc endoprosthesis. *Int Orthop* 13:173–176.

Büttner-Janz K., S. Hochschuler, and P. McAfee, Eds. 2003. *The artificial disc.* Berlin: Springer Verlag.

Calisse J., A. Rohlmann, and G. Bergmann. 1999. Estimation of trunk muscle forces using the finite element method and *in vivo* loads measured by telemeterized internal spinal fixation devices. *J Biomech* 32:727–731.

Cripton P.A., G.M. Jain, R.H. Wittenberg, and L.P. Nolte. 2000. Load-sharing characteristics of stabilized lumbar spine segments. *Spine* 25:170–179.

Cripton P.A., M. Sati, T.E. Orr, et al. 2001. Animation of *in vitro* biomechanical tests. *J Biomech* 34:1091–1096.

David T. 2003. Complications with the SB Charité artificial disc. In *The artificial disc.* K. Büttner-Janz, S. Hochschuler, and P. McAfee, Eds. Berlin: Springer Verlag.

Dooris A.P., V.K. Goel, N.M. Grosland, et al. 2001. Load-sharing between anterior and posterior elements in a lumbar motion segment implanted with an artificial disc. *Spine* 26:E122–129.

Eck J.C., S.C. Humphreys, and S.D. Hodges. 1999. Adjacent-segment degeneration after lumbar fusion: A review of clinical, biomechanical, and radiologic studies. *Am J Orthop* 28:336–340.

Eijkelkamp M.F., C.C. van Donkelaar, A.G. Veldhuizen, et al. 2001. Requirements for an artificial intervertebral disc. *Int J Artif Organs* 24:311–321.

Fernstrom U. 1966. Arthroplasty with intercorporal endoprosthesis in herniated disc and in painful disc. *Acta Chir Scand Suppl* 357:154–159.

Griffith S.L., A.P. Shelokov, K. Büttner-Janz, et al. 1994. A multicenter retrospective study of the clinical results of the LINK SB Charité intervertebral prosthesis. The initial European experience. *Spine* 19:1842–1849.

Guyer R.D., and D.D. Ohnmeiss. 2003. Intervertebral disc prostheses. *Spine* 28:S15–23.

Hedman T.P., J.P Kostuik, G.R. Fernie, and W.G. Hellier. 1991. Design of an interverte-bral disc prosthesis. *Spine* 16:S256–260.

Hellier W.G., T.P. Hedman, and J.P. Kostuik. 1992. Wear studies for development of an intervertebral disc prosthesis. *Spine* 17:S86–96.

Hochschuler S.H., D.D. Ohnmeiss, R.D. Guyer, and S.L. Blumenthal. 2002. Artificial disc: Preliminary results of a prospective study in the United States. *Eur Spine J* 11Suppl2: S106–110.

Huang R.C., F.P. Girardi, F.P. Cammisa, Jr., and T.M. Wright. 2003. The implications of constraint in lumbar total disc replacement. *Spine* 28Suppl:S412–417.

Kinzel G.L., A.S. Hall, and B.M. Hillberry. 1972. Measurement of the total motion between two body segments-I. Analytical development. *J Biomech* 5:93–105.

Kostuik J.P. 1998. Alternatives to spinal fusion. *Orthop Clin North Am* 29:701–715.

Kumaresan S., N. Yoganandan, F.A. Pintar, et al. 2001. Contribution of disc degeneration to osteophyte formation in the cervical spine: A biomechanical investigation. *J Orthop Res* 19:977–984.

Langrana N.A., J.R. Parsons, C.K. Lee, et al. 1994. Materials and design concepts for an intervertebral disc spacer. I. Fiber-reinforced composite design. *J Appl Biomater* 5:125–132.

LeHuec J.C., T. Kiaer, T. Friesem, et al. 2003. Shock absorption in lumbar disc prosthesis: A preliminary mechanical study. *Spine* 28:346–351.

Lemaire J-P. 2003. Mid-term (4 year) and long-term (10 year) results of the SB Charité prosthesis. In *The artificial disc.* K. Büttner-Janz, S. Hochschuler, and P. McAfee, Eds. Berlin: Springer Verlag.

Lemaire J.P., W. Skalli, F. Lavaste, et al. 1997. Intervertebral disc prosthesis. Results and prospects for the year 2000. *Clin Orthop* 337:64–76.

Link H.D. 2002. History, design and biomechanics of the LINK SB Charité artificial disc. *Eur Spine J* 11Suppl2:S98–105.

Link H.D., and A. Keller. 2003. Biomechanics of total disc replacement. In *The artificial disc.* K. Büttner-Janz, S. Hochschuler, and P. McAfee, Eds. Berlin: Springer Verlag.

Link S.B. 2001. Is fusion still today's gold standard? *Charité artificial disc, maintaining natural mobility.* Hamburg: Waldemar Link.

Linovitz R. 2001. PRODISC. Retrospective clinical study: 7 to 11 year follow up: Internal Report, Spine Solutions.

Lipman J., D. Campbell, F. Girardi, et al. 2003. Mechanical behavior of the PRODISC II intervertebral disc prosthesis in human cadaveric spines. 49th Annual Meeting of the Orthopedic Research Society. New Orleans, LA.

Marnay T. 2003. Lumbar disc replacement, 7–11 years results with PRODISC. *Eur Spine J* 11Suppl:S19.

McAfee P.C., B.W. Cunningham, G.A. Lee, et al. 1999. Revision strategies for salvaging or improving failed cylindrical cages. *Spine* 24:2147–2153.

McAfee P.C., I.L. Fedder, S. Saiedy, et al. 2003. SB Charité disc replacement: Report of 60 prospective randomized cases in a U.S. center. *Spine* 28Suppl:S424–433.

McGill S.M., and R.W. Norman. 1986. Partitioning of the L4-L5 dynamic moment into disc, ligamentous, and muscular components during lifting. *Spine* 11:666–678.

McGill S.M. 1990. Loads on the lumbar spine and associated tissues. In *Biomechanics of the spine, clinical and surgical perspective*. V.K. Goel and J.N. Weinstein, Eds. Boca Raton, FL: CRC Press.

Nachemson A.L. 1981. Disc pressure measurements. *Spine* 6:93–97.

Oxland T.R., M.M. Panjabi, and R.M. Lin. 1994. Axes of motion of thoracolumbar burst fractures. *J Spinal Disorders* 7:130–138.

Pointillart V. 2001. Cervical disc prosthesis in humans: First failure. *Spine* 26:E90–92.

Rohlmann A., G. Bergmann, F. Graichen. 1997. Loads on an internal spinal fixation device during walking. *J Biomech* 30:41–47.

Rohlmann A., G. Bergmann, F. Graichen, and G. Neff. 1999. Braces do not reduce loads on internal spinal fixation devices. *Clin Biomech* 14:97–102.

Rohlmann A., G. Bergmann, F. Graichen, and H.M. Mayer. 1998. Influence of muscle forces on loads in internal spinal fixation devices. *Spine* 23:537–542.

Spine Solutions. 2002a. I. PRODISC, company brochure. New York.

Spine Solutions 2002b. I. PRODISC, retrospective clinical study: 7 to 11 year follow-up. New York.

Suk K.S., H.M. Lee, S.H. Moon, and N.H. Kim. 2001. Recurrent lumbar disc herniation: Results of operative management. *Spine* 26:672–676.

Tropiano P., R.C. Huang, F.P. Girardi, and T. Marnay. 2003. Lumbar disc replacement: Preliminary results with ProDisc II after a minimum follow-up period of 1 year. *Spine* 28Suppl:S362–368.

Van Ooij A., F.C. Oner, and A.J. Verbout. 2003. Complications of artificial disc replacement: A report of 27 patients with the SB Charité disc. *Spine* 28Suppl:S369–383.

Wen N., F. Lavaste, J.J. Santin, and J.P. Lassau. 1993. Three-dimensional biomechanical properties of the human cervical spine *in vitro*, I. Analysis of normal motion. *European Spine Journal* 2:2–11.

White A.A., and M.M. Panjabi. 1990. *Clinical biomechanics of the spine*. 2nd ed. New York: J.B. Lippincott.

Whitecloud T.S., III. 1999. Modern alternatives and techniques for one-level discectomy and fusion. *Clin Orthop* 359:67–76.

Wilke H.J., P. Neef, M. Caimi, et al. 1999. New *in vivo* measurements of pressures in the intervertebral disc in daily life. *Spine* 24:755–762.

Yorimitsu E., K. Chiba, Y. Toyama, and K. Hirabayashi. 2001. Long-term outcomes of standard discectomy for lumbar disc herniation: A follow-up study of more than 10 years. *Spine* 26:652–657.

Zeegers W.S., L.M. Bohnen, M. Laaper, and M.J. Verhaegen. 1999. Artificial disc replacement with the modular type SB Charité III: 2-year results in 50 prospectively studied patients. *Eur Spine J* 8:210–217.

Zigler J.E. 2002. Early experience with PRODISC in the United States. *Spine*. A.G. Aesculap and K.G. Tuttlingen, Eds. Berlin: Springer-Verlag.

Zigler J.E., T.A. Burd, E.N. Vialle, et al. 2003. Lumbar spine arthroplasty: Early results using the ProDisc II: A prospective randomized trial of arthroplasty versus fusion. *Spine* 28Suppl:352–361.

Chapter 10. Reading Comprehension Questions

10.1. What is the primary motivation for TDR?
   a) To improve fusion of the degenerated disc
   b) To restore the motion of the natural disc
   c) To treat vertebral compression fractures
   d) To reduce osteolysis of the spine
   e) All of the above

10.2. When was TDR first clinically introduced?
   a) 1960s
   b) 1970s
   c) 1980s
   d) 1990s
   e) 2000s

10.3. Which of the following features is not included in the current design of the Charité TDR?
   a) Bone cement
   b) Metal backing
   c) Porous coating
   d) UHMWPE disc component
   e) None of the above

10.4. The TDR achieves fixation to which bony surfaces?
   a) Cancellous bone
   b) Pedicles
   c) Posterior elements
   d) Endplates
   e) All of the above

10.5. What major design feature of the PRODISC distinguishes it from the Charité disc?
   a) The inferior surface of the UHMWPE component in the PRODISC is totally constrained
   b) The superior surface of the UHMWPE component in the PRODISC is totally constrained
   c) The metallic components of the PRODISC are cemented
   d) The metallic components of the PRODISC are cobalt chrome alloy
   e) All of the above

Chapter 11

Mechanisms of Crosslinking and Oxidative Degradation of UHMWPE

Luigi Costa

Pierangiola Bracco
University of Torino
Torino, Italy

Introduction

The crosslinking and degradative reactions of ultra-high molecular weight polyethylene (UHMWPE) are governed by free radical reaction pathways. For these reactions to occur, free radicals must be induced in the polymer, for example by thermal decomposition of hydroperoxides or by high-energy radiation, which gives homolytic bond scissions with production of alkyl radicals. In previous chapters, we have often referred to crosslinking and degradation of UHMWPE, but we have not described in great detail the chemical mechanisms associated with these pathways. As noted previously, during irradiation, a cascade of chemical reactions may occur in UHMWPE depending on the environmental conditions. This chapter describes in detail the chemical reactions that take place in UHMWPE during irradiation and subsequent exposure to oxygen.

Mechanisms of Crosslinking

Crosslinking of a polymer is defined as the linking of two or more molecular chains by means of chemical (covalent) bonds. In this way, the molecular mass increases theoretically up to infinity. Thus, crosslinking results in one long, branched molecule with infinite molecular mass. Crosslinking can be achieved by chemical or by radiochemical reactions (Carlsson 1993, Clegg and Collyer 1991, Ivanov 1992, Shen, McKellop, and Salovey 1996).

In chemical crosslinking, one must either extend the polymerization process, or add suitable reactants and additives to induce the formation of chemical bonds between adjacent polymeric chains. One example of chemical crosslinking is rubber, which may be chemically crosslinked by addition of sulfur. With UHMWPE, peroxides such as dicumyl peroxide may be mixed with the resin powder to chemically crosslink the polymer during conversion to bulk rod or sheet (Shen, McKellop, and Salovey 1996). However, chemical crosslinking is not used to process UHMWPE for medical applications, and for this reason we will focus on radiation crosslinking for the remainder of this chapter.

Mechanism of Radical Formation during Irradiation

The degradation mechanisms of polyethylene in an inert atmosphere induced by high-energy radiation has been widely studied (Clegg and Collyer 1991, Ivanov 1992). Electron beam (e-beam) and gamma rays, employed for both sterilization and crosslinking processes, have a mean energy some orders of magnitude higher than that of chemical bonds. Their interaction with UHMWPE leads, through a complex energy transfer, to the scission of C–C and C–H bonds, giving H radicals and primary and secondary macroradicals (Scheme 1) (Clegg and Collyer 1991, Ivanov 1992). These macroradicals are dispersed throughout both the crystalline and the amorphous phases of the polymer.

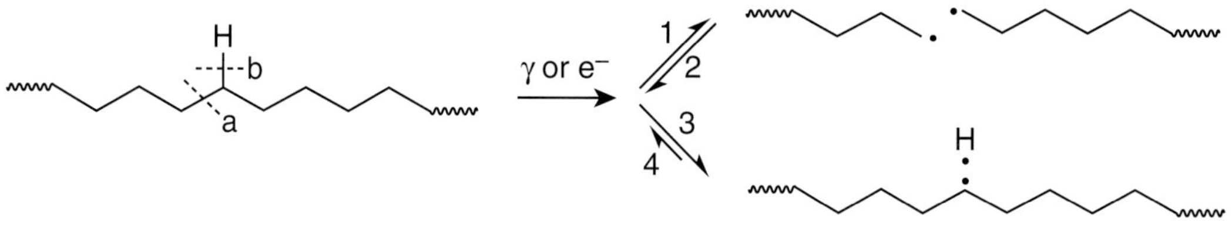

Scheme 1
Primary process of formation of primary and secondary macroalkyl radicals.

In previous studies, researchers have not detected primary alkyl radicals in irradiated UHMWPE (Lacoste and Carlsson 1992, Igarashi 1983, De Vries, Smith, and Franconi 1980). In addition, Nuclear Magnetic Resonance (NMR) and Fourier Transform InFrared (FTIR) spectroscopy studies have not revealed an appreciable increase in the concentration of terminal methyl units, which would be associated with radiolytic cleavage of the polymer chains. Because breaking of C–C is a random (stochastic) process, it can be assumed that the primary macroradicals resulting from Reaction 1 undergo recombination, in both amorphous and crystalline phase, giving back a C–C bond, with dissipation of energy in the polymer mass (Scheme 1, Reaction 2).

Orthopedic UHMWPE has a molecular mass of $2\cdot10^{-6}$ a.m.u. or higher. In this state, the polymer has a high viscosity, even in the molten state. Thus, macro-radicals have very low mobility, either in the molten or in the solid state, while the H radical, which has a diameter smaller than 1 Å, can migrate in the poly-mer mass, even in the crystalline phase, where distances between C atoms are in the order of 4 Å. H radicals resulting from Reaction 3 are very mobile and they can extract other H atoms intermolecularly or intramolecularly producing hydrogen, following Scheme 2.

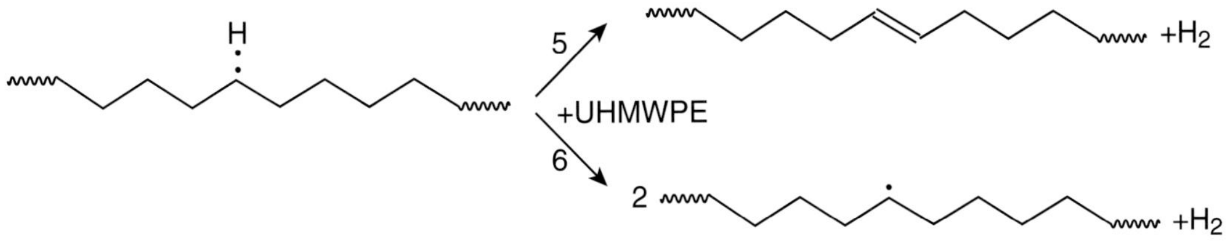

Scheme 2
Reactions of hydrogen formation.

Reaction 5 is extremely favored, being exothermal ($\Delta H = -288$ kJ/mol), with a very low entropy variation. The intramolecular process (Reaction 5) is extremely fast, and the secondary radicals decay, giving vinylene double bonds and molecular hydrogen (a gaseous product), which in turn can diffuse through the polymer mass.

On the other hand, intermolecular extraction (Reaction 6), is possible but less favored. Reaction 6 is exothermal ($\Delta H = -30$ kJ/mol) and associated to a very low entropy variation.

Among the vinylene double bonds, the transbonds are thermodynamically more stable. However, NMR studies have reported the presence of both cisvinylene and transvinylene in the amorphous phase and of transvinylene only in the crystalline phase (Perez and Vanderhart 1988, Randall 1988), formed according to a pseudozero order kinetic (Dole, Milner, and Williams 1958).

Reaction of Isolated Radicals

Some H radicals, not involved in Reaction 5, diffuse through the polymer bulk and extract H atoms from other macromolecules, resulting in the formation of isolated macroradicals in the polymer bulk (Scheme 2, Reaction 6). The proba-bility of extraction of the H atom decreases according to the following order: allyl, tertiary, secondary, primary, with a reactivity ratio between secondary and tertiary of 1 versus 9 (Arnaud, Moisan, and Lemaire 1984).

The concentration of vinyl double bonds in virgin (unirradiated) UHMWPE is approximately one double bond every 20,000 CH_2 groups and, for low irradiation dosages, the concentration of resulting vinylene double bonds is even lower. Then, statistically, the hydrogen atom extracted in Reaction 6 is likely to be a secondary hydrogen. Therefore, hydrogen radicals not involved in Reaction 5 decay, giving molecular hydrogen and secondary macroradicals, which in turn can give hydrogen transfer reactions, β scissions, or reactions with other reactive species. Reaction 6 leads to an increase of isolated radicals in the polymer mass (Scheme 2). The concentration of surviving radicals decreases with time after irradiation.

The radicals structure changes from secondary to allyl radical, more stable, which can survive even after a few years on the shelf at room temperature, being probably trapped in the crystalline phase (Scheme 3, Reaction 7) (Bhateja et al. 1995). The H transfer reaction in Scheme 3 is thermodynamically favored, because it leads to the formation of more stable allyl radicals. The activation energy for this process is only 40 kJ/mol, and thus the reaction can occur even at room temperature.

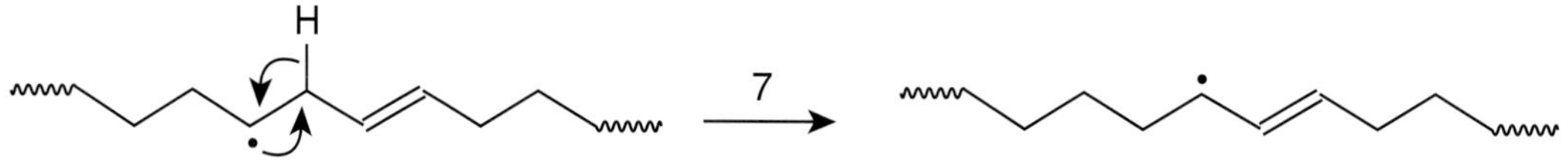

Scheme 3

Hydrogen transfer to allyl radical.

Evolution of secondary alkyl radicals via β scission, as shown in Scheme 4, rarely occurs at room temperature. Reaction 8 is the inverse reaction of polymerization and is endothermal ($\Delta H = 88$ kJ/mol), while Reaction 9 is more endothermal ($\Delta H = 146$ kJ/mol), because of the higher energy of the C–H bond, compared with the C–C one. Thus, both the reactions are extremely unlikely at room temperature, but may occur at higher temperatures (≈ 150–$250°C$) in the absence of oxygen.

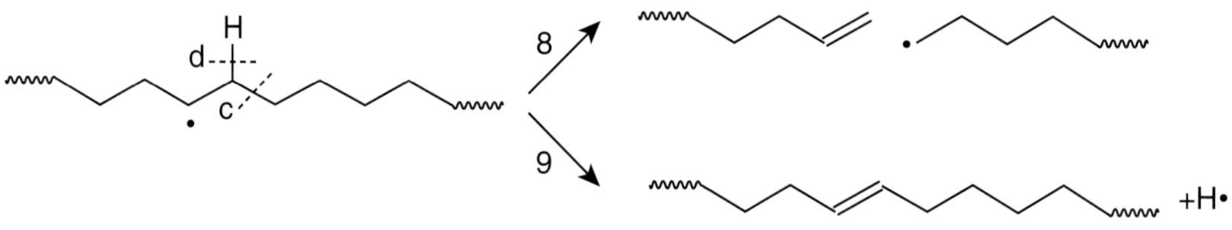

Scheme 4

β scission reaction of secondary radicals.

Y-Crosslinking Mechanism

UHMWPE is not a simple sequence of methylenes $-(CH_2)-$, but it also contains small but measurable concentrations of vinyl double bonds, tertiary carbons, and methyl groups. These short and long chain branches, as well as the residues of the catalyst, are incorporated in the less ordered amorphous phase, not in the crystalline lamellae. The amount of structural irregularities, as well as the degree of crystallinity, depends on the polymer synthesis method and on the subsequent processing conditions of the resin.

Vinyl and vinylene double bonds, the latter formed during irradiation, can react with macroradicals, in the same way that occurs during the polymerization process, resulting in the formation of Y-crosslinks, branching, and an increase in the molecular mass (Scheme 5). When a terminal vinyl double bond reacts with the macroradical on an adjacent polymer chain (Scheme 5, Reaction 10), a Y-crosslink is formed (Randall 1988). Reaction 10 is exothermal for $\Delta H =$ −88 kJ/mol, but it is controlled by inductive and steric effects. In UHMWPE, Reaction 10 occurs only with the terminal vinyl groups at the end of the polymer chains. There is experimental evidence from NMR studies to support the formation of Y-crosslinks from terminal vinyl species, as shown in Scheme 5. This crosslinking reaction predominates at room temperature and is the primary mechanism of crosslinking in gamma sterilization, as well as during elevated crosslinking with gamma or e-beam radiation when the UHMWPE is in the solid state (Perez and Vanderhart 1988).

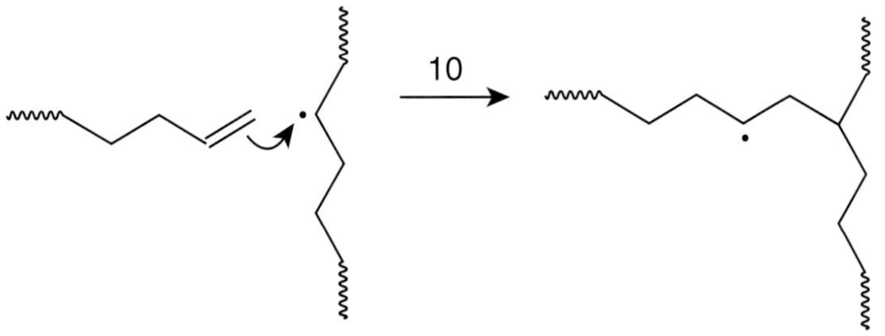

Scheme 5
Reaction of vinyl double bonds with secondary radicals.

Because of steric interference, the reaction shown in Scheme 5 is not observed with the vinylene double bonds, which are present within the polymer chain. In lower density polyethylenes, this reaction can occur with vinylidene species (Shinde and Salovey 1985, Smedberg, Hjertberg, and Gustafsson 1997). However, these types of vinylidene bonds are rarely observed in UHMWPE and are mentioned nonetheless for completeness.

H-Crosslinking Mechanism

Another crosslinking mechanism, leading to the formation of H-crosslinks, is shown in Scheme 6. Secondary and allyl macroradicals, decay via disproportion (Scheme 6, Reaction 11) or coupling (Reaction 12). Both reactions are exothermal ($\Delta H = −260$ kJ/mol and − 313 kJ/mol for Reactions 11 and 12, respectively).

Although thermodynamically feasible at room temperature, the formation of H-crosslinks is nonetheless blocked by steric hindrance when UHMWPE is in

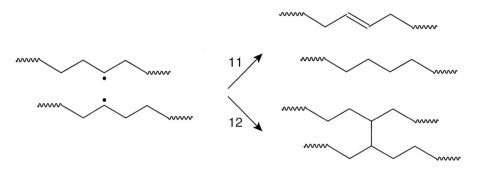

Scheme 6
Termination reactions: disproportion (11) and coupling (12).

the solid state. Radicals in the crystalline phase behave differently from radicals in the amorphous phase. Models of the crystalline phase in the solid state show that the minimum distance between two C atoms is 4 Å, thus much higher than the mean C–C bond length (1.54 Å). Therefore, the formation of H-crosslinks (Reaction 11) is sterically hindered in the crystalline phase. The ratio between the kinetic constants of disproportion and coupling for a secondary butyl radical in the liquid state at room temperature is about 1.5 Å, increasing by one order of magnitude for the same radical in the solid state (Tilman, Tilquin, and Claes 1982).

The presence of H-crosslinks in polyethylene irradiated in solid state has been ruled out in previous NMR studies. A different behavior can take place in the molten state, where the mobility of radicals is higher. Under these conditions, NMR studies confirm the presence of some H-crosslinks (Perez and Vanderhart 1988).

UHMWPE Oxidation

The mechanism of oxidation of short-chain hydrocarbons is widely known under the name of Bolland's cycle (Scheme 7) (Bolland 1949). Oxidation of polyethylene and polypropylene has been extensively studied (Lacoste and Carlsson 1992). More recently, the thermal and photo-oxidation of UHMWPE has been investigated (Costa, Luda, and Trossarelli 1997a). These processes are mostly similar to oxidation of high-density polyethylene (HDPE), with some exceptions, mainly because of the low mobility of the polymeric chains in UHMWPE.

Macroradicals can easily be found in the bulk of virgin UHMWPE. These radicals result from thermal decomposition of hydroperoxides (Lacoste et al. 1981) created during processing by compression molding or ram extrusion. These macroradicals, in the presence of oxygen, react readily giving hydroperoxyl radicals and the reaction is already favored at room temperature. We must

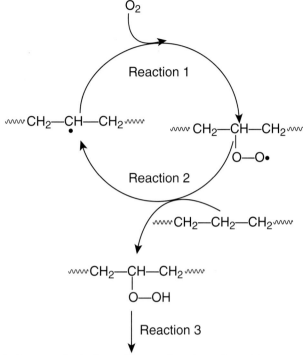

Scheme 7
Oxidation scheme of UHMWPE.

bear in mind that the oxygen is typically present during the conversion of UHMWPE powder, and it is, in practical terms, nearly impossible to obtain a totally oxygen-free polymer during processing, even when conducted in the presence of an inert gas, such as nitrogen.

Critical Products of the Oxidation Process

Hydroperoxides are the first products of oxidation (Carlsson et al. 1987a), which thermally and mechanically decompose and, over time, produce radicals. These continue the oxidation process, as was shown in Scheme 7. Further products of thermal, mechanical, or photolitic decomposition of hydroperoxides are ketones, carboxylic acids, and esters. The main products of thermal and photolitical decomposition of the secondary hydroperoxides formed in UHMWPE are ketones (Scheme 8).

Ketone formation occurs via a closed process, which does not create new radicals and then does not spread oxidation. Furthermore, because the process takes place without further chain scissions, it does not induce any variation of the molecular mass and only minimum changes of the mechanical properties of UHMWPE.

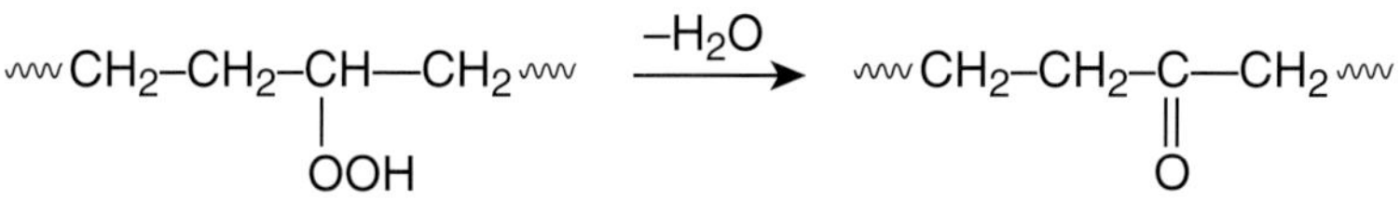

$$\text{wvw}\,CH_2-CH_2-CH-CH_2\,\text{wvw} \xrightarrow{-H_2O} \text{wvw}\,CH_2-CH_2-C-CH_2\,\text{wvw}$$

(with OOH below the left structure and O double-bonded below the right structure)

Scheme 8
Thermal decomposition of hydroperoxides to form ketones.

Acids are produced by scission of the polymeric chain, with a mechanism that has not yet been elucidated (Costa, Luda, and Trossarelli 1997b, Zaharadnickova, Sedlar, and Dastych 1991). Acid formation leads to a decrease in molecular mass and thus to a progressive deterioration of the mechanical properties of the UHMWPE. Acid determination, by using derivatization techniques and FTIR analyses (Lacoste and Carlsson 1992), makes it possible to quantify the molecular mass variation of the oxidized material in relation to the original molecular mass (Costa et al. 1998a).

Ester formation occurs during oxidative degradation, although the precise mechanism also is not completely clear. One hypothesis involves decomposition of primary peroxides, originating from the ending step of the polymerization process (Costa et al. 1998). Alcohols are formed by decomposition of hydroperoxides (Carlsson et al. 1987a) through alkoxy radicals, following Scheme 9.

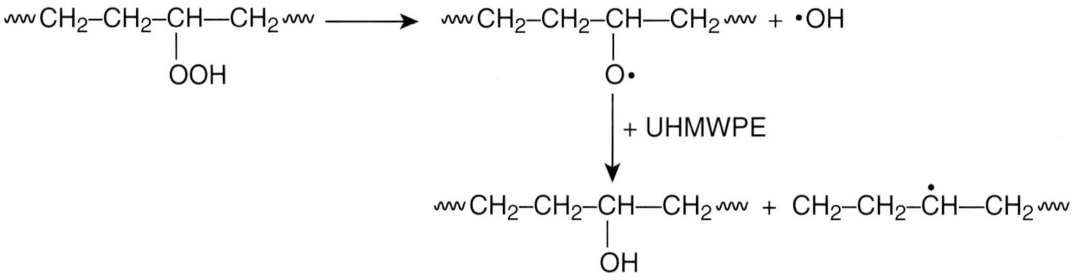

$$\text{wvw}\,CH_2-CH_2-CH-CH_2\,\text{wvw} \longrightarrow \text{wvw}\,CH_2-CH_2-CH-CH_2\,\text{wvw} + \cdot OH$$

(OOH below left; O· below right; then +UHMWPE arrow downward)

$$\text{wvw}\,CH_2-CH_2-CH-CH_2\,\text{wvw} + CH_2-CH_2-\overset{\bullet}{C}H-CH_2\,\text{wvw}$$

(OH below)

Scheme 9
Thermal decomposition of hydroperoxides to form secondary alcohols.

Hydroperoxides determination is the key factor to measure the level and the behavior of the oxidation (Carlsson et al. 1987b, Lacoste and Carlsson 1991, Shen, Yu, and McKellop 1999). Scientific studies and the American Society for Testing and Materials (ASTM) oxidation index standard (ASTM F2102) usually give the quantity of ketones and other carboxyl species present as an index of the oxidation degree. It must be pointed out that ketones, though a product of the oxidative process, do not produce polymeric chain scissions and so they do not result in substantial reduction of the UHMWPE mechanical properties. Quantification of ketones is reliable only if the ratio between ketones and carboxylic acids remains constant through the entire oxidative process.

UHMWPE Oxidation Induced by Gamma or E-Beam Irradiation

If UHMWPE is treated with high-energy radiation (either for sterilization or crosslinking) in the presence of oxygen, then oxidative degradation can take place. The radical produced by UHMWPE during gamma or e-beam irradiation reacts promptly with the oxygen present in the polymer, triggering an oxidation process reported in Scheme 7, which shows the possible reactions of the secondary radicals with oxygen.

During irradiation a number of radicals are formed at a very high rate. Consequently, oxidation proceeds much faster in irradiated UHMWPE than in unirradiated UHMWPE. The extent of oxidation within an UHMWPE hip or knee component, at a certain depth beneath the surface, depends on the actual concentration of alkyl radicals and on the amount of available oxygen. The actual concentration of alkyl radicals is a function of both the dispensed dosage rate and the temperature of the sterilization room, because further radicals can result from thermal decomposition of hydroperoxides (Costa et al. 2002, Yeom et al. 1998).

Whereas the distribution profile of the radicals in the bulk cannot be determined, the distribution of the products of their reaction with oxygen can. In other words, the degree and distribution of oxidative degradation in the bulk can be determined by measuring the distribution of the hydroperoxides and of their decomposition products, using FTIR (Costa et al. 2002). This distribution is a function of the rate at which gamma or e-beam radiation is supplied, the quantity of oxygen present within the polymer at the time of irradiation, the quantity of oxygen diffused afterward, and the temperature of the sample.

In the case of e-beam irradiation, the observed oxidative processes are qualitatively similar to those occurring with gamma ray irradiation. With e-beam irradiation, the dosage rate is orders of magnitude higher than with gamma radiation (10 kGy/second with e-beam versus 1–10 kGy/hour with gamma irradiation) and the irradiation time is on the order of seconds with e-beam, as opposed to hours with gamma irradiation. Comparing the two irradiation methods, under the same conditions and the same dispensed dosages, it has been observed that amount of oxidation was lower after e-beam irradiation than when gamma rays were used, because of the difference in dosage rates and the ability of oxygen to diffuse into the UHMWPE during the longer irradiation times used with gamma irradiation (Ikada et al. 1999).

Temperature Effects during Irradiation

Unsterilized UHMWPE components are typically stored at ambient temperature before being sent to a gamma irradiation plant. The temperature of the sterilization cell depends on the external temperature (sterilization occurred in winter or summer results in a different initial cell temperature) and of course on the intrinsic characteristics of the plant itself (Halls 1991). Gamma rays deposit energy into the sample, thereby increasing its temperature and the temperature of the cell. Reasonably, the cell temperature is in the range between 25 and 45°C (Ivanov 1992).

Orthopedic prostheses are sterilized with a dose of 25 to 40 kGy (Harrison 1991). Implants receive a dosage within this range, typically about 30 kGy (that is, 30 kJ/kg) (Costa et al. 1998b). The specific heat of UHMWPE is around 1.5 kJ/(K kg). It means that under adiabatic conditions, the temperature increase of UHMWPE component is approximately 20°C.

However, thermal energy absorbed by UHMWPE during irradiation is partially transferred to the surrounding environment. Like many polymers, UHMWPE has a low linear heat transfer coefficient (K_{UHMWPE} = 0.33 W/mK). As a consequence, the temperature of the UHMWPE component will be inhomogeneous and vary between that of the irradiation cell on the surface and that of the adiabatic limit near the center of the component. Based on these considerations, it is estimated that the temperature inside an UHMWPE component during gamma sterilization ranges between 45 and 65°C.

After sterilization, prostheses are stored at ambient temperature and the component has time to reequilibrate. Eventually, after implantation, the temperature of the UHMWPE component rises to body temperature (37°C).

Alkyl Macroradicals (R•)

Alkyl macroradicals (R•) originate either by direct interaction of the polymer with gamma rays (or e-beam) or by hydrogen abstraction from reactive radicals as ROO•, HOO•, RO•, OH•, and H•. The amount of R• is proportional to the absorbed dosage, and their actual concentration depends on the dosage rate.

Experimental studies have shown that vinylene concentration is constant with sample thickness (Costa et al. 1998). Alkyl macroradicals produced by gamma radiation are homogeneously distributed within the UHMWPE component because vinylene are the primary product of interaction between polyethylene and high-energy radiation. In e-beam irradiation, the radical distribution is a function of depth because of the cascade effect, which leads to a subsurface maximum in the absorbed dosage.

Peroxy Macroradicals (ROO•)

ROO• radicals are quite stable in UHMWPE, in spite of the large amount of available hydrogen atoms (Lazar et al. 1989). In fact, hydrogen abstraction has a high activation energy (60 kJ/mole) and proceeds very slowly at room temperature. ROO• decay is so slow that they have still been found after many weeks of storage at room temperature in air.

Hydroperoxide Decomposition (ROOH)

The O–O bond in ROOH is thermally unstable (bond energy 40 Kcal/mol) (Lazar et al. 1989), therefore it easily decomposes, producing very reactive OH•

and RO• radicals (Scheme 9). Thus, throughout the oxidative irradiation, a competition between ROOH formation and ROOH consumption occurs. It should be emphasized that an increase of 10°C in temperature generally doubles the rate of chemical reactions.

Oxygen Concentration in the UHMWPE Prosthetic Components

Fick's law governs the oxygen diffusion into the UHMWPE prosthetic component. Oxygen dissolves in the amorphous phase of polyethylene (Billingham 1990). As a result, oxidation occurs only in the amorphous domain, where oxygen is available. It has been reported that oxygen solubility in an HDPE, whose degree of crystallinity is similar to that of UHMWPE, is 1 mmol/kg at 25°C (Billingham 1990). Because UHMWPE components are typically up to 15 mm thick, the implants are saturated with oxygen before radiation sterilization. Following sterilization, oxygen may diffuse into the UHMWPE on the shelf if it is stored in air-permeable packaging.

According to Lacoste and Carlsson (1992) and Birkinshaw and colleagues (1989), about 2 mmols of alkyl radical (R•) per kilogram of UHMWPE are measured for every 10 kGy of irradiation. R• concentration is proportional to dosage, but the constant of proportionality is also a function of the dosage rate. However, with a typical sterilization dose of 30 kGy, the R• concentration is calculated to be 6 mmols/kg, which is much higher than the oxygen saturation concentration (1 mmol/kg). Thus, the reaction between alkyl radicals and oxygen will be oxygen diffusion controlled as soon as the dissolved oxygen is consumed.

The oxidation rate of UHMWPE components, resulting from high-energy radiation, depends on the following factors:

- Dosage rate of the irradiation source, which governs the duration of exposure to the environmental conditions at the sterilization facility
- Temperature of the sterilization facility
- Absorbed dosage, which controls the generation of primary alkyl radicals
- Sample thickness, which governs the oxygen concentration (O_2) distribution through the thickness of the implant

Typical conditions, to which gamma-sterilized prosthetic components may be subjected, are summarized as follows:

- Initial temperature of sterilization cell between 15 and 50°C
- Absorbed dosage between 28 and 40 kGy
- Dosage rate between 10 and 1 kGy/hour corresponding to 3 to 40 hours of irradiation
- Increase of temperature into the sample bulk of more than 20°C

In conclusion, both a dynamic gradient of O_2 and of temperature will be created through the thickness of an irradiated sample, whereas the dosage rate can be considered constant during irradiation. Local oxygen concentration and local temperature control the actual concentration of oxidized species, which varies along the thickness, provided that the R• is constant but depending on dosage rate.

Oxidative Degradation after Implant Manufacture

The oxidative process initiated during sterilization can continue during shelf storage and implantation. The rate and the extent of the process will depend on the storage temperature in the shelf and on the human body temperature, together with the amount of available oxygen *in vivo*.

In addition, oxidative degradation related to poor consolidation has been found in both *in vivo* and shelf-aged prostheses that were either gamma sterilized or ethylene oxide gas (EtO) sterilized (Costa et al. 2002). When oxidation related to consolidation occurs, the oxidation profile through the cups section is inhomogeneous and the maximum oxidation is observed near the center of the prosthesis. It is worth mentioning that this consolidation-related oxidation mechanism is often accompanied by whitening of the material, visible to the naked eye.

The origins of consolidation-related oxidation are not clear yet, but the processing conditions of the UHMWPE bar or sheet, together with the influence of the machining of the prosthetic component, have been proposed as contributing factors. Poor consolidation of the UHMWPE powder during processing,

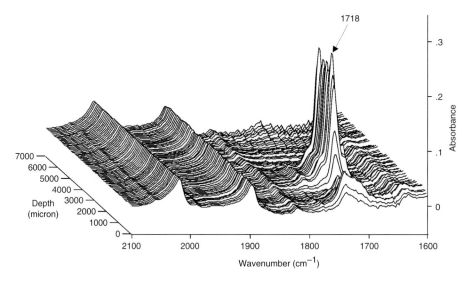

Figure 11.1
FTIR map collected along the cross-section of an EtO-sterilized, shelf-aged UHMWPE acetabular cup.

followed by internal stresses induced by machining, have been hypothesized to lead to the formation of free volumes between individual resin powder particles. These intergranular voids are more easily permeated by oxygen, so that facilitated oxidative degradation can take place at the grain boundaries, even if the initiating radical mechanisms are still unclear.

Figure 11.1 shows a collection of FTIR spectra through the thickness of a shelf-aged, EtO-sterilized cup where bulk oxidation was evident. Gamma irradiation of the UHMWPE was ruled out by examination of the transvinylene region of the FTIR spectra. The absorption at 1718 cm^{-1} is attributed to ketones. In this case, the oxidation was associated with poor consolidation of the UHMWPE, rather than the result of the sterilization process itself. These results suggest that prosthetic UHMWPE needs stabilization, as the totality of commercial polyolefines. Biocompatible stabilizers, such as vitamin E (an α-tocopherol), are easily available on the market and already employed in a number of different applications (Costa et al. 1998b, Costa et al. 2000).

In Vivo Absorption of Lipids

UHMWPE, as all the other polyethylenes, is a semicrystalline polymer, which means that amorphous, disordered regions coexist with crystalline, ordered domains. The crystalline portion is extremely compact. For this reason it is highly unlikely that a molecule may penetrate the space (about 4.5 Å) between the polymeric chains, except for the tiny hydrogen and helium molecules whose diameter is about 2 Å. The amorphous portion is less compact and has continuous free volume among the polymeric chains. These voids may allow diffusion of not only gaseous molecules, but also of molecules of larger size, having flexible linear chains. Obviously, only substances with a certain affinity for polyethylene will be able to diffuse. In other words, apolar, long-chained molecules, which have a fairly good solubility in polyolefines, are the most likely diffusants. The two Fick's laws govern the diffusion process. Diffusion rate is a function of many factors, such as temperature, pressure, and molecular mass of the diffusant. UHMWPE prosthetic components *in vivo* are in intimate contact with joint fluid, basically a filtered component of blood's plasma, rich in lipids and triglycerides.

Figure 11.2 shows a collection of FTIR spectra along the cross-section of an EtO-sterilized cup, retrieved after 4 years *in vivo*. The absorption at 1740 cm^{-1} is attributed to ester groups. Note the typical profile, decreasing from the surfaces toward the bulk. Recent studies revealed the composition of the diffused fraction by extraction and gas chromatography/mass spectrometry analysis (Costa et al. 2001). It has been pointed out that the diffused products are mostly squalene (a cholesterol precursor), cholesterol, cholesteryl esters with fatty acids, and triglycerides, whose relative amount is highly variable depending on the patient. Figure 11.3 shows a typical gas chromatogram of the extracted fraction from an UHMWPE component.

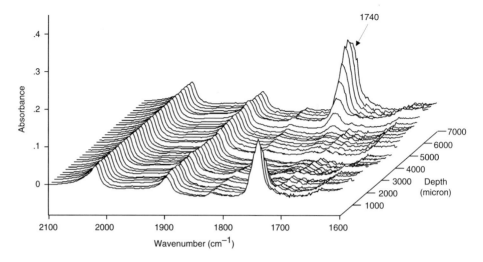

Figure 11.2
FTIR map collected along the cross-section of an EtO-sterilized acetabular cup, explanted after 4 years *in vivo*.

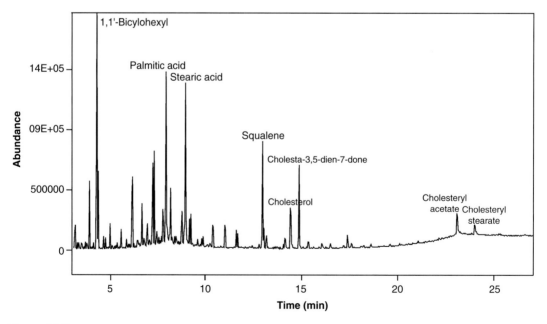

Figure 11.3
Typical GC/MS analysis of the soluble fraction extracted from a retrieved UHMWPE prosthetic component.

One question is whether the diffusion phenomenon can affect the mechanical properties of the polymer and to what extent. Recent experimental data, obtained *in vitro* with model compounds, suggests that modifications of mechanical properties induced by diffusion are quite limited (Turner et al. 2003).

Acknowledgments

We are very grateful to Professors Elena Brach del Prever, Maurizio Crova, and Paolo Gallinaro of the Traumatology and Orthopaedics Hospital (CTO) of Torino, Italy, whose interest in research and continuous encouragement has been the foundation of our work on UHMWPE.

References

Arnaud R., J.Y. Moisan, and J. Lemaire. 1984. Primary hydroperoxidation in low density polyethylene. *Macromolecules* 17:332–336.

Bhateja S.K., R.W. Duerst, E.B. Aus, and E.H. Andrews. 1995. Free radicals trapped in polyethylene crystals. *J Macromol Sci Phys* B34(3):263–272.

Billingham N.C. 1990. Physical phenomena in the oxidation and stabilisation of polymers. In *Oxidation inhibition in organic materials* Vol. II J. Pospisíl and P.P. Klemchuk, Eds. Boca Raton: CRC.

Birkinshaw C., M. Buggy, S. Daly, and M.J. O'Neill. 1989. *Appl Polym Sci* 38: 1967–1973.

Bolland J.L., 1949. *Quart. Rep.*, 3, 1.

Carlsson D.J. 1993. Degradation and stabilisation of polymers subjected to high energy radiation. In *Atmospheric oxidation and antioxidants,* Vol. II, Scott G. Ed, Ed. Amsterdam: Elsevier.

Carlsson D.J., R. Brousseau, C. Zhang, and D.M. Wiles. 1987a. Polyolefin oxidation: Quantification of alcohol and hydroperoxide products by nitric oxide reactions. *Polym Degr Stab* 17:303–323.

Carlsson D.J., R. Brousseau, C. Zhang, and D.M. Wiles. 1987b. Polyolefin oxidation: Quantification of alcohol and hydroperoxide products by nitric oxide reactions. *Polym Degr Stab* 17:303–323.

Clegg D., and A. Collyer. 1991. Irradiation effects on polymers. New York: Elsevier.

Costa L., M.P. Luda, and L. Trossarelli. 1997a. UHMWPE. 2. Thermally and photo initiated oxidation. *Polym Degr Stab* 58:41–54.

Costa L., M.P. Luda, and L. Trossarelli. 1997b. Ultra high molecular weight polyethylene: 1. Mechano-oxidative degradation. *Polym Degr Stab* 55:329–338.

Costa L., M.P. Luda, L. Trossarelli, et al. 1998a. Oxidation in orthopaedic UHMWPE sterilized by gamma-radiation and ethylene oxide. *Biomaterials* 19:659–668.

Costa L., M.P. Luda, L. Trossarelli, et al. 1998b. *In vivo* UHMWPE biodegradation of retrieved prosthesis. *Biomaterials* 19:1371–1385.

Costa L., M.P. Luda, P. Bracco, and L. Trossarelli. 1998. Oxidation of UHMWPE during sterilisation by gamma and beta ray. Macro 98 37th International Symposium on Macromolecules Gold Coast, Australia. 12. July 17.

Costa L., and E.M. Brach del Prever. 2000. *UHMWPE for arthroplasty.* Torino, Italy: Edizioni Minerva Medica.

Costa L., P. Bracco, E. Brach del Prever, and M.P. Luda. 2002. Oxidation in prosthetic UHMWPE, 224th ACS National Meeting, Boston, MA, August 18–22.

Costa L., K. Jacobson, P. Bracco, and E.M. Brach del Prever. 2002. Oxidation in ethylene oxide sterilised UHMWPE. *Biomaterials* 23:1613–1624.

Costa L., P. Bracco, E.M. Brach del Prever, et al. 2001. Analysis of products diffused in UHMWPE prosthesis components *in vivo*. *Biomaterials* 22:307–315.

De Vries K.L., R.H. Smith, and B.M. Franconi. 1980. Free radicals and new end groups resulting from chain scission: 1. Gamma-irradiation of polyethylene. *Polymer* 21:949–956.

Dole M., D.C. Milner, and T.F. Williams. 1958. Irradiation of polyethylene. II Kinetics of insaturation effects. *J Am Chem Soc* 80:1580–1588.

Halls N.A. 1991. Gamma-irradiation processing. In *Irradiation effects on polymers.* D. Clegg and A. Collyer, Eds. New York: Elsevier.

Harrison N. 1991. Radiation sterilization and food packaging. In *Irradiation effects on polymers.* D. Clegg and A. Collyer, Eds. New York: Elsevier.

Igarashi M. 1983. Free radical identification by ESR in polyethylene and nylon. *J Poly Sci Poly Chem* 21:2405–2425.

Ikada Y., K. Nakamura, S. Ogata, et al. 1999. Characterization of ultrahigh molecular weight polyethylene irradiated with gamma-rays and electron beams to high doses. *J Poly Sci Part A Poly Chem* 37:159–168.

Ivanov V.S. 1992. Radiation chemistry of polymers. Utrech The Netherlands: VSP.

Lacoste J., D.J. Carlsson, S. Falicki, and D.M. Wiles. 1981. Polyethylene hydroperoxide decomposition products. *Polym Degr Stab* 34:309–323.

Lacoste J., and D.J. Carlsson. 1991. A critical comparison of methods for hydroperoxide measurement in oxidized polyolefins. *Polym Degr Stab* 32:377–386.

Lacoste J., and D.J. Carlsson. 1992. Gamma-, photo, and thermally-initiated oxidation of linear low density polyethylene: A quantitative comparison of oxidation products. *J Poly Sci Part A Poly Chem* 30:493–500.

Lazar M., J. Rychly, W. Klimo, et al. 1989. Free radicals in chemistry and biology. Baton Rouge: CRC.

Perez E., and Vanderhart. 1988. A ^{13}C Cp-Mas NMR study of irradiated polyethylene. *J Poly Sci Poly Phys* 26:1979–1993.

Randall J.C. 1988. Carbon^{13} NMR gamma-irradiated polyethylene. In *Crosslinking and scission in polymers NATO ASI series C.* Vol. 292, O. Guven, Ed. Boston: Kluwer Academic.

Shen F.W., H.A. McKellop, and R. Salovey. 1996. Irradiation of chemically crosslinked UHMWPE. *J Poly Sci Poly Phys* 34:1063–1077.

Shen F.W., Y.J. Yu, and H. McKellop. 1999. Potential errors in FTIR measurements of oxidation in ultrahigh molecular weight polyethylene implants. *J Biomed Mater Res* 48:203–210.

Shinde A., and R. Salovey. 1985. Irradiation of polyethylene. *J Poly Sci Poly Phys Ed* 23:1681–1689.

Smedberg A., T. Hjertberg, and B. Gustafsson. 1997. Crosslinking reactions in an unsaturated low density polyethylene. *Polymer* 38:4127–4138.

Tilman P., B. Tilquin, and P. Claes. 1982. Estimation du rapport Kd/Kc pour des radicuax alkyles en phase solide. *J de Chimie Physique* 79:629–632.

Turner J.L., S.M., Kurtz, P. Bracco, and L. Costa. 2003. The effect of cholesteryl acetate absorption on the mechanical behavior of crosslinked and conventional UHMWPE. 49th ORS, New Orleans, February 2–5.

Yeom B., Y. Yu, H.A. McKellop, and R. Salovey. 1998. Profile of oxidation in irradiated polyethylene. *J Poly Sci Part A Poly Chem* 36:329–339.

Zaharadnickova A., J. Sedlar, and D. Dastych. 1991. Peroxy acids in photo-oxidised polypropylene. *Polym Deg Stab* 32:155–174.

Chapter 11. Reading Comprehension and Group Discussion Questions

11.1. How does high-energy radiation produce allyl macroradicals in UHMWPE? Explain the most relevant chemical reactions starting with the formation of alkyl macroradicals.

11.2. Why are allyl macroradicals important in the overall oxidation scheme of UHMWPE?

11.3. How do H- and Y-crosslinks in UHMWPE differ, both in terms of their molecular structure as well as in their mechanism of formation?

11.4. What are hydroperoxides and how are they formed?

11.5. In what way are hydroperoxides significant in the overall oxidation pathway for UHMWPE?

11.6. Why are some carbonyl species in UHMWPE more damaging than others?

11.7. Which carbonyl species are the most damaging?

11.8. Explain the mechanism of lipid contamination of UHMWPE in the body. Does lipid contamination damage UHMWPE? Why or why not?

Chapter *12*

Characterization of Physical, Chemical, and Mechanical Properties of UHMWPE

Stephen Spiegelberg
Cambridge Polymer Group, Inc.
Boston, MA

Introduction

There are many properties to consider when developing a new ultra-high molecular weight polyethylene (UHMWPE) material for implant applications. The characteristics of the UHMWPE powder may be measured to verify that there are not variations in different manufacturing lots. After consolidation, where the polyethylene powder is compressed into a solid slab of material by ram extrusion, compression molding, or other techniques including isostatic pressing, manufacturers will test the solid polyethylene slab to determine the quality of the consolidation, and to determine if any deleterious effects have occurred. Highly crosslinked UHMWPE is commonly used now, whereby the polyethylene slab is exposed to radiation, which forms chemical bonds between polymer chains. Several analytical techniques have been developed to assess if the material has been exposed to enough radiation to improve the desirable properties of the UHMWPE, while at the same time ensuring that other properties are not diminished by the radiation.

This chapter summarizes the common test techniques used in the orthopedic community for analyzing UHMWPE during the various stages of processing. The techniques described in this chapter can also be used to characterize

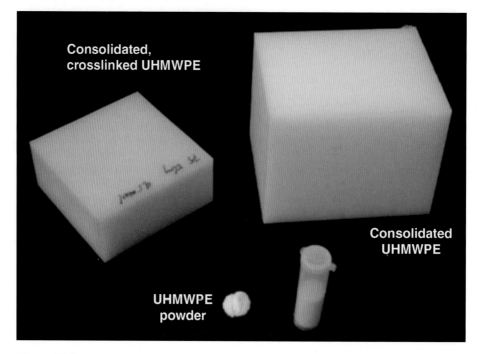

Figure 12.1

UHMWPE powder, consolidated stock, and irradiated stock. The irradiated stock has not been thermally treated and shows the characteristic yellow color caused by free radicals generated from electron beam.

explanted (retrieved) components in an effort to better understand the reasons for revision. The tests can be applied to UHMWPE powder, consolidated stock, or irradiated stock, as shown in Figure 12.1.

What Does the Food and Drug Administration Require?

The Food and Drug Administration (FDA) has set guidelines for the battery of tests a researcher should conduct on a new UHMWPE material before it receives approval for implantation into humans. This list represents the minimum number of tests that should be conducted; individuals may want to conduct more tests on their own, and the FDA may require other tests depending on the nature of the material being considered for FDA clearance.

Table 12.1 summarizes some of the tests discussed in the FDA guidance document *Data Requirements for Ultra-High Molecular Weight Polyethylene (UHMWPE) used in Orthopedic Devices* (FDA 1995). If the new polyethylene has almost identical properties to a polyethylene already on the market after stage 1 testing, no further testing may be required. If it does differ in properties from an existing

Table 12.1
Tests Required by the Food and Drug Administration for a New Polyethylene Submission

Stage 1 tests	Stage 2 tests	Stage 3 test
Ultimate tensile strength	Creep	Biocompatibility
Yield strength	Wear	
Young's modulus	Fatigue	
Poisson's ratio	Crack propagation	
Percentage of elongation	J-integral	
Molecular weight	Thin sectioned photomicrograph	
Density and porosity	Infrared spectra and chemical structure	
Percentage of crystallinity		
Glass transition temperature, T_g		
Crystallization temperature range, T_c		
Melting temperature, T_m		
Oxidation temperature, T_o		

FDA-approved polyethylene, stage 2 testing will be necessary. If the stage 2 test results are also different from existing FDA-approved polyethylenes, stage 3 tests may be required.

The American Standards for Testing and Materials (ASTM) has compiled a list of the recommended properties that polyethylene components must meet or exceed if they are to be used in orthopedic devices. This list is summarized in ASTM F648 and D4020-01a. The FDA usually refers to these standards when considering a new UHMWPE for orthopedic implants. A summary of all possible tests, including their ASTM standard numbers, where applicable, is summarized in Table 12.2, and indicates where the individual tests are appropriate for UHMWPE powder, consolidated stock, or irradiated stock.

Physical Property Characterization

The physical properties of UHMWPE include its morphology, which encompasses both microscopic and macroscopic analysis of the material, as well as thermal properties and molecular weight.

Table 12.2
Common Test Techniques Used to Assess the Various Properties of UHMWPE

Physical	Chemical	Mechanical	In vitro
Transmission electron microscopy[a, b]	Fourier transform infrared spectroscopy (ASTM 1421, F2102)[a, b, c]	Small punch (Kurtz et al. 1999)[a, b]	Accelerated aging (ASTM F2003-02)[a, b]
Scanning electron microscopy[a, b, c]	Electron spin resonance spectroscopy[b]	Compression (ASTM D2990)[a, b]	Wear testing (ASTM F732-00)[a, b]
Density (ASTM D1505)[a, b]	Gel permeation chromatography (ASTM D6474)[a, c]	Tensile (ASTM D2990, D638)[a, b]	
Differential scanning calorimetry (ASTM D3417)[a, b, c]	Dilute solution viscometry (ASTM D2857, F4020)[b, c]	Fatigue (ASTM E647)[a, b]	
Oxidation induction time (ASTM F3895)[a, b]	Swelling analysis (ASTM D2765, F2214)[b]	J-Integral (ASTM E813)[a, b]	
Fusion assessment[a]	Sol-Gel (ASTM D2765)[a, b]	Creep (ASTM 2990)[a, b]	
	Trace element (ASTM F648)[c]		

[a]Consolidated; [b]post-irradiation; [c]powder state.
The superscripts indicate which tests are suitable for the polyethylene in a powder state, after consolidation, and after crosslinking via irradiation.

Differential Scanning Calorimetry

Differential scanning calorimetry (DSC) is the most common technique used to measure the thermal properties of the UHMWPE, and provides the melting point and degree of crystallinity, as briefly summarized in Chapter 1. Given that the glass transition of polyethylene is approximately −160°C, most researchers will not measure this parameter. A picture of a DSC is shown in Figure 12.2. This unit is equipped with an autoloader, which facilitates testing of multiple samples. In DSC, a small amount of the UHMWPE, approximately 5–10 mg, is measured on an accurate microbalance and then sealed in an aluminum sample pan. The sample pan, along with an empty reference pan, is placed in the DSC chamber, which houses two heaters, one for each pan. The DSC heats the two pans at a known rate, usually 10°C/minute, from 0 to 200°C, and monitors the heat flow. A DSC trace for GUR 1050 powder is shown in Figure 12.3. The melting endotherm at 141°C is clearly visible. When the endotherm is integrated from 20 to 160°C ($\Delta H_{\text{endotherm}}$) and normalized with the heat of fusion of pure UHMWPE ($\Delta H_f = 291$ J/g), the calculated degree of crystallinity is $X = 81\%$.

$$\%X = \frac{\Delta H_{\text{endotherm}}}{\Delta H_f}\%$$

After consolidation, the melting point and the degree of crystallinity decrease, usually down to 140°C and 60%, respectively. For a DSC trace of consolidated

Figure 12.2
DSC with autoloader.

UHMWPE, see Figure 1.8. The crystallinity further decreases after radiation if melt annealing is performed following radiation. Consequently, DSC should be performed after each of these processing steps.

Oxidation induction time (OIT) measurements are less commonly performed, but they do provide some information about the oxidative stability of the UHMWPE. In this test, the DSC is heated to 200°C, then it is held isothermally. The purge gas, which is normally nitrogen, is then switched to pure oxygen. The time to the start of the degradation exotherm caused by oxidation is reported as the OIT. Open sample pans are used in this test technique.

Scanning Electron Microscopy

Scanning electron microscopy (SEM) is used to examine the morphology of the powder before consolidation. The procedure is straightforward: two-sided carbon tape is fixed to an SEM sample stub, and the UHMWPE powder is sprinkled onto the surface. A light gold, platinum, or carbon coating is applied ($\approx$ 100 Å), and the sample is examined in an SEM chamber. A typical SEM micrograph of UHMWPE powder is shown in Figure 12.4. The flakes are 50–100 nm in diameter. SEM can also be used to examine the consolidated resin, as shown in

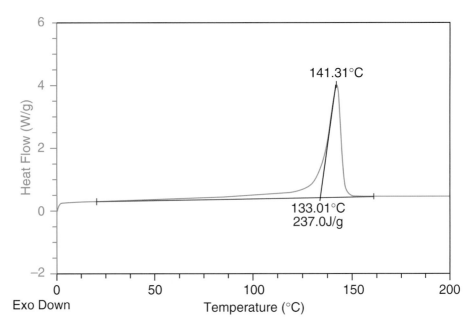

Figure 12.3
DSC trace of GUR 1050 powder (first heat). The crystallinity, based on the melting endotherm and a heat of fusion of 291 J/g, was 81% based on an integration range of 20–160°C.

Figure 12.4
Scanning electron micrograph of UHMWPE powder.

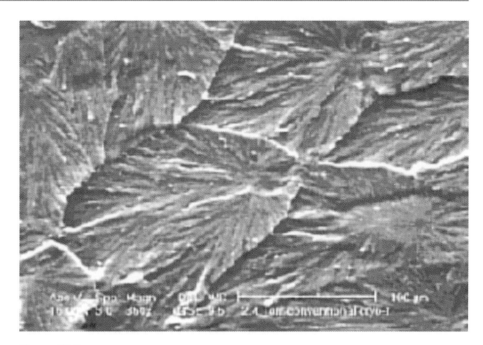

Figure 12.5
Scanning electron micrograph of consolidated UHMWPE, showing freeze-fractured surface.

Figure 12.5. This sample was freeze fractured, and clearly shows the polyethylene flakes on the fracture surface. Voids in the sample caused by poor consolidation would be readily visible.

Intrinsic Viscosity

The intrinsic viscosity (IV) is related to the molecular weight of the polymer through the Mark-Houwink relationship (Fox and Flory 1948). The technique used to perform IV measurements is described in ASTM D2857 and ASTM D4020. A Ubbelohde No. 1 viscometer, as shown in Figure 12.6, is placed in an oil bath at 135°C. The polymer powder is dissolved in decahydronapthalene at 150°C with a volume of solvent in milliliters equal to 1.8 times the mass of polyethylene in milligrams. After mixing for 1 hour, the solution is transferred to the viscometer, and then allowed to equilibrate at 135°C. The elution time through the capillary of the viscometer is measured, t_s, and compared with the elution time of the pure solvent, t_o, and the constant for the viscometer, k. The relative viscosity, η_r, specific viscosity, η_{sp}, intrinsic viscosity $[\eta]$, and viscosity-average molecular weight are computed as follows:

$$\eta_r = \frac{t_s - k / t_s}{t_o - k / t_o} = \eta_{sp} + 1$$

Figure 12.6
Ubbelohde viscometer.

$$[\eta] = (2\eta_{sp} - 2\ln\eta_r)^{1/2}/c$$

$$M_v = 5.37 \times 10^4 [\eta]^{1.37}$$

Gel Permeation Chromatography

An alternative technique to measure the molecular weight uses gel permeation chromatography (GPC). In GPC, the polymer is dissolved in a suitable solvent, which for UHMWPE is usually trichlorobenzene. The dilute solution is injected in a flow stream that passes through columns containing a rigid gel with different porosities. The smaller polymer chains are temporarily entrained by the pores in the gel, while the larger polymer chains pass through the column with less pore interaction. As a result, the chains become separated according to size. As they leave the gel columns, a refractometer detector measures the relative concentration of material. By calibrating the columns with polymers of known molecular weight, the molecular weight distribution of the UHMWPE can be determined. A typical distribution for UHMWPE is shown in Figure 12.7, in which polyethylene standards were used. The moments of molecular weight were calculated from the distribution and are plotted on the curve (Rempp and

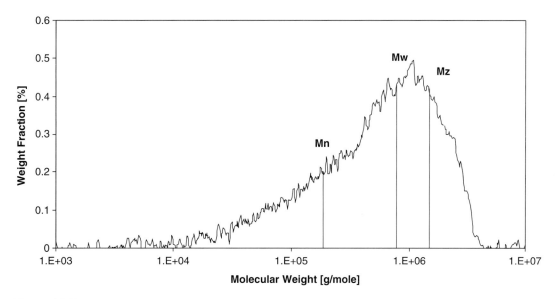

Figure 12.7

Molecular weight distribution for UHMWPE, relative to polyethylene standards. The number-average, weight-average, and z-average molecular weights are shown as M_n, M_w, M_z, respectively.

Merrill 1986). This technique, like any technique that requires solvating the UHMWPE, will measure only the molecular weight of the soluble portion of the material.

Fusion Assessment

The morphology of the consolidated UHMWPE should be analyzed to ensure that proper fusion has occurred during the molding process. Transmission optical microscopy is an easy approach to examining this morphology. A microtomed slice of material is placed under a microscope and examined at 10× to 40×. According to the morphology evaluation protocol described in ASTM F648 (Annex 2), UHMWPE films of at least 100 μm thickness are viewed under transmitted light or dark field illumination at no less than 40× magnification. The images shown in Figure 12.8 show a sample with poor consolidation, indicated by pronounced grain boundaries, and a sample with good consolidation.

Transmission Electron Microscopy

The morphology of the consolidated UHMWPE can be examined with transmission electron microscopy (TEM) to a very high degree of magnification.

Figure 12.8

Optical micrograph of microtomed polyethylene films prepared by compression molding. The image on the left shows clear boundaries between the original UHMWPE flakes, indicating poor consolidation. The image on the right shows good consolidation. The parallel lines in this micrograph are knife marks caused during microtoming (magnification: 15×).

Structures down to 100 nm can be readily viewed with this technique. Changes in crystalline structure with consolidation and radiation treatment can also be viewed. As opposed to the surface analysis technique of SEM, TEM takes a spatial average of the polymer morphology, relying on differences in electron density to yield structure. Consequently, microtoming is required to yield 50–200 nm thick slices, which are then stained with chlorosulfonic acid to enhance contrast. An example of the ribbonlike lamellae found in UHMWPE is shown in Figure 12.9. This micrograph shows primary large lamellae, with smaller lamellae interspersed.

Density Measurements

The density of the polyethylene is related to both its crystallinity and any porosity that may be present. One of the more accurate techniques for measuring density involves the use of a density gradient column, as described in ASTM D1505-99e1. In this technique, a vertical cylindrical tube is filled with two or more miscible liquids that have a density range that spans the density of the material to be investigated. The tube is temperature controlled because the density of the fluids will change with temperature. The resulting liquid mixture will have a linearly varying density with respect to height, with the lowest density at the top of the column. Glass floats of accurately known densities are placed at the top of the column, and then fall to the height in the column where their density matches that of the fluid. These floats are used to calibrate the column. The density of the test sample is determined by placing a piece in the column, measuring the height in the column where it settles, and

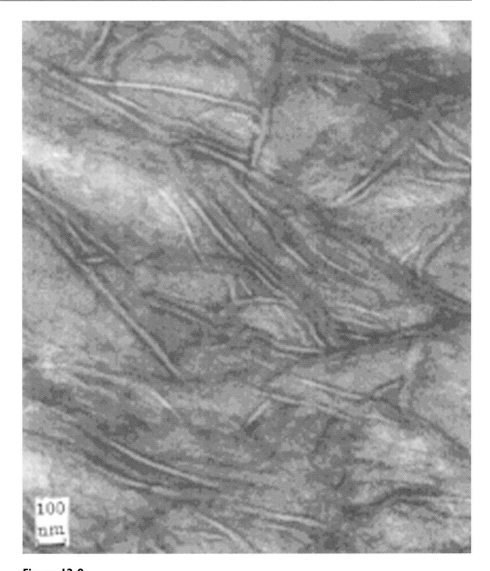

Figure 12.9
Transmission electron micrograph of consolidated UHMWPE, showing lamellae.

comparing this height with the calibration curve. For UHMWPE, an ethanol-water column will provide the necessary range of 0.72–1 g/ml density range.

Alternatively, the material can be determined by the water-displacement technique. The sample is weighed carefully in air with a sensitive analytical balance. The same sample is then suspended from a wire attached to the balance, then placed in water (or any liquid that will cause it to sink). The weight of the sample in the liquid, less the weight of the wire, is used to determine the mass of liquid displaced by the sample, which provides the volume of the sample. The density can then be calculated.

Chemical Property Characterization

Chemical characterization of the UHMWPE powder is typically performed at the powder-manufacturing site as part of their quality control protocol. However, researchers may want to periodically verify the purity and composition of the powder. ASTM 648 outlines several tests that can be used to analyze the powder. With regards to chemical characterization, trace element analysis and Fourier transform infrared spectroscopy (FTIR) are the two common approaches.

Trace Element Analysis

In trace element analysis, one is typically looking for elements that could have been inadvertently introduced during the powder-manufacturing process. The most common trace elements found in polyethylene powder include titanium, calcium, chlorine, and aluminum. Quantification of these elements is usually carried out by weighing known amounts of the powder, pyrrolizing the polyethylene through heat or microwave digestion, and performing mass spectroscopy to quantify the yields of these materials, reporting results in parts per million. Sample preparation is important in order to get good quantitative results. Samples must be sealed in containers before pyrrolization or microwave digestion of the material; if not, erroneously low concentrations can be measured because of airborne loss of materials. Chlorine has to be measured through titrimetric methods.

Fourier Transform Infrared Spectroscopy

FTIR is performed to measure the chemical structure of the polyethylene. FTIR can be performed either on the polyethylene powder, using the potassium bromide (KBr) pellet technique, or on the consolidated polyethylene. An FTIR with a microscope attachment, which is useful for examining the chemical structure as a function of position in a sample, is shown in Figure 12.10. The polyethylene powder is blended in a 1:20 ratio with KBr powder that has been carefully dried in an oven to remove all absorbed water. The two powders should be ground together with a mortar and pestle, and then are placed in a KBr pellet anvil. One should place enough powder to cover the bottom of the mold, then press. If done correctly, the KBr pellet should be transparent, with semiopaque regions where the polyethylene powder is sitting. One should collect and average at least 32 scans in transmission mode with an FTIR bench. The background spectrum should be collected on a pure KBr pellet. Any unusual peaks may indicate the presence of contaminants or oxidation. This technique is semiquantitative.

After consolidation of the polyethylene powder into solid stock, few if any chemical changes should have occurred. Degradation, or chain scission, caused by heat or mechanical deformation may sometimes appear after consolidation in isolated cases. FTIR can be used to see if bulk degradation has occurred in the sample.

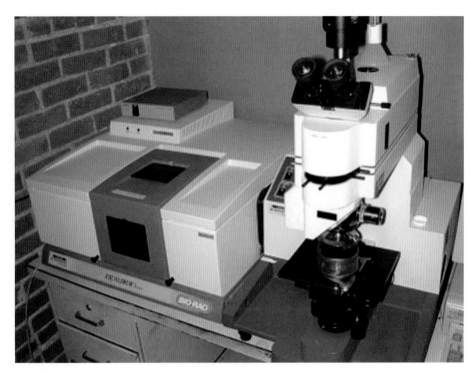

Figure 12.10
FTIR spectrometer with microscope attachment.

To perform FTIR on consolidated material, researchers will typically prepare a thin film of the material, and then perform the infrared analysis in transmission mode. A thin film can be prepared by microtoming. A researcher will cut the consolidated polyethylene slab with a band saw into a piece small enough to fit into the chuck of a microtome, shown in Figure 12.11. They will then microtome several millimeters away from the cut surface to avoid sampling material than may have degraded in the band saw. A 100–200 μm thick sample is then microtomed, and placed in the FTIR bench or on the microscope stage of a micro-FTIR system. A minimum of 24 scans should be conducted, and the spectra is then rationed against a blank background. Typical spectra are shown in Figure 12.12 and Figure 12.13.

To look for evidence of degradation, researchers will usually examine the terminal vinyl group at 910 cm^{-1}, or a double bond at the end of a polymer chain, as shown in the following equation.

$$-CH=CH_2$$

The presence of a terminal vinyl group usually means that the polymer chain has broken, which will leave behind a vinyl group on each end.

Oxidation can also sometimes occur, which leads to carbonyl formation. Carbonyls, which include ketones, esters, and ethers, have a principle absorption peak of approximately 1700 cm^{-1}. ASTM F2102 describes the technique to

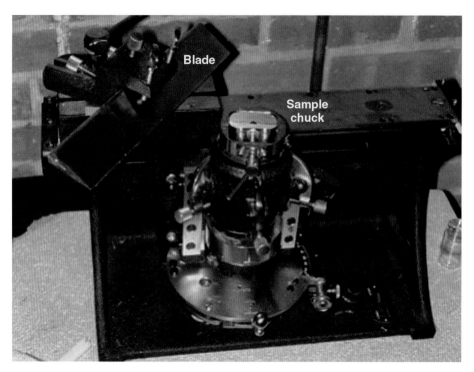

Figure 12.11

Sliding microtome, which is used for preparing thin films of polyethylene.

measure the oxidation index, or the area of the carbonyl absorption peak as a ratio against the methylene stretch at 1396 cm^{-1} (Costa et al. 1998, Kurtz et al. 2000, Spiegelberg and Schaffner 1999).

In addition to the previously mentioned peaks, the uniformity of the received radiation dose can be determined by examining the transvinylene peak at 965 cm^{-1} (Johnson and Lyons 1995, Lyons and Johnson 1993). Transvinylene groups appear in polyethylene as a radiolytic product during exposure to ionizing radiation with a yield related to the number of crosslinks formed. Monitoring this group as a function of location in the sample allows one to determine the degree of uniformity of the received radiation dose.

Electron Spin Resonance

Residual free radicals trapped in the polyethylene are known to cause long-term stability problems in implanted polyethylene prosthesis. The free radicals can react with oxygen to form unstable hydroperoxides, which can decompose to carbonyls. The latter are readily susceptible to chain scission and subsequent loss of mechanical properties. The only known method of directly measuring residual free radicals is by electron spin resonance (ESR) spectroscopy (also

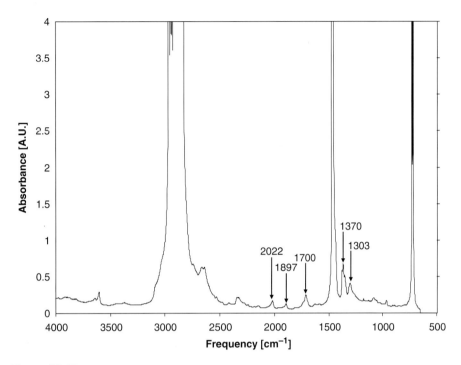

Figure 12.12

FTIR trace of UHMWPE irradiated to 15 Mrads and gamma sterilized in air, with key peak locations indicated.

known as electron pair resonance, or EPR) (Jahan et al. 1991). Figure 12.14 shows an ESR spectroscopy trace of unirradiated and irradiated, or sterilized, UHMWPE. The multiple peaks in the irradiated sample indicate the presence of residual free radicals. The number of free radicals per gram can be determined by performing a double integration on this curve.

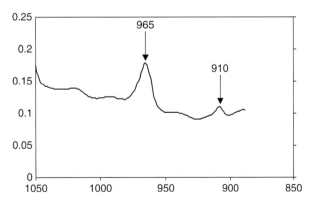

Figure 12.13

Peaks of interest in the footprint region of the spectra shown in Figure 12.12.

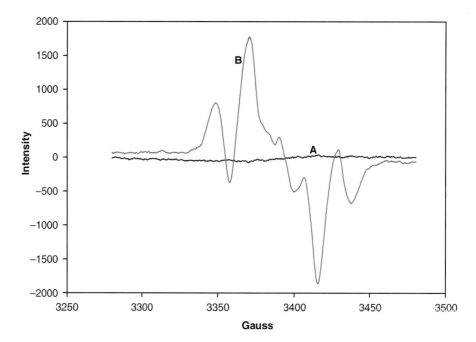

Figure 12.14
ESR trace of unsterilized (A) and gamma-sterilized (B) UHMWPE.

Swell Ratio Testing

When a crosslinked UHMWPE is placed in a good solvent, it cannot dissolve, but it will swell to a certain degree. The degree of crosslinking can be indirectly determined by measuring the swell ratio of the crosslinked polyethylene. In ASTM D2765, a crosslinked specimen is weighed, allowed to swell in a chamber containing xylene at 130°C, and is then reweighed. The swell ratio, q, is determined from the weight gain and the densities of the polyethylene and xylene.

An alternate approach is discussed in ASTM 2214. In this technique, the change in sample height is monitored with a probe that is in light contact with the sample. A swelling apparatus is shown in Figure 12.15, which uses a laser micrometer to monitor the height change of the sample. An example of the transient response of the swelling of crosslinked UHMWPE is shown in Figure 12.16. Based on the steady state swell ratio, q, the interaction parameter, χ_1, for polyethylene and xylene at 130°C, and the molar volume of xylene, V_1, the crosslink density, v_x (moles of crosslinks/unit volume), can be calculated from Flory network theory (Flory 1953) from the following expression:

$$v_x = -\frac{\ln(1 - q^{-1}) + q^{-1} + \chi_1 q^{-2}}{V_1(q^{-1/3})}$$

The previous expression assumes a network with tetrafunctional crosslinking (i.e., no branching).

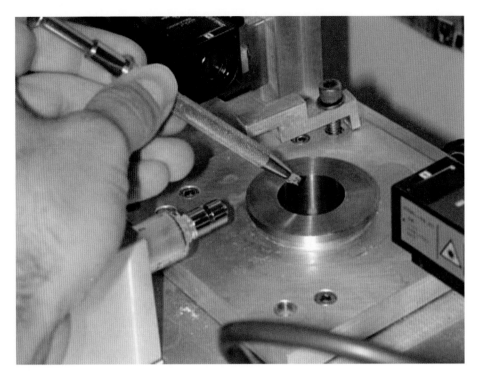

Figure 12.15

Polyethylene cube (4 × 4 × 4 mm) being inserted into test chamber of swelling apparatus. The laser micrometer is visible.

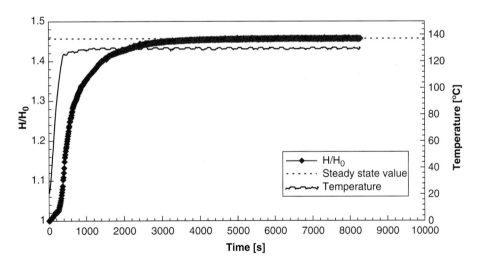

Figure 12.16

Transient swelling data for UHMWPE in xylene at 130°C. Steady state was achieved after approximately 1 hour.

Similar to the ASTM method for swelling analysis, sol-gel measurements involve weighing methods to determine the degree of crosslinking in polymer systems. However, instead of weighing the solvent uptake in the polymer matrix, one measures the amount of material extracted (i.e., soluble) from the network. In this manner, one measures the percentage of gel, or fully crosslinked, material in the sample. ASTM D2765 describes several methods for making these measurements. In one approach, the sample is carefully weighed, packaged in a preweighed porous stainless steel mesh bag, then placed in a flask containing boiling xylene and equipped with a condensor. After 24 hours, the bag and remaining sample are dried in a vacuum oven and reweighed. This approach can be used effectively for polymer films and unusually shaped structures. However, it is less effective in computing a crosslink density than the swelling method previously described. Antioxidant must be used in both these methods to prevent thermally induced chain scission.

Mechanical Property Characterization

Mechanical properties are critical for UHMWPE that is to be used in an orthopedic device. The tests indicated next are commonly used by researchers in preparing an FDA application.

Poisson's Ratio

Poisson's ratio, v, indicates the variation of the strain ($\varepsilon = \Delta L / L_0$) of a sample in the transverse direction, ε_t, and the longitudinal direction, ε_L, when the sample is subjected to either tension or compression, and is defined as:

$$v = -\frac{\varepsilon_t}{\varepsilon_L}$$

To measure Poisson's ratio, a sample must be instrumented with strain gauges in both transverse and longitudinal directions, which are monitored while the sample is deformed below its elastic limit in either tension or compression.

J-Integral Testing

J-integral tests indicate the resistance of a material to crack propagation under a steady tensile deformation, and are a means of reporting the toughness of the material in geometries that contain notches or flaws. ASTM J813 describes the testing technique for metals, which can be applied to some plastics. In this test, a tensile specimen, usually in the form of a compact tensile specimen, is notched with an initial flaw. It is pulled apart at a constant crosshead speed, monitoring the tensile load along with the crack length Δa, which is usually measured optically. The *J* parameter is computed from the load-displacement curves and the sample geometry, to generate a plot like that shown in Figure 12.17. Researchers report J_Q, which is the intersection of the *J-Δa* line with the 0.2 mm offset line. This technique provides criteria to determine if the test was conducted in plane strain conditions (i.e., the test results are independent of sample geometry). If the criteria determine the test results were in plane stress conditions, researchers should be cautious in reporting data.

Tensile Testing

Tensile testing is perhaps the most common of the mechanical tests conducted on UHMWPE, and is described in detail in ASTM D638. In this test, a dogbone specimen is machined or punched out, as shown in Figure 12.18A, and pulled at a user-specified crosshead speed in a tensile load frame. There are several standard size dogbones described in this standard. The load, *F*, and displacement, *ΔL*, are monitored, and are converted to engineering stress and strain, as shown in Figure 12.18B. The area *A* is the initial cross-sectional area of the sample. Young's Modulus, *E*, the yield stress, σ_Y, ultimate tensile stress, σ_{UTS}, and percentage of elongation, %ε are all determined from this curve.

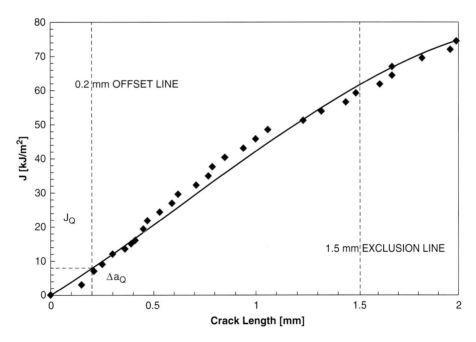

Figure 12.17

J-integral plot for UHMWPE compact tensile specimen.

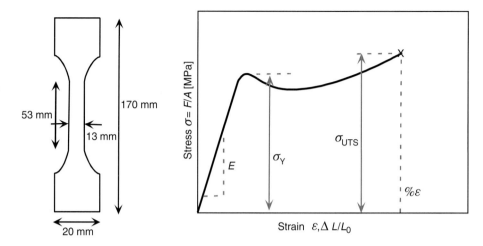

Figure 12.18

(**A**) Type I dogbone tensile specimen. (**B**) A typical tensile test plot for UHMWPE, showing the various parameters that can be determined.

Fatigue Testing

Fatigue testing, as described in ASTM E647, monitors the crack propagation in a specimen as it is subjected to an oscillating tensile load. The number of oscillations, N, used in these tests are typically in the millions of cycles. The purpose of the test is to monitor the crack propagation resistance under cyclical loading, a type of deformation to which most implants will be subjected (Baker, Bellare, and Pruitt 2003).

In this test a specimen, most often a compact tensile specimen, as used in the J-integral tests and shown in Figure 12.19, is placed in a hydraulic load frame, and then oscillated at either a fixed displacement (displacement control) or between fixed loads (load control). The choice of test depends in part on the application for the material. During the test, the crack length, Δa, is periodically measured. The stress intensity factor, K, which depends on the load range used,

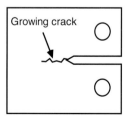

Figure 12.19

Compact tensile specimen for fatigue crack growth studies.

sample geometry, and crack length, is computed from the raw data (*a* versus *N*), which is then used to prepare a curve of log d*a*/d*N* versus log Δ*K*.

Creep

Creep, or room-temperature flow upon application of a stress such as body weight, is an issue in orthopedic devices containing a UHMWPE (Kang and Nho 2001). These components are usually machined to tight tolerances, and long-term stability of the implant depends on maintaining these tolerance dimensions in the body. If an acetabular liner or tibial tray undergoes excessive creep, the leg can shorten or dislocations can occur. Minimization of creep is thus desired.

Creep is measured by application of a constant load to a sample, then measuring the height displacement of the load as a function of time, and reporting the data in terms of strain, *ε*, versus time, as shown in Figure 12.20. The data is often converted to compliance, *J*(*t*), which is inversely proportional to the modulus, as expressed relative to the applied compressive stress, *σ*(*t*), as:

$$\varepsilon(t) = J(t)\sigma(t)$$

Small Punch

A newer mechanical analysis has emerged that makes use of the biaxial deformation of small discs of UHMWPE, yielding a stress-strain curve (Kurtz et al. 1999). Researchers have demonstrated a dependence on the area under the load-displacement curve to wear testing, and suggest that the biaxial nature of wearing is the reason for the good comparison (Edidin and Kurtz 1999). The obvious benefit of this technique is the small sample size, allowing spatial measurement of mechanical properties on retrieved components. This technique is described in further detail in ASTM F2183 and Chapter 13.

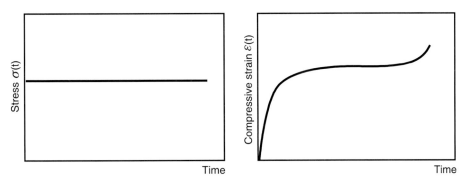

Figure 12.20
Applied load history and resultant strain history for creep analysis of polyethylene.

Other Testing

Finally, two standard methods have emerged to anticipate the response of implanted UHMWPE components before and during implantation. Accelerated aging, through increased temperatures and oxygen content, is used to try to predict oxidation levels over many years of shelf storage (Kurtz et al. 2001, Sun 1996). Hip simulators are used to monitor wear behavior under a physiologically accurate wearing pattern (McKellop 1978, Ramamurti et al. 1998, Schroeder and Pozorski 1996). These two approaches are discussed extensively in other manuscripts, and are beyond the scope of this chapter.

Conclusion

There is a large battery of tests available to the researcher to examine the effects of radiation on the properties of UHMWPE. These effects include evolution of morphological structure, production of chemical species, and changes in mechanical properties. Caution must be used in sample preparation, test conditions, and data interpretation. This list of analytical tests is not all encompassing. As noted in previous sections, some regulatory agencies may require additional tests, depending on the nature of the material and the final application. Additionally, researchers may want to perform other tests to explore additional benefits of their material.

References

Baker D.A., A. Bellare, and L. Pruitt. 2003. The effects of degree of crosslinking on the fatigue crack initiation and propagation resistance of orthopedic-grade polyethylene. *J Biomed Mater Res Part A* 66A:146–154.

Costa L., M.P. Luda, L. Trossarelli, et al. 1998. Oxidation in orthopaedic UHMWPE sterilized by gamma radiation and ethylene oxide. *Biomaterials* 19:659–668.

Edidin A.A., and S.M. Kurtz. 1999. Validation of a modern hip simulator using four clinically-applied polymeric biomaterials. In *Society for Biomaterials, 25th annual meeting*. Providence, RI: Society for Biomaterials.

Federal Drug Administration. 1995. *Data requirements for ultra-high molecular weight polyethylene (UHMWPE) used in orthopedic devices*. Document FOD 180. Rockville, MD: FDA.

Flory P.J. 1953. *Principles of polymer chemistry*. Ithaca and London: Cornell University Press.

Fox T.G., and P.J. Flory. 1948. Viscosity-molecular weight and viscosity-temperature relationships for polystyrene and polyisobutylene. *J Am Chem Soc* 70:2384–2395.

Jahan M.S., C-H Wang, G. Schwartz, and J.A. Davidson. 1991. Combined chemical and mechanical effects on free radicals in UHMWPE joints during implantation. *J Biomed Mat Res* 25:1005–1017.

Johnson W.C., and B.J. Lyons. 1995. Radiolytic formation and decay of trans-vinylene unsaturation in polyethylene: Fournier transform infra-red measurements. *Radiat Phys Chem* 46(4–6):829–832.

Kang P.H., and Y.C. Nho. 2001. The effect of gamma-irradiation on ultra-high molecular weight polyethylene recrystallized under different cooling conditions. *Radiat Phys Chem* 60(1–2):79–87.

Kurtz S.M., C.W. Jewett, J.R. Foulds, and A.A. Edidin. 1999. Miniature specimen mechanical testing technique scaled to articulating surface of polyethylene components for total joint arthroplasty. *J Biomed Mater Res* 48(1):75–81.

Kurtz S.M., O.K. Muratoglu, F. Buchanan, et al. 2001. Interlaboratory reproducibility of standard accelerated aging methods of oxidation of UHMWPE. *Biomaterials* 22:1731–1737.

Kurtz S.M., O. Muratoglu, M. Evans, and A.A. Edidin. 2000. Advances in the processing, sterilization, and crosslinking of ultra-high molecular weight polyethylene for total hip arthroplasty. *Biomaterials* 20(18):1659–1688.

Lyons B.J., and W.C. Johnson. 1993. *Radiolytic formation and decay of trans-vinylene unsaturation in polyethylene*. In *Irradiation of polymeric materials: Processes, mechanisms, and applications*. E. Reichmanis, C.W. Frank, and J.H. O'Donnell, Eds. Washington, DC: American Chemical Society.

McKellop H. 1978. Wear characteristics of UHMW polyethylene: A method for accurately measuring extremely low wear rates. *J Biomed Mat Res* 12:895.

Ramamurti B.S., D.M. Estok, M. Jasty, and W.H. Harris. 1998. Analysis of the kinematics of different hip simulators used to study wear of candidate materials for the articulation of total hip arthroplasties. *J Bone Joint Surg* 16:365–369.

Rempp P., and E.W. Merrill. 1986. *Polymer synthesis*. New York: Huethig & Wepf.

Schroeder D.W., and K.M. Pozorski. 1996. Hip simulator wear testing of isostatically molded UHMWPE effect of EtO and gamma irradiation. In *42nd annual meeting, Orthopaedic Research Society*. Atlanta: Orthopaedic Research Society.

Spiegelberg S.H., and S.R. Schaffner. 1999. Oxidation profiles in shelf-stored ultra-high molecular weight polyethylene samples. In *Society for Biomaterials, 25th annual meeting*. Providence, RI: Society for Biomaterials.

Sun, D.C. 1996. Simple accelerated aging method for long-term post-radiation effects in UHMWPE implants. In *Fifth world biomaterials congress*. Toronto: Society for Biomaterials.

Chapter 12. Reading Comprehension and Group Discussion Questions

12.1. What mechanical property data are currently recommended by the FDA for a new UHMWPE formulation?

12.2. What type of testing would you perform to measure the degree of crystallinity of UHMWPE?

12.3. You have synthesized a novel UHMWPE powder resin. How would you go about characterizing its molecular weight?

12.4. What chemical characteristics of UHMWPE can be measured using FTIR?

12.5. Which of the test methods outlined in this chapter characterize the mechanical failure properties of UHMWPE?

Chapter 13

Development and Application of the Small Punch Test to UHMWPE

Avram Allan Edidin

Drexel University

Philadelphia, PA

Introduction

Mechanical testing of miniature specimens—as opposed to bulk large specimens—is attractive when the need arises to directly probe the properties of a specific component rather than a generic material. Testing of miniature specimens obtained directly from a specific structure becomes paramount when evidence suggests that the properties of the material under consideration are known to change over time. In such cases, there can be no assurance that assays of supposedly equivalent material, aged in a similar manner, will have any fidelity to individual components of interest.

Nowhere does this dilemma between a direct testing of a component and testing of an analogue appear more forcibly in biomechanics than in the determination of mechanical properties from ultra-high molecular weight polyethylene (UHMWPE) orthopedic bearings. These components are found in total hip, knee, and shoulder arthroplasties, contain between 40 and 200 g of material, and the designs typically have characteristic length scales of between 4 and 60 mm. It may readily be appreciated that fabrication of any mechanical test specimen, even one greatly reduced in size, will be quite challenging and that obtaining replicates from a single specimen will be even more so.

Unfortunately, the failure modes of UHMWPE bearings are predominantly mechanical at one or multiple length scales, making paramount accurate determination of mechanical properties through time. Furthermore, the properties of UHMWPE are known to vary with starting resin grade, processing methodology, irradiation regime, and storage environment (Kurtz et al. 1999). The number of interdependent variables suggest that attempts to replicate the eventual condition of a particular UHMWPE component will be an inexact approximation at best.

This chapter reviews the development and application of miniature specimen mechanical testing techniques, based on the small punch test, to the characterization of UHMWPE components for total joint replacement. The development of the small punch test as applied to UHMWPE was motivated by two clinically relevant and related problems: the relationship between process and system variables and polymer degradation, and the relationship between mechanical properties and the ensuing wear of the arthroplasty bearing. To a first order approximation, the polymer degradation problem was motivated by mechanical failure of tibial plateau bearings, and the wear problem was motivated by the clinical need to reduce the prevalence of small-particle mediated late-onset osteolysis in total hip arthroplasty systems.

Overview and Metrics of the Small Punch Test

The small punch test methodology (also referred to as the disk bend test) was originally proposed by Manahan and colleagues (1981) for the characterization of metallic components from the power industry. Small punch testing was adapted for UHMWPE by Kurtz and associates (1997, 1999) for direct characterization of the mechanical behavior of UHMWPE components. In the small punch test for UHMWPE, a disc of material 0.0200 inches thick by 0.250 inches diameter is indented by a hemispherical punch thereby creating a biaxial state of tension in the material. Furthermore, because of the enclosure of the specimen in the punch-and-die apparatus, the thickness tolerance is only 0.0005 inches. Specimen preparation may be performed by any conventional machining operations performed so as to minimize anisotropy in the finished specimens. A schematic of specimen preparation and testing apparatus is shown in Figure 13.1. Additional details of the small punch test for UHMWPE are provided in ASTM F2183-02 (2002).

Naturally, specimens of this size are particularly attractive to the study of UHMWPE bearing components, because multiple specimens can be obtained from almost any single component. Further suitability is noted in that multiple specimens may often be prepared through the *thickness* of the bearing, which is of interest in the study of oxidative degradation, wear-induced chain alignment, or any other mechanism that leads to a material inhomogeneity through the thickness of a bearing.

Four basic metrics are obtained from a load-displacement curve of a material subjected to the small punch test. These include the initial stiffness, peak load, ultimate load, and the maximum displacement at failure. A fifth derived metric, integral work to failure (WTF), is found by integrating the area under the

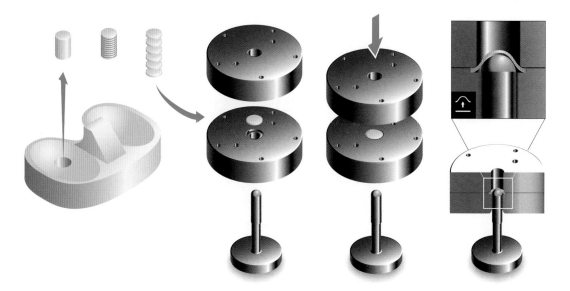

Figure 13.1

Schematic of specimen preparation and small punch test. Typically, cores of net diameter are taken with a custom coring tool set to 0.250 ± 0.001 inches and then sliced into specimens on a lathe. Following facing operations, the final specimens are [19.7–20.0 mils] thick. Specimens are then inserted into a capturing die apparatus through which the hemispherical punch passes. The closed die is mounted to the crosshead of the testing machining and forced over the fixed punch. Dimensions provided here in English units because the test was originally developed in the United States.

load-displacement curve and represents the total energy required to fail the specimen. Many tests have distinct membrane bend and membrane stretch portions, roughly represented by the portion of the curve before and after the peak load is reached respectively. A typical load-displacement curve for GUR 1020 UHMWPE is shown in Figure 13.2.

Although quantitative metrics form the basis for performing statistical comparisons, a fair amount of information about a material may be gleaned by observing the shape of load-displacement curves. In particular, the degree of geometric strain hardening, so called because it is observed on only the load-displacement curve rather than a true stress-strain curve, often differentiates pristine from degraded UHMWPE or ordinary from crosslinked UHMWPE. An example comparing virgin (unirradiated) to crosslinked UHMWPE is shown in Figure 13.3.

Linear polyethylenes with molecular weights below approximately 1 million typically display geometric strain softening. Degradation of UHMWPE is often first observed through a reduction or roll-off of the load-displacement curve as compared with the nondegraded material. As will be discussed shortly, this observation represents a mechanical means of detecting chain scission arising from oxidation.

During the development of the small punch test for UHMWPE, a validation analysis was performed comparing the results obtained from small punch

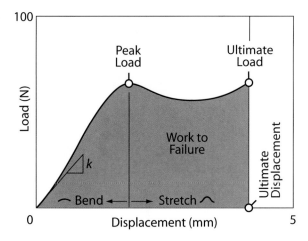

Figure 13.2

Representative load-displacement curve developed by testing of a small punch specimen in equib-iaxial tension; curve here reflects typical behavior of GUR 1020. Primary metrics including initial stiffness, peak load, ultimate load, and ultimate displacement. WTF shown in grey. Unirradiated UHMWPE exhibits a bend and a stretch region as shown.

testing with properties measured during conventional, large-specimen uniaxial mechanical tests (Kurtz et al. 1997). Historically, mechanical tests of UHMWPE were performed in uniaxial tension using specimens such as those described in ASTM D648. Kurtz and colleagues describe an experiment comparing the uniaxial response of 4150- and 1120-based UHMWPE specimens with the response

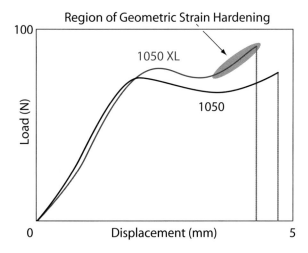

Figure 13.3

Load-displacement curves developed from representative testing of Hoechst 1050 resin in both as-converted, labeled 1050, and crosslinked, labeled 1050 XL, condition. Crosslinking increases the amount of geometric strain hardening in the region shown in grey. In general, substantial infor-mation can be gleaned from the overall form of the load-displacement curve as well as from the quantitative metrics themselves.

of the same materials observed under equibiaxial tension (1997). The researchers availed themselves of the ability to test the same exact material in both arms of their experiment because of the small specimen size required for the small punch testing. They observed, by an inverse approach using a finite element model of the small punch set-up to simulate the test itself, that the initial stiffness in the small punch test could be related linearly to the material's modulus by the relationship

$$E = 13.5\,k$$

where k is the stiffness (in N/mm) observed in the membrane bend portion of the test and E is the conventional Young's modulus (in MPa).

Accelerated and Natural Aging of UHMWPE

Following the validation of the small punch test for virgin UHMWPE and the reconciliation of the reported metrics between the conventional large-specimen uniaxial and the novel miniature specimen biaxial tests, investigators began to apply the small punch test to characterize the effects of oxidative degradation that occurs during accelerated and natural aging on the mechanical behavior of UHMWPE (Kurtz et al. 1999). Kurtz and associates (1999) used the small punch test to investigate the mechanical signature of oxidative degradation by comparing the mechanical behavior of GUR 4150 and GUR 1120 small punch specimens before and after aging in different atmospheres. They observed that the miniature small punch specimens were resistant to subsequent degradation following accelerated aging in an air oven if the specimens had been irradiated in a nitrogen atmosphere. This finding mirrored that of Clough and Gillen (1981) who observed similar protective benefits of nitrogen atmosphere irradiation when similar experiments were performed in linear polyethylenes with lower molecular weights. Representative load-displacement curves from the experiment are shown in Figure 13.4.

Later work by Edidin and colleagues (2002) suggests that the protective benefit of irradiating UHMWPE in an inert environment such as nitrogen gas is not as apparent when artificial aging is performed on the bulk specimens as opposed to the prepared small punch test specimens. More specifically, the degree to which changing the irradiation atmosphere protects from future degradation appears to be a function of heating rate, whereby very fast rates quench the surface layer, thereby stopping further free–radical–mediated oxidative degradation (Kurtz et al. 1999). How this interplay of variables might affect *in vivo* oxidation of irradiated UHMWPE is discussed later in this chapter.

As previously noted, an important motivator in the development of the small punch test for UHMWPE was its ability to assess the mechanical performance of retrieved arthroplasty components, whether retrieved from the shelf or from the body. During the mid-1990s manufacturers began to switch the irradiation and storage atmosphere from air to either nitrogen or argon. By 2000 some initial data describing the mechanical performance of inert atmosphere irradiated

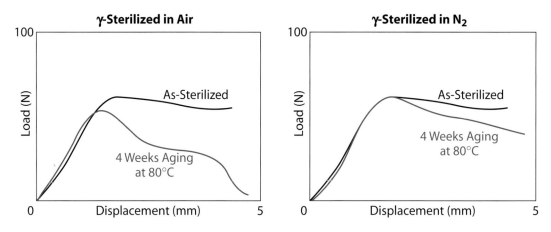

Figure 13.4

Load-displacement curves observed by testing specimens prepared from GUR 4150 resin. At left, the specimens irradiated in air and aged in an air atmosphere oven are seen to have degraded mechanical properties. At right, specimens irradiated in nitrogen but also aged in an air atmosphere oven are seen to degrade to a markedly lower degree. (Adapted from Kurtz et al. 1999.)

components could be obtained using the small punch test to assay the components. These data were then compared with data obtained from similar components irradiated in air stored for the same period (Edidin et al. 2000). The results of this comparison were the first to show that insert gas packaging systems were capable of preventing oxidation during shelf storage. Figure 13.5 compares

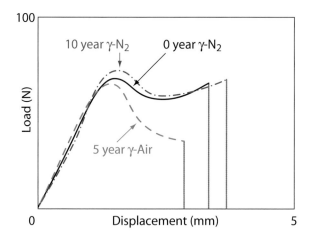

Figure 13.5

Representative load-displacement curve obtained from bearing surfaces of three acetabular components irradiated and stored in either nitrogen or air. After 10 years storage in nitrogen, load-displacement curve is nearly identical to that obtained immediately poststerilization in nitrogen. Component irradiated and stored in air is substantially degraded after only 5 years. Base resin was GUR 1020 compression molded in all cases. (Adapted from Edidin et al. 2000.)

specimens made from GUR 1020 components 10 years after irradiation and storage in either air or nitrogen atmosphere packaging.

Testing of naturally and artificially aged UHMWPE arthroplasty components repeatedly showed that the first mechanical signature of degradation was a decrease in the load carrying capability of the specimens at displacements beyond that recorded at peak load. Oxidative degradation has been previously shown to reduce effective molecular weight of polyethylene by cleaving the long backbone chains of the semicrystalline polymer matrix. Edidin and associates proposed the hypothesis that the load-displacement curve of degraded UHMWPE should have certain similarity to the load-displacement response of nondegraded linear polyethylene with intrinsically lower molecular weight. The hypothesis was tested by comparing the load-displacement curves and resultant metrics obtained from GUR 1050, GUR 1020, GHR 8110, and LM 6007.00 high-density polyethylene (HDPE). These materials have molecular weights of approximately $3–5 \times 10^6$, 1×10^6, 500,000, and 120,000, respectively. As can be seen in Figure 13.6, lower molecular weight polyethylenes exhibit load-displacement behaviors akin to those exhibited by degraded UHMWPE (Edidin and Kurtz 2000, Kurtz et al. 1999).

Clearly oxidative chain scission has an effect on mechanical performance similar to reduction of molecular weight *ab initio*. This experiment provided the strongest mechanical evidence of oxidation to date, in that a mechanistic relationship between oxidation and molecular weight was displayed.

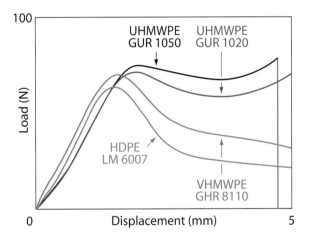

Figure 13.6

Comparison of representative load-displacement curves obtained from testing of four linear polyethylenes of varying molecular weights. The roll-off in load sustained by the specimen in the lower molecular weight materials is similar to that exhibited by higher molecular weight materials that have undergone oxidative degradation (see example in Figure 13.5). (Adapted from Kurtz et al. 1999 and Edidin et al. 2000.)

In Vivo Changes in Mechanical Behavior of UHMWPE

The efficacy of barrier inert atmosphere packaging, described previously, gives some assurance that a just-opened UHMWPE bearing component will have minimal oxidative degradation regardless of time spent on the shelf. However, the question remains as to whether oxidative degradation occurs *in vivo*. Premnath and colleagues (1996) suggest that *in vivo* oxidation is likely to be a very slow process based on an argument that oxygen dissolved in water saturates at one-eighth the concentration in air. Earlier examination of retrieved components had indirectly suggested that in-service time and oxidation level were not linked (Gomez-Barrena et al. 1998), but this study and others suffered from unknown shelf and postretrieval aging times; only *in vivo* times were known.

Recently Kurtz and colleagues presented data from retrieval analyses based on examination of well-characterized components obtained from previously consented patients (2003). By judicious use of patient records, coupled with the use of manufacturing records, Kurtz and associates were able to track each bearing from the time of manufacture, through implantation, and eventually to the research laboratory. Chemical and mechanical small punch testing analysis of the retrieved acetabular bearings was performed in both the articulating and nonarticulating regions. The study was designed such that the time interval between explantation and testing was minimized, thereby decreasing the opportunity for degradation post-*vivo*. Furthermore, components were stored in nitrogen during this interval. The authors reported on 16 retrieved components originally irradiated and stored in air, with a mean shelf aging time of 0.4 years, a mean implantation time of 11.5 years, and a mean postimplantation time of 0.7 years while stored at −5°C. Because of the relatively short *ex-vivo* aging times compared with the lengthy *in vivo* times, a case was made that any degradation observed would be attributable to *in vivo* oxidative degradation.

Small punch testing of the retrieved specimens showed that the unworn region of the articulating surface was significantly degraded compared with the worn region of the articulating surface. A summary of the differences observed is shown in Figure 13.7 for all 75 small punch tests performed on the acetabular liners retrieved from 16 patients. The finding of maximal degradation in the unworn region was novel; the experiment was not designed to determine whether the difference in degradation rates between the worn and unworn regions reflected a loss of material at the articulating surface, or whether loading affected the degradation pathway itself. These data were among the first to conclusively show that *in vivo* degradation occurs in the absence of confounding variables associated with some earlier retrieval studies such as unknown shelf and postexplantation times. The data also suggest that the assumptions of Premnath and colleagues (1996), related to the concentration in water at room temperature, do not necessarily pertain to the conditions *in vivo*, in which the joint fluids are maintained at body temperature.

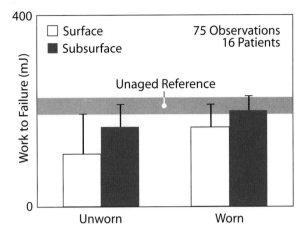

Figure 13.7
WTF versus implantation time for entire cohort of 16 retrieved acetabular components. Data obtained from specimens machined from the surface and subsurface of the worn and unworn region of the articulating surface. (Adapted from Kurtz et al. 2003.)

Effect of Crosslinking on Mechanical Behavior and Wear

In the late 1990s problems associated with submicron UHMWPE wear debris, such as increased incidence of osteolysis with increasing wear (Dumbleton, Manley, and Edidin 2002), grew to the point that concerted efforts to reduce the flux of particulate emanating from the bearing couples *in vivo* were made. The outcome of these research efforts was the development of various highly crosslinked materials that exhibited markedly reduced wear in laboratory simulators. If in fact the primary wear mechanism in a hip arthroplasty is in some way related to the mechanical performance of the bearing, then the small punch test would likely be able to detect differences between the mechanical signature of conventional and highly crosslinked UHMWPEs. Mechanical assessment of several proprietary highly crosslinked UHMWPEs was performed in 2001 by Edidin and Kurtz who repeatedly observed that highly crosslinked UHMWPE components exhibited marked geometric strain hardening after attainment of the initial maximum load on the load-displacement curve. Representative material curves are shown in Figure 13.8.

When assayed using the small punch test, the load-displacement curve for gamma irradiated GUR 1050 exhibits greater geometric strain hardening and greater change from its nonirradiated condition than does the same data obtained using virgin compared with irradiated GUR 1020, suggesting that all other things being equal, the efficiency of crosslinking is lower in lower Mw resins. Figure 13.9 illustrates this finding by comparing the load-displacement curve of GUR 1020 and GUR 1050 resins converted in the same manner (compression molding) before and after irradiation. The increased efficiency of crosslinking observed in higher molecular weight resins has been corroborated in a chemical manner by

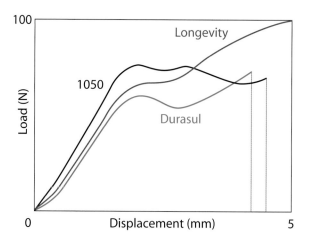

Figure 13.8

Comparison of GUR 1050 in three conditions as described by the respective load-displacement curves. Longevity and DURASUL are commercially available formulations of highly crosslinked UHMWPE, which exhibit the characteristic geometric strain hardening in the drawing phase of the small punch test. Free radical quenching by remelting UHMWPE above its peak melt transition typically reduces the observed peak load as exhibited here (Edidin and Kurtz 2001).

Spiegelberg and associates (2001), who noted a greater crosslink density was observed in UHMWPEs with lower polydispersion indices, with GUR 1050 having the lowest polydispersion index of the materials tested.

An alternative explanation for the increased wear resistance of GUR 1050 is based on an energy argument arising from the observation that crosslinked

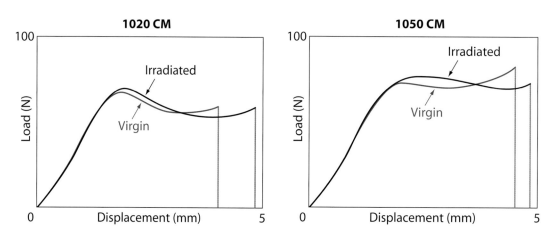

Figure 13.9

Comparison of representative load-displacement curves obtained using GUR 1020 and GUR 1050 resins converted by compression molding before and after irradiation. A more substantial change following irradiation of the degree of geometric strain hardening is observed for the GUR 1050 material than for the GUR 1020 material, suggesting that the higher molecular weight–based resin is more efficiently crosslinked. A single irradiation dose of 25 to 40 kGy was absorbed by both materials that were irradiated simultaneously. (Adapted from data in Edidin et al. 2002.)

materials typically exhibit higher integral WTF than noncrosslinked UHMWPEs. However, because crosslinking may also influence other metrics that change wear rates, such as hindering chain mobility and/or changing surface lubricity, Edidin and Kurtz designed an experiment to investigate solely the effect of changing WTF on observed wear rates (2000). They used a spectrum of semicrystalline polymers with previous clinical application to test the hypothesis that wear rate was proportional to the WTF of a material when tested under equibiaxial tension. The materials chosen were UHMWPE, HDPE, polyacetal, and polytetrafluoroethylene. The basic design of the experiment consisted of fabricating identical acetabular liners on production tooling from the four materials, measuring gravimetric wear on a hip simulator using commonly accepted protocols (Wang et al. 1998), and measuring the articulating surface WTF using the small punch test. All materials were tested in the native, nonsterile and nonirradiated condition to limit confounding variables. Analysis of variance revealed correlations between gravimetric wear rate and several of the small punch metrics, but the best correlates, with r^2 of 0.96, were found between the WTF and wear rate. This correlation, shown in Figure 13.10, displayed a highly linear relationship between each material's WTF and the ensuing wear rate, suggesting that intrinsic equibiaxial toughness or WTF plays a fundamental role in the generation of wear.

Crosslinked UHMWPEs are typically tougher than their noncrosslinked equivalents, as may be observed in Figure 13.9. However, the data shown in Figure 13.9 was obtained using materials with complicated and proprietary thermal annealing and/or quenching steps following gamma irradiation. Because annealing tends to substantially increase the elongation failure, it is not guaranteed that the low wear exhibited by highly crosslinked UHMWPE in the simulator is necessarily a continuation of the curve in Figure 13.10 extrapolated

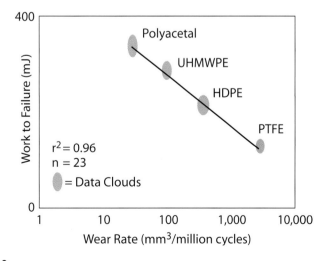

Figure 13.10

Relationship between WTF as calculated from equibiaxial loading to failure for materials with historical clinical application. Data suggests that adhesive-abrasive wear rates are correlated with the innate WTF of the bearing material (Edidin et al. 2000).

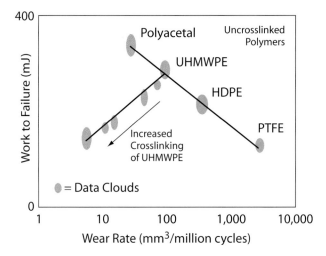

Figure 13.11

Comparison of WTF versus wear rate for crosslinked and noncrosslinked acetabular components fabricated from a variety of polyolefins tested in a multistation wear simulator. Increasing the crosslink density of UHMWPE without thermal annealing still leads to a substantial decrease in wear rate. (Adapted from Edidin et al. 1999, 2000.)

out to even tougher materials. In an experiment, Kurtz and Edidin combined the data in Figure 13.10 with data obtained from sequentially irradiated acetabular components with no thermal treatment postirradiation to produce the curve shown in Figure 13.11 (Edidin and Kurtz 2000, Edidin et al. 1999).

In sum, the experiments relating integral toughness to gravimetric wear metrics suggest that although marked geometric strain hardening is observed in all crosslinked UHMWPEs tested to date, the causality of integral toughness to the wear performance of such components remains unclear.

Shear Punch Testing of UHMWPE

The prominence of geometric strain hardening in highly crosslinked UHMWPE, coupled with uncertainty as to the overall role of integral toughness on wear resistance, begged the question of whether the presumed resistance to chain mobility in equibiaxial tension evidenced by geometric strain hardening is solely responsible for decreased wear. Wear at the acetabular bearing surface, in particular, has been attributed to the cross-shear phenomenon wherein debris is liberated following two orthogonal passes of the femoral head's contact point across some portion of the contact patch of the bearing (Wang et al. 1998). If these orthogonal passes induce some sort of shear-related damage accumulation, then modification of the mechanical test apparatus to permit shear loading might be able to differentiate wear resistance better than the tensile loading apparatus.

Kurtz and colleagues (2003) developed a modified punch that loaded the specimen in shear for a substantial portion of the range of travel of the punch in a typical test. Rather than a domed head, the modified punch used a flat head and was mated to a die without a radius at the pinch point between the specimen and the upper die face. In keeping with recommendations in the literature, the annular clearance between the punch and the die was set at 0.033 mm, yielding a length-to-width ratio greater than 15, previously shown to minimize bending during shear testing (G'Sell, Boni, and Shrivastava 1983, Gul 1997). No particular thickness-to-width ratios were designed into the apparatus because it was assumed that the disk-shaped specimens would provide sufficient lateral restraint to eliminate out-of-plane deformations previously associated with shear tests. Correlation between *in vitro* data and finite element analyses was demanded to yield a relationship between the initial stiffness of the UHMWPE and its Young's modulus as

$$E = 0.283 \, k$$

where k is the initial stiffness (in N/mm) observed on the load-displacement curve and E is Young's modulus (in MPa). The consistently observed features of the shear punch test are shown in Figure 13.12. Complete shear-through of the specimen occurs when the punch has traversed the initial thickness of the specimen (0.5 mm), but pure shear conditions do not exist beyond the displacement resulting in the peak load.

Testing using the shear punch apparatus of conventional and highly cross-linked UHMWPE strongly suggested the representative behaviors in shear

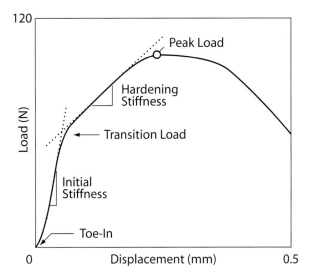

Figure 13.12

Characteristic features observed on load-displacement curves obtained using the shear punch apparatus. An initial and hardening stiffness in addition to the peak load are the primary metrics used to perform analytical comparisons. Like the conventional small punch test, the shear punch test also may be interpreted qualitatively. (Adapted from Kurtz et al. 2002.)

were in fact quite similar, as shown in Figure 13.13. This finding suggests that the cross-shear motion pattern by the femoral head on the bearing surface likely still results in tensile final rupture of the wear debris fibrils, and that therefore the term "cross-shear" is a misnomer.

Kurtz and coworkers also examined the shear behavior of shelf-aged tibial bearing components to investigate the role played by shear failure in the breakdown of degraded components (2003). The surface and subsurface mechanics of components sterilized and stored in air were measured. Because degradation occurs in a spatially highly nonuniform manner, comparisons were made between the heavily degraded subsurface region and the substantially less degraded surface region. This comparison revealed substantial differences on the overall appearance of the two curves, with the more heavily degraded subsurface specimens exhibiting a complete lack of geometric shear strain hardening and a higher initial stiffness, as shown in Figure 13.14. Bartel and colleagues (1986) suggested that the peak stresses in a loaded tibial bearing occur approximately 1 mm below the articulating surface, or in the same region as the subsurface samples tested by Kurtz and associates (2003). Thus, clinical failure modes of UHMWPE bearings observed in retrieved tibial bearings, consisting primarily of pitting and delamination, may arise from the substantially lower resistance to both tensile and shear strain hardening postyield. Lowered integral toughness may also play a role, as suggested by Kurtz and colleagues (2000) who observed a trend relating lower toughness with higher qualitative damage scores in a pilot retrieval study. Overall, the tensile and shear small punch test findings have proven key tools in understanding the relationship

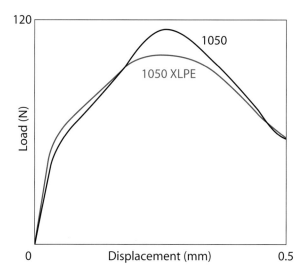

Figure 13.13

Load-displacement curves observed by testing virgin and crosslinked UHMWPE in shear punch apparatus. Curves are quite similar, suggesting that changes in shear resistance are unlikely to explain the increased resistance to wear exhibited by crosslinked UHMWPEs. Materials tested were virgin GUR 1050 and GUR 1050 irradiated to 75 kGy and annealed at temperatures below the melt point. (Adapted from Kurtz et al. 2002.)

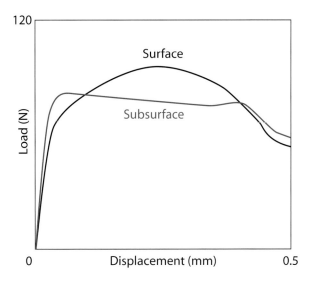

Figure 13.14

Comparison of load-displacement curves from specimens obtained at the surface and at approximately 1500 microns subsurface from a tibial bearing component originally made from GUR 1120 sheet, irradiated with 25 to 40 kGy of gamma irradiation and aged on the shelf in air for 8.5 years. Oxidative degradation in shelf-aged components is known to peak subsurface, and the data suggests that resistance to shear flow is highly compromised by degradation. (Adapted from Kurtz et al. 2002.)

between mechanics of UHMWPE to both adhesive/abrasive wear, and frank mechanical breakdown of arthroplasty bearing components.

Fatigue Punch Testing of UHMWPE

Seminal work first applied the equibiaxial tension small punch test to the investigation of degradation and wear mechanisms, because these problems were manifest at the time of the test's development. However as discussed briefly in the review of the shear punch test development earlier, bearing components fail not just from continual adhesive/abrasive wear mechanisms, but also from fatigue mechanisms, such as delamination and frank failure associated with crack initiation and propagation. The means of determining resistance to failure in fatigue of aged or crosslinked materials could provide useful information on the suitability of various bearing materials to their proposed application. For example, although some highly crosslinked materials are remelted, others are annealed, and thus the relative fatigue performance of such materials may be as important as their resistance to *in vitro* or *in vivo* oxidation. In addition, the suitability of crosslinked materials in general to applications undergoing cyclic loading such as tibial plateau bearings requires the development of new testing methods. Therefore, methodological changes were made to the tensile small punch apparatus, as described by Villarraga and associates (2002), to permit cyclic loading of the test specimen at body temperature. These changes included the addition of a boroscope of some sort, typically a medical arthroscope, to permit direct visualization of the specimen during the crack initiation and propagation phases of specimen failure. An environmental chamber was used to encase the punch and die apparatus to permit testing of specimens at the body temperature. Lastly, the fatigue punch test was conducted in a servohydraulic testing machine as opposed to the electro-mechanical testing machine typically used under displacement control in the previously described quasistatic punch tests.

Testing of materials under cyclic loading to determine resistance to cracking may be performed with or without a preexisting flaw in the test specimen. The fatigue small punch test is performed using the same specimen as the static test. Failure is presumed to occur once a critical level of damage accumulation has occurred in the absence of a created flaw or defect; this methodology follows a total-life or damage-accumulation model. Because damage accumulation models are primarily useful to make comparisons among the performance of different materials, as opposed to providing quantitative estimates of a material's absolute performance in the presence of an existing flaw, it is desirable to have specimens fail in a reasonable (fairly low) number of cycles. To this end, Villarraga and associates (2002) describe first testing each material in a conventional static small punch test to determine the mean peak load (P_{max}) exhibited by that material when tested at a constant loading rate of 200 N/s. Because the peak load represents the inflection between membrane bending and membrane stretch deformation modes with its associated chain realignment, it is a natural choice for the limit load in the fatigue punch test. The fatigue punch test itself is then performed under cyclic loading conditions to a maximum load of 0.55 to 0.90 P_{max} depending on the number of cycles to failure expected or desired, with higher percentages shortening the test. Loading rate during the cyclic portion of the test is also 200 N/s. The overall experimental schematic is show in Figure 13.15.

In their work to develop the fatigue punch test, Villarraga and colleagues (2002) tested four conditions of GUR 1050 UHMWPE used in contemporaneous clinical applications. These conditions included virgin (representing ethylene oxide gas or gas plasma–sterilized bearings), gamma sterilized (2.5 to 4.0 Mrad in nitrogen), annealed highly crosslinked UHMWPE (100 kgy and 110°C), and remelted highly crosslinked UHMWPE (100 kgy and 150°C). All testing was conducted at 37°C using an environmental chamber. Two metrics comparing the relative fatigue resistance of a given material were developed. These were a classic S–N curve relating maximum peak load to survival, and a hysteresis metric giving some indication of the relative energy retained in the test specimen during each cyclic excursion.

Substantial differences were observed between the control (virgin GUR 1050) material and both crosslinked materials in their displacement to failure, their hysteresis curves, and in the overall appearance of the specimens at failure. The control material exhibited a dramatic increase in displacement just prior to failure not exhibited by either of the crosslinked materials. Lower loads lead to a greater number of cycles prior to failure at larger displacements in all materials tested. Representative examples of the displacement versus cycles relationships obtained at both relatively low and somewhat higher number of cycles to failure are shown in Figure 13.16. The corresponding appearance of the kinds

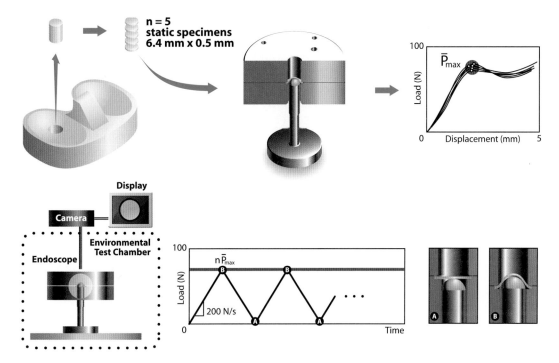

Figure 13.15

Schematic of fatigue punch methodology as described by Villarraga and colleagues (2002). Specimens are prepared conventionally and then tested to failure at a constant load rate of 200 N/s at 37°C to determine peak load. Subsequently additional specimens from the same material are tested in under cyclic loading using a triangular waveform ramping again at 200 N/s to a percentage of the peak load previously determined. Percentage is varied across specimen group to cause failure under a greater or lower number of cycles.

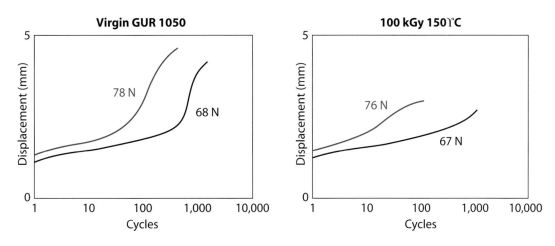

Figure 13.16

Displacement versus cycles curves obtained following testing of virgin and remelted highly crosslinked GUR 1050 specimens at differing peak loads. Crosslinked specimens exhibited substantially lower displacements at failure than control specimens. (Adapted from Villarraga et al. 2002.)

of specimens tested to generate the curves in Figure 13.16 are shown following fracture in Figure 13.17.

The maximum hysteresis behavior was relatively similar across the materials with the exception of that exhibited by the XLPE annealed material, which appeared to require lower loads to generate the same amount of stored energy in the specimen. The cumulative hysteresis plot for each material is shown in Figure 13.18. How the ability of a candidate bearing material to dissipate energy during a load–unload cycle relates to its suitability as a bearing component material will require further study in conjunction with carefully monitored clinical application.

Figure 13.17

Representative low-power magnification SEM photographs of samples of the same materials tested in Figure 13.16. At top, virgin GUR 1050 specimen and at bottom, a specimen of remelted (150°C) highly crosslinked GUR 1050. Specimen at top failed at 7289 cycles; specimen on the bottom failed at 7994 cycles. (Adapted from Villarraga et al. 2002.)

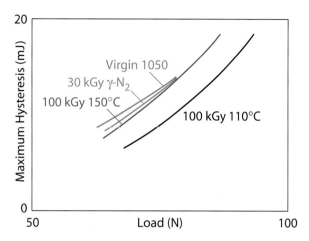

Figure 13.18

Maximum hysteresis exhibited as a function of load for each of the materials tested by Villarraga and colleagues during the development of cyclic small punch test. (Adapted from Villarraga et al. 2002.)

Conclusion

In summary, since 1998 the small punch test has been extensively developed to permit direct assay of the mechanical properties of orthopedic UHMWPE bearings. Originally adapted strictly as a miniature-specimen, biaxial tension test, it has since been extended to permit testing in both shear and cyclic modalities. As such, it has proven extremely useful in the detection of the mechanical signature of oxidative degradation and the signature associated with elevated levels of irradiation-induced crosslinking. Further testing has enabled associations between wear and mechanical properties to be made from the selfsame components, although a quandary remains as to which mechanical property drives wear reduction. Most recently, application of the small punch specimen geometry to a cyclic loading regime has permitted initial investigation into the suitability of highly crosslinked UHMWPE materials to applications wherein they will be subjected to cyclic loading. The greatest strength of the small punch testing methodology is its minimal requirement for material, thereby permitting mechanical assays to be made from almost any experimental specimen configuration, and one would expect this strength to continue to be leveraged across the mechanical, chemical, and tribological spectra that describe the orthopedic bearing world.

References

ASTM F 2183-02. 2002. Standard test method for small punch testing of ultra-high molecular weight polyethylene used in surgical implants. West Conshohocken, PA: American Society for Testing and Materials.

Bartel D.L., V.L. Bicknell, and T.M. Wright. 1986. The effect of conformity, thickness, and material on stresses in ultra-high molecular weight components for total joint replacement. *J Bone Joint Surg* 68:1041–1051.

Clough R.L., and K.T. Gillen. 1981. Radiation-thermal degradation of PE and PVC: Mechanism of synergism and dose rate effects. *Radiat Phys Chem* 18:661–669.

Dumbleton J.H., M.T. Manley, and A.A. Edidin. 2002. A literature review of the association between wear rate and osteolysis in total hip arthroplasty. *J Arthroplasty* 17(5): 649–661.

Edidin A.A., L. Pruitt, C.W. Jewett, et al. 1999. Plasticity-induced damage layer is a precursor to wear in radiation- cross-linked UHMWPE acetabular components for total hip replacement. Ultra-high-molecular-weight polyethylene. *J Arthroplasty* 14:616–627.

Edidin A.A., and S.M. Kurtz. 2000. The influence of mechanical behavior on the wear of four clinically relevant polymeric biomaterials in a hip simulator. *J Arthroplasty* 15:321–331.

Edidin A.A., J. Muth, S. Spiegelberg, and S.R. Schaffner. 2000. Sterilization of UHMWPE in nitrogen prevents oxidative degradation for more than ten years. *Transactions of the 46th Orthopedic Research Society* 25:1.

Edidin A.A., and S.M. Kurtz. 2001. Development and validation of the small punch test for UHMWPE used in total joint replacements. In *Functional biomaterials.* N. Katsube, W. Soboyejo, and M. Sacks, Eds. Winterthur, Switzerland: Trans Tech Publications.

Edidin A.A., M.P. Herr, M.L. Villarraga, et al. 2002. Accelerated aging studies of UHMWPE. I. Effect of resin, processing, and radiation environment on resistance to mechanical degradation. *J Biomed Mater Res* 61:312–322.

Gomez-Barrena E., S. Li, B.S. Furman, et al. 1998. Role of polyethylene oxidation and consolidation defects in cup performance. *Clin Orthop* 352:105–117.

G'Sell C., S. Boni, and S. Shrivastava. 1983. Application of the plane simple shear test for determination of the plastic deformation of solid polymers at large strains. *J Materials Sci* 18:903–918.

Gul R. 1997. Improved UHMWPE for use in total joint replacement. Ph.D. Diss., Boston: Massachusetts Institute of Technology.

Kurtz S.M., J.R. Foulds, C.W. Jewett, et al. 1997. Validation of a small punch testing technique to characterize the mechanical behavior of ultra-high molecular weight polyethylene. *Biomaterials* 18:1659–1663.

Kurtz S.M., C.W. Jewett, J.R. Foulds, and A.A. Edidin. 1999. A miniature-specimen mechanical testing technique scaled to the articulating surface of polyethylene components for total joint arthroplasty. *J Biomed Mater Res (Appl Biomater)* 48:75–81.

Kurtz S.M., O.K. Muratoglu, M. Evans, and A.A. Edidin. 1999. Advances in the processing, sterilization, and crosslinking of ultra-high molecular weight polyethylene for total joint arthroplasty. *Biomaterials* 20:1659–1688.

Kurtz S.M., L.A. Pruitt, D.J. Crane, and A.A. Edidin. 1999. Evolution of morphology in UHMWPE following accelerated aging: The effect of heating rates. *J Biomed Mater Res* 46:112–120.

Kurtz S.M., C.M. Rimnac, L. Pruitt, et al. 2000. The relationship between the clinical performance and large deformation mechanical behavior of retrieved UHMWPE tibial inserts. *Biomaterials* 21:283–291.

Kurtz S.M., W. Hozack, M. Marcolongo, J. Turner, C. Rimnac, and A. Edidin. 2003. Degradation of mechanical properties of UHMWPE acetabular liners following long-term implantation. *J Arthroplasty* 18(7 Suppl 1):S68–78.

Kurtz S.M., C.W. Jewett, J.S. Bergstrom, et al. 2003. Miniature specimen shear punch test for UHMWPE used in total joint replacements. *Biomaterials* 23:1907–1919.

Manahan M.P., A.S. Argon, and O.K. Harling. 1981. The development of a miniaturized disk bend test for the determination of postirradiation mechanical properties. *J Nucl Mater* 104:1545–1550.

Premnath V., W.H. Harris, M. Jasty, and E.W. Merrill. 1996. Gamma sterilization of UHMWPE articular implants: An analysis of the oxidation problem. *Biomaterials* 17:1741–1753.

Spiegelberg S.H., S.M. Kurtz, and A.A. Edidin. 2001. Effects of molecular weight distribution on the network properties of radiation- and chemically crosslinked ultra-high molecular weight polyethylene. *Trans. 25th Society for Biomaterials* 215.

Villarraga M.L., S.M. Kurtz, M.P. Herr, and A.A. Edidin. 2002. Multiaxial fatigue behavior of conventional and highly crosslinked UHMWPE during small punch testing. *J Biomed Mater Res* in press.

Wang A., A. Essner, V.K. Polineni, et al. 1998. Lubrication and wear of ultra-high molecular weight polyethylene in total joint replacements. *Tribology International* 31:17–33.

Chapter 13. Reading Comprehension and Group Discussion Questions

13.1. Which of the metrics of the small punch test characterize mechanical failure properties of UHMWPE?

13.2. What are the advantages of the small punch test over standard mechanical test methods described in Chapter 12?

13.3. How do the features of the small punch test load-displacement curve shift as a function of molecular weight, oxidation, and crosslinking?

13.4. What is the stress state in the center of the UHMWPE specimen during the small punch test? How does the stress state evolve as the test progresses?

13.5. How does the shear punch test differ from the small punch test? Explain how the stress state in the deformed region evolves as the test progresses.

Chapter 14

Computer Modeling and Simulation of UHMWPE

Jörgen Bergström

Exponent, Inc.
Natick, MA

Introduction

Computer models to simulate the mechanical behavior of ultra-high molecular weight polyethylene (UHMWPE) used in joint arthroplasty components continue to evolve and increase in sophistication. The strategies for predicting the mechanical response of UHMWPE are typically phenomenologically or physically based, and a particular implementation can vary from very simple to extremely complex. This chapter introduces existing methodologies to predict the mechanical response of UHMWPE, with the goal of outlining the value and range of application for available analytical tools for UHMWPE.

A thorough understanding of the mechanics of UHMWPE is important for efforts to improve the performance of orthopedic components. Elastic properties, resistance to plastic deformation, stress and strain at failure, fatigue behavior, and wear resistance of UHMWPE are believed to play roles in the life expectancy of an UHMWPE bearing. There exists a fundamental relationship between a material's intrinsic mechanical properties, akin to state variables, and how a structure made of the material will respond under mechanical stimuli. This material-specific fundamental relationship is referred to as a constitutive model. A validated constitutive model is a required input to a finite element (FE)–based simulation of a structure made of the material in question.

In this chapter we review different candidate constitutive models for UHMWPE, list the strengths and limitations of each model, and show how they can be used in simulations of UHMWPE orthopedic components. One of the most important uses of a validated constitutive model is to predict failure of the structure at hand on some length scale. Modeling of UHMWPE orthopedic bearings is usually performed to examine questions related to macroscopic failure, such as in a knee bearing insert, or microscopic failure, such as occurs concomitant to the generation of wear particles in a hip bearing insert. This chapter does not attempt to review this class of failure theories, but instead focuses on the use and application of constitutive models for conventional and highly crosslinked UHMWPE.

Overview of Available Modeling and Simulation Techniques

Many techniques can be used to model the mechanical behavior of UHMWPE. The two main classes consist of analytical and computational techniques, each of which can employ material models of varying sophistication. Table 14.1 provides a brief description of different modeling approaches with their associated advantages and disadvantages. The analytical closed-form solution methods by necessity require greatly simplified material model descriptions. The FE-based simulation methods are more flexible in that they can allow for more sophisticated models for polymer behavior, ranging from linear elastic to very advanced viscoplastic constitutive models. Though FE methods are computationally relatively expensive to implement, the decrease in the cost of computing power has enabled them to rise to the fore in the analysis of the response of mechanical systems in general and the validation of material models in particular.

FE analysis involves the discretization of a bounding geometry into smaller discrete elements, typically triangles or rectangles and their three-dimensional analogues. This discretization enables the solution of complicated boundary-valued problems involving both complex geometries and complex material behavior by imposing a weak solution of the ensuing Raleigh-Ritz problem. The size and scope of the simulations performed have increased tremendously following a number of numerical performance improvements. These improvements include faster computers permitting execution of more complicated analyses encompassing large deformations and complex constitutive models; the development of robust FE solvers that can reliably accommodate geometric and material nonlinearity, contacting surfaces, and time dependence; automated meshing software that can mesh difficult initial geometries and remesh deformed geometries if required for numerical stability; and, most importantly from the perspective of this discussion, the development of physically based constitutive models that accurately capture the experimentally determined behavior of UHMWPE.

Any FE analysis requires the input of appropriate geometry, loading conditions, and material behavior. Key to the development of any material model is good data from the laboratory describing the candidate material's behavior under some set

Table 14.1
Summary of Available Modeling Approaches and Techniques for Predicting the Behavior of UHMWPE

Modeling approach	Advantages	Disadvantages
Handbook solution (using a linear elastic material model)[a]	Fast and easy to perform	Simple geometries only Does not consider nonlinearity material behavior Only for very small strains May give inaccurate results
Closed-form analytical solution (using an isotropic plastic material model)[b]	Fast and easy to perform More accurate than a linear elastic solution	Handbook solutions are available only for very few geometries and loading conditions May give inaccurate results
FE analysis using a simple material model (e.g., linear elastic, hyperelastic, linear viscoelastic, isotropic plasticity)	Can account for complex geometries Relatively easy to perform	Does not consider the true material behavior in general deformation states Typically valid only for small–intermediate deformations May give inaccurate results
FE analysis using an advanced material model for polymers	Can account for complex geometries Can account for the experimentally determined characteristic response Enables simulations of complex thermomechanical, composite responses	The advanced material models are currently only available as user-supplied subroutines for the FE packages Requires expertise both to calibrate and use

[a]Roark, Budynas, and Young 2001, [b]Lubliner 1998.

of known and understood loading conditions. Validation of a model is performed both by comparison to the original data, but more importantly, by comparison to additional data obtained in a different loading condition.

Characteristic Material Behavior of UHMWPE

Conventional and highly crosslinked UHMWPE materials exhibit similar qualitative behavior as a result of the underlying similarities of the material microstructures. In this section we examine the behavior of four different UHMWPE materials to illustrate the characteristic material response of UHMWPE. This data will also be used as the basis for the development of material models in the rest of the review. Two of the four materials were conventional UHMWPE and two were highly crosslinked UHMWPE. All materials were created from the same lot of ram-extruded GUR 1050 (Figure 14.1).

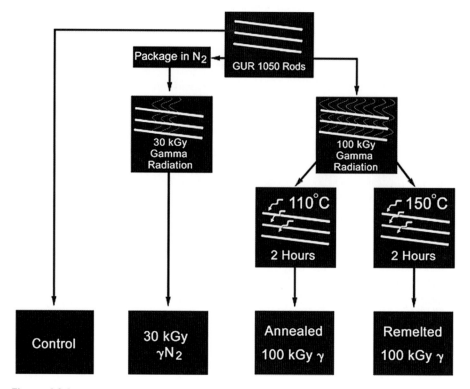

Figure 14.1
UHMWPE materials used in the experimental investigation.

The first conventional UHMWPE material was a control, virgin, unirradiated material. The second conventional UHMWPE material was gamma-radiation sterilized in nitrogen with a dose of 30 kGy (referred to as 30 kGy, γ-N$_2$). The two highly crosslinked UHMWPEs were both gamma irradiated with an absorbed dose of 100 kGy. One of the crosslinked materials was heat treated until the specimen center reached 110°C for 2 hours (referred to as 100 kGy, 110°C). The second was heat treated at 150°C for 2 hours (referred to as 100 kGy, 150°C). The degree of crystallinity of the four material types was determined by differential scanning calorimetry (DSC), see Table 14.2.

Table 14.2
Degree of Crystallinity of the Four Different Types of GUR 1050

Material	Degrees of crystallinity
Unirradiated	0.50
30 kGy, γ-N$_2$	0.51
100 kGy, 110°C	0.61
100 kGy, 150°C	0.46

The uniaxial tension response of the four materials was examined at room temperature using specimens with a diameter of 10 mm and a gauge length of 25 mm. The specimens were monotonically loaded to failure at a displacement rate of 75 mm/minute (corresponding to a true strain rate of about 0.014/ second. A noncontacting video extensometer was used to measure the axial strain. Representative true stress–true strain curves for the four types of UHMWPE are shown in Figure 14.2. It is clear that the highly crosslinked materials fail at lower stress and strain than the two conventional materials. Figure 14.2 also shows that the four materials behave rather similarly up to a strain of approximately 0.8, and that the crosslinked material that was remelted at 150°C has a slightly lower flow stress.

The uniaxial compressive response of the four materials was examined using cylindrical specimens with a diameter of 10 mm and a height of 15 mm. The compression testing was performed at room temperature using an electromechanical load frame, and the displacement of the load platens was used to measure the applied deformation. The resulting true stress–true strain response (Figure 14.3) was calculated from the measured force–displacement response assuming a constant volume deformation. The compressive behavior of the four materials follows the same trends as observed in the tension experiments: there are only small differences between the materials, and the crosslinked material that was remelted (100 kGy, 150°C) has a slightly lower yield stress. The representative behavior shown in Figures 14.2 and 14.3 further indicate

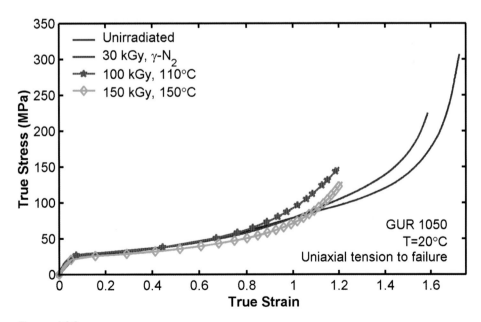

Figure 14.2

Representative experimental uniaxial tension data for the four UHMWPE materials. The tension experiments were performed at a displacement rate of 75 mm/minute.

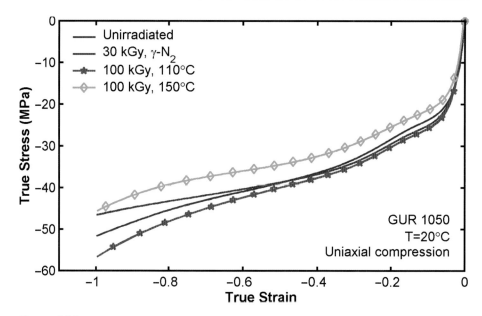

Figure 14.3

Representative experimental uniaxial compression data for the four UHMWPE materials. The compression experiments were performed at an engineering strain rate of −0.02/second.

that the stress is higher in tension than in compression at the same magnitude of applied true strain.

The influence of different strain rates on the compressive behavior of the highly crosslinked GUR 1050 that was heat treated at 110°C (100 kGy, 110°C) is shown in Figure 14.4. As shown in Figure 14.4, the yield strength is strongly dependent on the applied strain rate, and Young's modulus is weakly dependent on the applied strain rate. The significant amount of nonlinear recovery during the unloading is also noteworthy.

The influence of temperature on the uniaxial compressive behavior of the annealed material (100 kGy, 110°C) is shown in Figure 14.5. It shows that the yield stress and flow behavior is highly dependent on the applied temperature.

To evaluate the performance of different constitutive models it is important to perform not only uniaxial tension and compression experiments, but also multiaxial testing. It is known that some constitutive models can be made to agree well with one simple test but are not good predictors of general multiaxial deformation histories. In this work, we have probed the multiaxial response by using a set of experiments performed on miniaturized disc specimens, measuring 6.4 mm in diameter and 0.5 mm in thickness. The testing was performed using a closed-loop servohydraulic test system. The specimens were tested by indentation with a custom-built, hemispherical punch head at a constant displacement rate of 0.5 mm/minute at room temperature, following ASTM 2183-02 (ASTM F 2183-02 2002). A schematic of the experimental setup is shown in Figure 14.6 and the resulting force-displacement results are shown

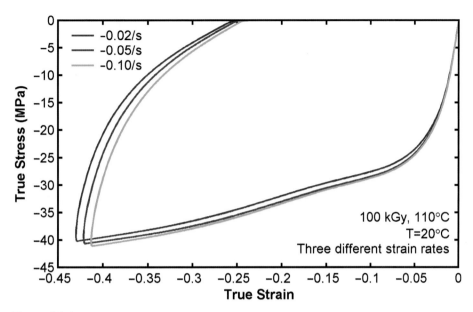

Figure 14.4
Influence of different strain rates on the cyclic compression behavior of GUR 1050 (100 kGy, 110°C).

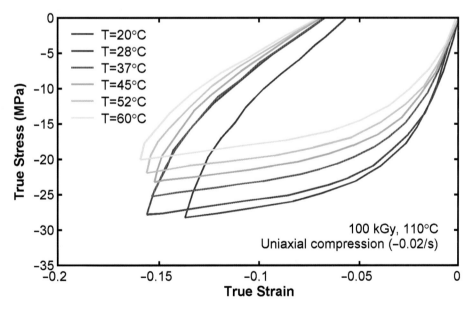

Figure 14.5
Influence of temperature on the compressive stress–strain response of GUR 1050 (100 kGy, 110°C).

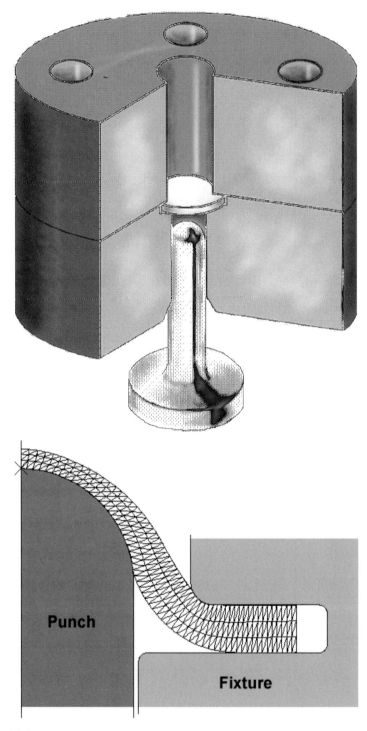

Figure 14.6

Schematic representation of the small punch set-up.

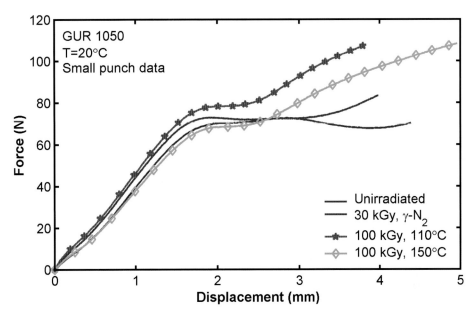

Figure 14.7
Representative experimental equibiaxial small punch data for the four UHMWPE materials.
The punch test was performed at a displacement rate of 0.5 mm/min.

in Figure 14.7. The punch force response is initially almost linear with the applied punch displacement, then reaches a plateau, and then finally starts to increase until final failure. It is also interesting to note that in the small punch tests the highly crosslinked materials fail at higher force levels than the conventional materials, which differs from their characteristic behavior in uniaxial tension (Figure 14.2).

Material Models for UHMWPE

UHMWPE, as well as other thermoplastics, exhibit a complicated nonlinear response when subjected to external loads. Their behavior, as demonstrated previously, is characterized by initial linear viscoelasticity at small deformations, followed by distributed yielding, viscoplastic flow, and material stiffening at large deformations until ultimate failure occurs. The response is further complicated by a dependence on strain rate and temperature. It is clear that higher deformation rates and lower temperatures increase the stiffness of the material.

It is also clear that thermoplastics are very different and typically exhibit a broader range of behavior in comparison with other structural materials, such as metals. The observed behavior is a manifestation of the different microstructures of the two types of materials and the different micromechanisms controlling the deformation resistance. It is therefore not surprising that different material models should be used when simulating UHMWPE compared with metals.

Specifically, and as will be shown later, traditional material models for metals (e.g., the J_2-plasticity model [Lubliner 1998]), although convenient and often familiar to the simulation engineer, should be used with great caution. They are rarely, if ever, a good choice for thermoplastics such as UHMWPE. Recall that, as shown in Figures 14.2 and 14.3, that the magnitudes of the true stress in tension and compression for a given magnitude of applied strain are often very different for UHMWPE. However, stress predictions based on the J_2-plasticity model are always symmetrical in the strain, clearly in disagreement with the experimental data. As will be discussed later, there are models that can more accurately capture the difference between tensile and compressive behavior of polymers.

Several candidate material models are available to predict the behavior of UHMWPE. Because the models have varying degrees of complexity, computational expense, and difficulty in extracting the material parameters from experimental data, it is a good idea to use the simplest material model that captures the necessary material characteristics for the application and situation at hand. Unfortunately, it is often difficult to determine the required conditions on the material model. Hence, it is recommended that a more advanced model be used in order to ensure accuracy and reliability of the predicted data. At a later stage, a less advanced model can be attempted if the computational expense is too great. At that time the accuracy of the different model predictions can also be tested and validated.

Given their obvious advantages, development of advanced constitutive models for UHMWPE and other thermoplastics is an active area of research that is continuously evolving and improving. In the last few years, models attempting to predict the majority of the experimentally observed characteristics of thermoplastics have been developed (Arruda and Boyce 1995, Bergström, Rimnac, and Kurtz 2003, Bergström et al. 2002a, Bergström et al. 2002b, Hasan and Boyce 1995, Khan and Zhang 2001, Kletschkowski, Schomburg, and Bertram 2002). Some of the modeling approaches are summarized later in the chapter. A number of traditional models can also be used to predict different aspects of the UHMWPE behavior. These models are often easier to use but have a limited domain of applicability. The three main models of this category are linear elasticity, hyperelasticity, and linear viscoelasticity. These models also have the added benefit of being directly available in all commercial FE packages.

The next few sections present these traditional models and how they apply to UHMWPE. Then, as an example of a more advanced material model, the Hybrid model (Bergström, Rimnac, and Kurtz 2003, Bergström et al. 2002a, Bergström et al. 2002b) is presented.

Linear Elasticity

Linear elasticity is the most basic of all material models. Only two material parameters need to be determined experimentally: Young's modulus (E) and Poisson's ratio (ν). Young's modulus can be obtained directly from uniaxial tension or compression experiments; typical values (Kurtz et al. 2002) for a few select UHMWPEs at room temperature are presented in Table 14.3.

Table 14.3

Representative Values of Young's Modulus at Room Temperature for a Few Different Types of UHMWPE[a]

Materials	Young's modulus (MPa)
Unirradiated	830
30 kGy γ-N$_2$	930
100 kGy γ 110°C	990
100 kGy γ 150°C	780

[a]Kurtz, et al. 2002.

   Poisson's ratio can be determined by measuring the transverse strain during uniaxial tension or compression experiments. Because of the small magnitude of the transverse strain it is often difficult to accurately determine Poisson's ratio. Instead, it is often sufficient to assume a value for Poisson's ratio of about 0.4. Note that unless the UHMWPE component is highly confined, Poisson's ratio has only very weak influence on the predicted material response.

   Only in special circumstances can UHMWPE be represented with a high degree of accuracy using linear elasticity (Figure 14.8). As shown in Figure 14.8, linear elasticity is a reasonable model only if the effective (von Mises) strains in the material are less than approximately 1%. Although a linear elastic representation of UHMWPE is accurate only for small strains, such a model can be

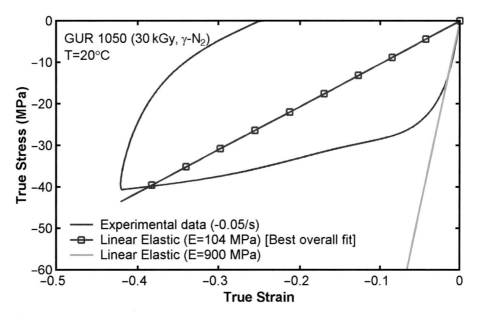

Figure 14.8

Comparison between experimental data for UHMWPE (GUR 1050, 30 kGy, γ-N$_2$) and predictions from linear elasticity.

of value if the UHMWPE component is part of a larger system and the response of the UHMWPE component is not the focus of the study and has little influence on overall response.

Hyperelasticity

A natural extension of linear elasticity is hyperelasticity (Ogden 1997). Hyperelasticity is a collective term for a family of models that all have a strain energy density that depends only on the currently applied deformation state (and not on the history of deformations). This class of material models is characterized by a nonlinear elastic response, and does not capture yielding, viscoplasticity, or time dependence. Strain energy density is the energy that is stored in the material as it is deformed, and is typically represented either in terms of invariants of the deformation gradient (F): $\hat{I}_1$, $\hat{I}_2$, and J, where

$$\hat{I}_1 = \hat{\lambda}_1^2 + \hat{\lambda}_2^2 + \hat{\lambda}_3^2, \tag{14.1}$$

$$\hat{I}_2 = \hat{\lambda}_1^{-2} + \hat{\lambda}_2^{-2} + \hat{\lambda}_3^{-2}, \tag{14.2}$$

$$J = det\ [F] = \lambda_1 \lambda_2 \lambda_3, \tag{14.3}$$

or expressed directly in terms of the distortional principal stretches: $\hat{\lambda}_1$, $\hat{\lambda}_2$, and $\hat{\lambda}_3$. The distortional stretches can be obtained from the applied principal stretches by $\hat{\lambda}_i = J^{-1/3}\lambda_i$, $I = 1, 2, 3$. The two main types of hyperelastic models are the polynomial model and the Ogden model. In the polynomial model the strain energy density is given by:

$$W = \sum_{i+j=1}^{N} C_{ij}\left(\hat{I}_1 - 3\right)^i\left(\hat{I}_2 - 3\right)^j + \sum_{i=1}^{N} D_i(J - 1)^{2i}, \tag{14.4}$$

and in the Ogden model the strain energy density is given by:

$$W = \sum_{i=1}^{N} C_i\left[\hat{\lambda}_1^{\alpha_i} + \hat{\lambda}_2^{\alpha_i} + \hat{\lambda}_3^{\alpha_i} - 3\right] + \sum_{i=1}^{N} D_i[J - 1]^{2i}, \tag{14.5}$$

where C_i and D_i are scalar material parameters. Commercial FE packages typically contain a number of other hyperelastic representations as well.

Hyperelastic models are often used to represent the behavior of crosslinked elastomers, where the viscoelastic response can sometimes be neglected compared with the nonlinear elastic response. Because UHMWPE behaves differently than do elastomers, there are only a few specific cases when a hyperelastic representation is appropriate for UHMWPE simulations. One such case is when the loading is purely monotonic and at one single loading rate. Under these conditions it is not possible to distinguish between nonlinear elastic and viscoplastic behavior, and a hyperelastic representation might be considered. Note that if a hyperelastic model is used in an attempt to capture the stress–strain response that is observed at large strains, including yielding, then there is a significant risk that the model will not be unconditionally stable or Drucker-stable (Lubliner 1998). A lack of Drucker-stability implies that even if a hyperelastic model fits the uniaxial experimental data, there is a risk that predictions of multiaxial deformation states will be in error. It is therefore often safer to use a more sophisticated model cable of capturing yielding and flow behavior in a more robust and accurate way. Figure 14.9 shows the best possible fit of an Ogden 1-term hyperelastic model to the experimental data for GUR 1050 (30 kGy, γ-N_2). From Figure 14.9 it is clear that this model is not able to reproduce the observed experimental data.

Linear Viscoelasticity

Linear viscoelasticity is an extension of linear elasticity and hyperelasticity that enables predictions of time dependence and viscoelastic flow. Linear viscoelasticity has been extensively studied both mathematically (Christensen 2003) and experimentally (Ward and Hadley 1993), and can be very useful when applied under the appropriate conditions. Linear viscoelasticity models are available in all major commercial FE packages and are relatively easy to use. The basic foundation of linear viscoelasticity theory is the Boltzmann's superposition principle, which states, "Every loading step makes an independent contribution to the final state."

   This idea can be used to formulate an integral representation of linear viscoelasticity. The strategy is to perform a thought experiment in which a step function in strain is applied, $\varepsilon(t)=\varepsilon_0 \cdot H(t)$, where $H(t)$ is the Heaviside step

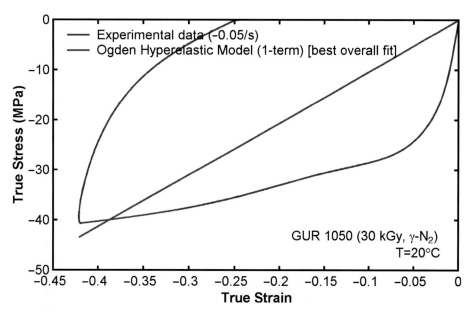

Figure 14.9

Uniaxial compression prediction using the Ogden hyperelasticity model.

function[1], and the stress response, $\sigma(t)$, is measured. Then a stress relaxation modulus can be defined by $E(t) = \sigma(t)/\varepsilon_0$. Note that ε_0 does not have to be infinitesimal because of the assumed superposition principle. To develop a model capable of predicting the stress response from an arbitrary strain history, one begins by decomposing the strain history into a sum of infinitesimal strain increments, $\Delta\varepsilon_i$ at τ_i:

$$\varepsilon(t) = \sum_{i=1}^{N} \Delta\varepsilon_i H(t - \tau_i), \tag{14.6}$$

The stress response can then be written as follows.

$$\sigma(t) = \sum_{i=1}^{N} \Delta\varepsilon_i E(t - \tau_i). \tag{14.7}$$

In the limit as the number of strain increments goes to infinity, the stress response becomes

$$\sigma(t) = \int_{-\infty}^{t} E(t - \tau)d\varepsilon(t) = \int_{-\infty}^{t} E(t - \tau)\frac{d\varepsilon(t)}{d\tau}d\tau. \tag{14.8}$$

This equation can be generalized to a three-dimensional deformation state for an isotropic material as follows.

$$\mathbf{T}(t) = \int_{0}^{t} 2G(t - \tau)\dot{e}d\tau + \mathbf{I}\int_{0}^{t} K(t - \tau)\dot{\phi}d\tau, \tag{14.9}$$

where $G(t)$ is the stress relaxation shear modulus, $\dot{e}$ the rate of change of deviatoric strains, $K(t)$ the stress relaxation bulk modulus, and $\dot{\phi}$ the rate of change of volumetric strains. Only two relaxation moduli need to be determined to predict any arbitrary deformation. The relaxation moduli can be determined from stress relaxation tests and are typically specified in FE packages as a power series of exponential functions (Prony series):

$$G(t) = G_\infty + \sum_{i=1}^{N} G_i e^{-t/\tau_i}, \tag{14.10}$$

where, G_∞, G_i, τ_i are material parameters.

The theory behind linear viscoelasticity is simple and appealing. It is important to realize, however, that the applicability of the model for UHMWPE is restricted to strains below the yield strain. One example comparing predictions based on linear viscoelasticity for experimental data for UHMWPE past yield is shown in Figures 14.10 and 14.11. Figure 14.10 shows the best predictive

---

[1]The Heaviside step function $H(t)$ is a function which takes the value 1 for $t > 0$, and the value 0 for $t = 0$.

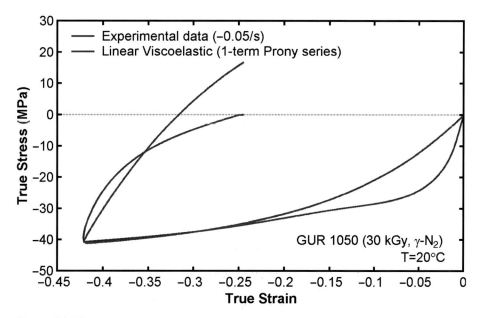

Figure 14.10
Comparison between linear viscoelasticity (one-term Prony series) and experimental data for GUR 1050 (30 kGy, γ-N₂).

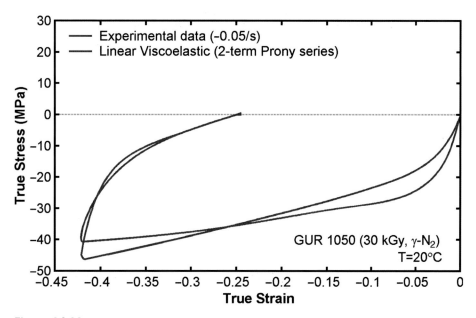

Figure 14.11
Comparison between experimental data for GUR 1050 (30 kGy, γ-N₂) and predictions from linear viscoelasticity theory with a two-term Prony series.

response of a linear viscoelasticity model based on a one-term Prony series, and Figure 14.11 shows the best response of a two-term Prony series viscoelasticity model. The figures show that adding terms to the Prony series typically improve the predictive response.

Although the predictions in Figure 14.11 are relatively good, these results are deceiving. The danger of using linear viscoelasticity for strains past yielding is illustrated in Figure 14.12. It shows that although linear viscoelasticity theory can be made to fit one experimental tension or compression test at small to moderate strains, it gives very poor predictions even at slightly different deformation rates, in serious contradiction to the experimental data.

Another interesting aspect of linear viscoelasticity is that it can be extended to enable predictions at different temperatures. The basis for this approach is based on a time-temperature superposition principle (Ward and Hadley 1993). This approach has been shown to work well in a restricted temperature range, but it does not change the requirement of infinitesimal strains.

Isotropic J_2-Plasticity

An example of a material model based on the physics of material behavior is classical metals plasticity theory. This theory, often referred to as J_2-flow theory, is based on a Mises yield surface with an associated flow rule, followed by rate-independent isotropic hardening (Khan and Huang 1995). Physically, plastic flow in metals is a result of dislocation motion, a mechanism known to be driven by shear stresses and to be insensitive to hydrostatic pressure.

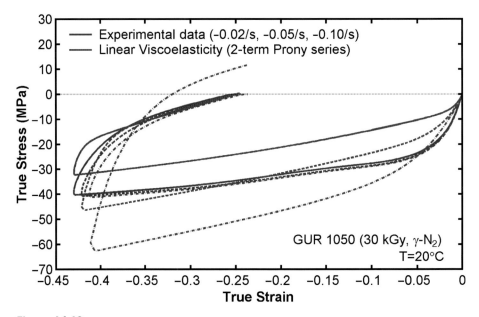

Figure 14.12

Comparison between experimental data for GUR 1050 (30 kGy, γ-N₂) and predictions from linear viscoelasticity theory with a two-term Prony series.

A consequence of the latter mechanism is that metals are observed to be nearly incompressible during plastic deformation.

For materials in which the basic physics of deformation is different from that described earlier, the J_2-plasticity model often performs quite poorly when applied to complex loading conditions. For example, material systems that develop texture or other forms of anisotropy during deformation, or materials that behave differently in tension and in compression are poor candidates for a J_2-plasticity model. Because the mechanisms governing plastic deformation in UHMWPE are quite different from those in metals, a more robust material model may be appropriate.

The deformation resistance of amorphous polymers in the glassy regime is determined by a combination of thermally activated segmental rearrangements (Argon 1973) (primarily responsible for the plastic behavior at small to moderate strains) and orientation-induced strain hardening caused by polymer chain stretching and alignment (primarily responsible for the deformation resistance at large strains [Haward 1973]). In addition, for semicrystalline polymers, such as UHMWPE, mechanisms associated with crystallographic slip, twinning, martensitic transformations, and crystallite/lamellar rotation also play a role (Lin and Argon 1994).

Finding the material parameters for the J_2-plasticity theory is straightforward and can be obtained from a simple uniaxial tension experiment. The material model, in fact, is a piecewise linear model (Figure 14.13) in which the material parameters specify the vertices of the stress–strain curve. The J_2-plasticity model can be made to fit monotonic, constant strain-rate, constant temperature test data well (Figure 14.14). Two of the main limitations of the theory are

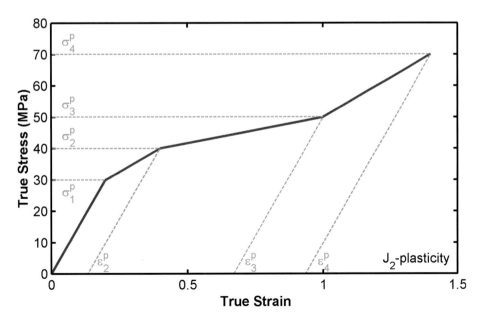

Figure 14.13

Definition of material parameters used in the J_2-plasticity theory.

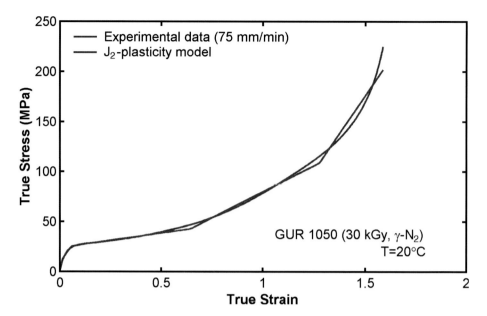

Figure 14.14

Comparison between experimental uniaxial tension data for the four UHMWPE materials and predictions from the J_2-plasticity model.

1) that it always predicts a linear unloading behavior, and 2) that it is incapable of considering the influence of different deformation rates. As shown in Figure 14.15, this makes the model inappropriate for many loading situations. Despite these limitations, the J_2-plasticity model has been the most widely used approach in the orthopedic research community for simulating the behavior of UHMWPE. As shown here, however, the J_2-plasticity model is not an accurate general tool for predicting the large-deformation-to-failure behavior of UHMWPE. In addition, the J_2-plasticity model does not accurately predict cyclic loading of UHMWPE. These are serious limitations because UHMWPE joint components undergo large deformations at the articulating surface while being subjected to cyclically applied loads.

To address these limitations, a new constitutive model was developed for conventional and highly crosslinked UHMWPEs (Bergström, Rimnac, and Kurtz 2003). This model, which is inspired by the physical micromechanisms governing the deformation resistance of polymeric materials, is an extension of specialized constitutive theories for glassy polymers that have been developed during the last 10 years, is discussed later.

The Hybrid Model

A number of more advanced and general models for predicting the yielding, viscoplastic flow, time-dependence, and large strain behavior of UHMWPE and other thermoplastics have recently been developed (Bergström, Rimnac,

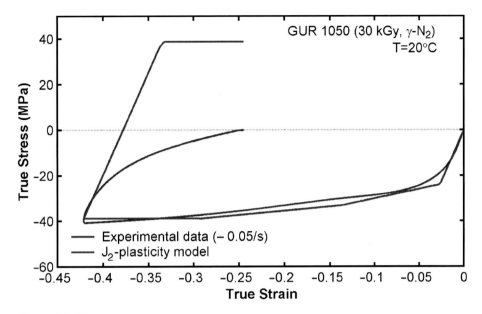

Figure 14.15

Comparison between experimental cyclic compression data and predictions from the J_2-plasticity model.

and Kurtz 2003, Bergström et al. 2002a, Bergström et al. 2002b). In this section one of these models, the hybrid model (HM)—specifically developed for UHMWPE—is discussed.

The HM is a constitutive model aimed at predicting the large strain time-dependent behavior of both crosslinked and uncrosslinked UHMWPE. The kinematic framework used in the HM is based on a decomposition of the applied deformation gradient into elastic and viscoplastic components: $F = F^e F^p$ (Figure 14.16). The spring and dashpot representation shown in Figure 14.16A is a one-dimensional embodiment of the model framework used to capture the viscoplastic flow characteristics. With the exception of the top spring (E), all spring and dashpot elements are highly nonlinear (described in detail later). Figure 14.16B depicts a map of the decomposition of a given material deformation state. This decomposition specifies how the three-dimensionality of the deformation gradients and stress tensors are connected and evolve during an applied deformation history.

In the HM, the deformation state is decomposed into elastic, backstress, and viscoplastic components. The backstress network incorporates time-dependent viscoplasticity to capture the nonlinear unloading response. The rationale for this decomposition is based on a presumption of continuous interaction between the amorphous and crystalline UHMWPE domains enhanced by the intrinsic entanglements arising from the material's very high molecular weight. Additional interaction between the amorphous and crystalline domains may also arise from covalent crosslinks between the polymer chains. At large deformations, however, the underlying molecular deformation resistance, the

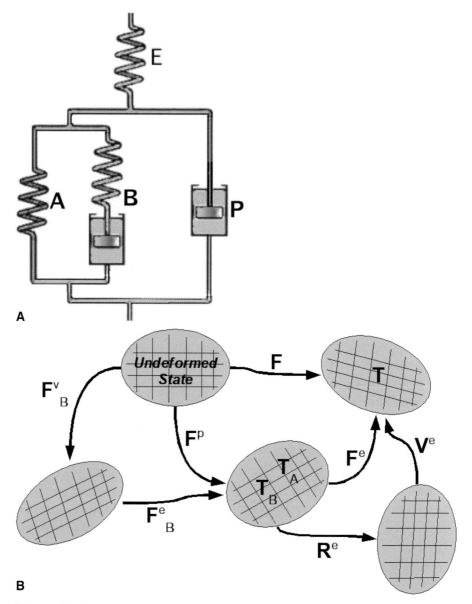

Figure 14.16
(**A**) Rheological representation of the augmented HM. (**B**) Deformation map showing the kinematics and stress tensors used in the augmented HM. These figures illustrate how the model represents the viscoplastic flow, and how the deformation state is generalized into three dimensions.

backstress network of molecular chains, has the ability to undergo viscoplastic flow. This flow behavior is caused by the absence of an isotropic crosslinked microstate in the material, which creates both regions with highly stretched molecular chains and regions that are less stretched. The flow behavior is a function of the highly deformed material state and the interaction between the amorphous and crystalline domains, and can be accurately captured using an energy activation representation. The kinematics of the viscoplastic flow of the backstress network is captured by decomposing the deformation gradient acting on part B of the backstress network (Figure 14.16A) into elastic and viscoelastic components: $F^p = F^e_B F^v_B$.

The Cauchy stress in the system is given by the isotropic linear elastic relationship:

$$\mathbf{T} = \frac{1}{J^e}\left(2\mu_e\,\mathbf{E}^e + \lambda_e\mathrm{tr}\left[\mathbf{E}^e\right]\mathbf{I}\right), \tag{14.11}$$

where μ_e and λ_e are Lame's constants, which can be obtained from Young's modulus and Poisson's ratio by $\mu_e = E_e/(2\,(1 + v_e))$ and $\lambda_e = E_e v_e/((1 + v_e)(1 - 2v_e))$, $J^e = \det[F^e]$ is the relative volume change of the elastic deformation, F^e is the deformation gradient, $\mathbf{E}^e = \ln[\mathbf{V}^e]$ is the logarithmic true strain, and $\mathbf{V}^e$ is the left stretch tensor (Gurtin 1981), which can be obtained from the polar decomposition of F^e.

The stress acting on the equilibrium portion of the backstress network is given by the following expression:

$$\mathbf{T}_A = \frac{1}{1 + q_A}\left[\mathbf{T}_{8chain}\left(\mathbf{F}^p;\mu_A,\lambda^{lock}_A,\kappa_A\right) + q_A\mathbf{T}_{I_2}\left(\mathbf{F}^p;\mu_A\right)\right], \tag{14.12}$$

$$\mathbf{T}_{I2} = \mu_A\left[I^{\mathbf{p}^*}_1\mathbf{B}^{\mathbf{p}^*} - \frac{2I^{\mathbf{p}^*}_2}{3}\mathbf{1} - \left(\mathbf{B}^{\mathbf{p}^*}\right)^2\right], \tag{14.13}$$

where $\mathbf{T}_A$ is a tensor-valued function of the viscoplastic deformation gradient, F^p, and the material parameters $\{\mu_A, \lambda^{lock}_A, \kappa_A, q_A\}$, where μ_A is the shear modulus, λ_A^{lock} is the locking stretch, κ_A is the bulk modulus, and q_A is a material parameter specifying the relative magnitudes of $\mathbf{T}_{8chain}$ and $\mathbf{T}_{I2}$, and $\mathbf{B}^{p^*}$ is the left Cauchy-Green deformation tensor. This hyperelastic stress representation is based on the eight-chain model (Arruda and Boyce 1993), and a term containing I_2 dependence of the strain energy density. The I_2 dependence is introduced by the crystalline domains and is manifested by the asymmetry in the response between tension and compression (Bergström, Rimnac, and Kurtz 2003).

The stress driving the viscoplastic flow of the backstress network is obtained from the same hyperelastic representation that was used to calculate the backstress, and has a similar framework as used in the Bergström-Boyce representation of crosslinked polymers at high temperatures (Bergström and Boyce 1998, 2000):

$$\mathbf{T}_B = s_B \cdot \mathbf{T}_A\left(\mathbf{F}^e_B\right), \tag{14.14}$$

where s_B is a dimensionless material parameter specifying the relative stiffness of the backstress network. At small deformations, the stiffness of the backstress network is constant and the material response is linear elastic. At larger applied deformations, viscoplastic flow caused by molecular chain sliding is initiated. With increasing viscoplastic flow, the crystalline domains become distorted and provide additional molecular material to the backstress network. This is manifested by an initial reduction in the effective stiffness of the backstress network with imposed viscoplastic deformation and is captured in the model by allowing the parameter, s_B, to evolve with the plastic deformation. The parameter, s_B, evolves with imposed plastic deformation to capture the distributed yielding:

$$\dot{s}_B = -p_B \cdot \left(s_B - s_{Bf}\right) \cdot \dot{\gamma}_C ,$$ (14.15)

where p_B is a material parameter specifying the transition rate of the distributed yielding event, s_{Bf} is the final value of s_B reached at fully developed plastic flow, and $\dot{\gamma}_C$ is the magnitude of the viscoplastic flow rate:

$$\dot{\gamma}_B^v = \dot{\gamma}_0 \left(\frac{\tau_B}{\tau_B^{base}}\right)^{m_B} .$$ (14.16)

The velocity gradient of the viscoelastic flow of the backstress network is given by the following equation

$$\mathbf{L}_B^v = \dot{\gamma}_B^v \left(\mathbf{F}_B^e\right)^{-1} \frac{dev[\mathbf{T}_B]}{\tau_B} \mathbf{F}_B^e ,$$ (14.17)

where $\dot{\gamma}_B^v$ is the rate of viscoplastic flow of the time-dependent network B, $\tau_B = \left\|dev[\mathbf{T}_B]\right\|_F$, τ_B^{base} and m_B are material parameters, and $\dot{\gamma}_0$ is a constant coefficient with a value of $1/s$.

The yielding and plastic flow of the material is captured using the tensorial flow rule (Bergström et al. 2002b, Bergström, Rimnac, and Kurtz 2003):

$$\mathbf{L}^p = \dot{\gamma}_C \left[\left(\mathbf{R}^e\right)^T \frac{dev[\mathbf{T}_C]}{\tau_C} \mathbf{R}^e\right] ,$$ (14.18)

where $\mathbf{L}^p = \dot{\mathbf{F}}^p \mathbf{F}^{p-1}$, $\mathbf{T}_C = \mathbf{T} - [\mathbf{F}^e(\mathbf{T}_A + \mathbf{T}_B)\mathbf{F}^{eT}] / J^e$ is the stress acting on the relaxed configuration convected to the current configuration, $\tau_C = \left\|dev[\mathbf{T}_C]\right\|_F$ is the effective shear stress (calculated using the Frobenius norm) driving the viscoplastic flow,

$$\dot{\gamma}_C = \dot{\gamma}_0 \cdot \left(\tau_C / \tau_C^{base}\right)^{m_C}$$ (14.19)

is the magnitude of the viscoplastic flow, τ_C^{base} and m_C are material parameters, and $\dot{\gamma}_0$ a constant coefficient with a value of $1/s$.

In total, the HM contains 13 material parameters: 2 small strain elastic constants (E_e, v_e), 4 hyperelastic constants for the backstress network (μ_A, λ_A^{lock}, κ_A, q_A), 5 flow constants of the backstress network (s_{Bi}, s_{Bf}, p_B, τ_B^{base}, m_B), and 2 yield and viscoplastic flow parameters (τ_C^{base}, m_C). These parameters can readily be determined from a few select experiments.

The capability of the HM to predict the response of UHMWPE was evaluated by comparing the model predictions with the aforementioned experimental data. The first step in this effort was to calibrate the HM to the uniaxial tensile and cyclic experimental data, for the material of interest. The first step in this procedure (Bergström, Rimnac, and Kurtz 2003), the bootstrapping step, was to find an initial estimation of the material parameters. This set of material parameters can be obtained, for example, from previous known values for a similar material (Bergström et al. 2002a). Then, a specialized computer program based on the Nelder-Mead simplex minimization algorithm was used to iteratively improve the correlation between the predicted data sets and the experimental data. The quality of a theoretical prediction, and therefore of the chosen material parameters, was evaluated by calculating the coefficient of determination, r^2. The reported material parameters for each material are from the set having the highest r^2-value. The resulting material parameters obtained from this procedure are listed in Table 14.4. It is clear that between the four types of UHMWPE materials, only four material parameters were different, the remaining parameters

Table 14.4
Hybrid Model Material Parameters for the Three Different Types of GUR 1050[a]

Material Parameter	30 kGy γ-N$_2$	100 kGy γ 110°C	100 kGy γ 150°C
E_e (MPa)	2020	2009	1270
v_e	0.46	0.46	0.46
μ_A (MPa)	**8.22**	**10.15**	**8.14**
λ_A^{lock}	**4.40**	**2.80**	**2.52**
κ_A (MPa)	2000	2000	2000
q_A	0.20	0.20	0.20
s_{Bi}	40.0	40.0	40.0
s_{Bf}	10.0	10.0	10.0
p_B	27.0	27.0	27.0
$\tau_B{}^{base}$ (MPa)	**25.0**	**26.2**	**20.7**
m_B	9.50	9.50	9.50
$\tau_C{}^{base}$ (MPa)	8.00	8.00	8.00
m_C	3.30	3.30	3.30

[a]E_e, is the Young's modulus; v_e, Poisson's ratio; v_A, shear modulus of network A; λ_A^{lock}, locking stretch of network A; κ_A, bulk modulus of network A; q_A, parameter specifying the asymmetry between tension and compression; s_{Bi}, parameter that controls the initial flow resistance; s_{Bf}, parameter that controls the final flow resistance; p_B, parameter that controls the distributed yielding; τ_B^{base}, parameter that controls the yield strength of network B; m_B, parameter controlling the rate dependence of network B; τ_C^{base}, parameter that controls the yield strength of network C; and m_C, parameter that controls the rate dependence of network B. Parameters that are unique for each material are written in bold text.

do not vary between different types of UHMWPE, and are in that sense independent of the thermal and radiation treatment.

The predicted results for the uniaxial tension, and the tension-compression data are shown in Figures 14.17 and 14.18, respectively. It is clear that the HM accurately captures both the large strain time-dependent response and the intermediate strain cyclic response.

After the optimal set of material parameters was found, the same parameters were then used to simulate the small punch test. This validation simulation was performed to check the capability of the HM to predict a multiaxial deformation history. It is well known that many constitutive models can predict uniaxial deformation histories relatively well, but that it is significantly more difficult to accurately predict multiaxial deformation states. It has been shown, for example, that the J_2-plasticity model can accurately predict monotonic uniaxial tension or compression data for UHMWPE, but is very poor at predicting cyclic or multiaxial deformation states (Bergström, Rimnac, and Kurtz 2003).

The small punch validation simulations were performed using the ABAQUS (HKS Inc., RI) FE package. The simulations used an axisymmetric representation with 360 quadratic triangular elements (CAX8H) to represent the small punch geometry (Figure 14.6B). In the simulations the friction coefficient between the specimen and the punch, and between the specimen and the die, was taken as 0.1 (Bergström et al. 2002a). The validation simulation data (Figure 14.19) indicate good agreement the large strain multiaxial deformation history in the small punch test, suggesting that the HM model is fairly robust.

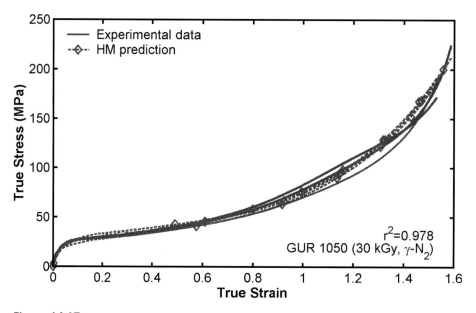

Figure 14.17

Comparison between experimental uniaxial compression data and predictions from the HM for GUR 1050 (30 kGy, γ-N₂). The three data sets are for true strain rates of 0.007/second, 0.018/second, and 0.035/second.

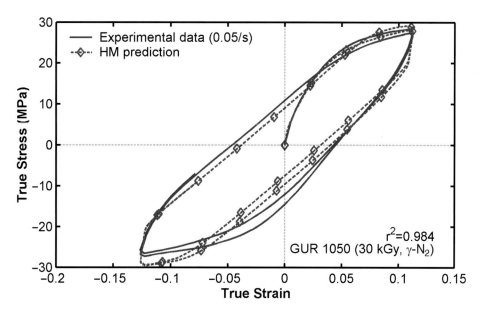

Figure 14.18

Comparison between experimental uniaxial cyclic tension and compression data and predictions from the HM for GUR 1050 (30 kGy, γ-N$_2$). The experimental data correspond to a true strain rate of 0.05/second.

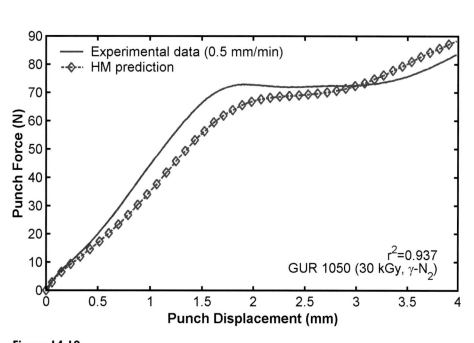

Figure 14.19

Comparison between experimental small punch data and predictions from the HM for GUR 1050 (30 kGy, γ-N$_2$). The experimental data correspond to a punch rate of 0.5 mm/minute.

Discussion

The mechanical response of UHMWPE at large deformations is very complex and encompasses nonlinear behavior during both loading and unloading. Initially, at small strains, the response is linear elastic. With increasing deformation, localized yielding is initiated at sites where the flow resistance is the lowest. The flow resistance then evolves and becomes more homogeneous in both the crystalline and the amorphous domains. Finally, at large deformations the imposed molecular chain stretching and alignment continually stiffens the response until ultimate failure is reached. To model these events is challenging, but necessary to develop a better understanding of the deformation behavior of UHMWPE and to predict the fatigue, fracture, and wear response of the material.

The current state of the art in mechanical modeling is based on FE simulations. FE modeling is a well-developed field of mechanical engineering that since the 1980s has reached a significant level of maturity. There are three types of input needed to simulate an UHMWPE component using an FE simulation: the component geometry, the applied loads and the boundary conditions, and the material model representation. Of these the material model representation is the most challenging, and was reviewed in this chapter. As discussed earlier, there are a number of different classes of material models that can be used, ranging from very simple but inaccurate, to very sophisticated and accurate.

The most advanced material model presently available for UHMWPE is the HM. This model focuses on creating a mathematical representation of the deformation resistance and flow characteristics for conventional and highly crosslinked UHMWPE at the molecular level. The physics of the deformation mechanisms establish the framework and equations necessary to model the behavior on the macroscale. As already mentioned, to use the constitutive model for a given material requires a calibration step where material-specific parameters are determined. A variety of numerical methods may be used to determine the material-specific parameters for a constitutive theory. In the previous section we employed a numerical optimization technique to identify the material parameters for the constitutive theory.

Of greater importance is how well the physics-inspired model framework represents the governing micromechanisms, and ultimately, how well the model can predict the behavior of a given material under different loading conditions than that for which the model was originally calibrated. The simulations of the small punch test demonstrate that the HM provides satisfactory and valid predictions of large-deformation multiaxial behavior of conventional and highly crosslinked UHMWPEs.

In summary, there are a number of different constitutive models that can be used to predict different aspects of UHMWPE behavior. The most advanced of the currently available models is the HM, which has been shown to be able to predict the behavior of both conventional and highly crosslinked UHMWPE used in total joint replacements. The HM is currently limited to isothermal deformation histories, although research is ongoing to enable to arbitrary thermomechanical deformation states. In addition, fatigue, fracture, and wear are targets for current and future studies.

Acknowledgments

Special thanks to Steven Kurtz, Exponent, Inc., and Clare Rimnac, Case Western University, for the experimental test characterization, and for the collaborative efforts resulting in the development of the Hybrid Model. Thanks also to Anton Bowden, Exponent Inc., and Avram Edidin, Drexel University, for editorial assistance with this chapter. This research was supported by NIH R01 47192.

References

Argon A.S. 1973. A theory for the low temperature plastic deformation of glassy polymers. *Philos Mag* 28:839–865.

Arruda E.M., and M.C. Boyce. 1993. A three-dimensional constitutive model for the large stretch behavior of rubber elastic materials. *J Mech Phys Solids* 41:389–412.

Arruda E.M., and M.C. Boyce. 1995. Effects of strain rate, temperature and thermomechanical coupling on the finite strain deformation of glassy polymers. *Mech Mater* 19:193–212.

ASTM F 2183-02. 2002. Standard test method for small punch testing of ultra-high molecular weight polyethylene used in surgical implants. West Conshohocken, PA: American Society for Testing and Materials.

Bergström J.S, and M.C. Boyce. 1998. Constitutive modeling of the large strain time-dependent behavior of elastomers. *J Mech Phys Solids* 46:931–954.

Bergström J.S., and M.C. Boyce. 2000. Large strain time-dependent behavior of filled elastomers. *Mech Mater* 32:627–644.

Bergström J.S., S.M. Kurtz, C.M. Rimnac, and A.A. Edidin. 2002a. Constitutive modeling of crosslinked UHMWPE for joint replacements. *Transactions of the 48th Orthopedic Research Society* 59: paper no. 0050.

Bergström J.S., S.M. Kurtz, C.M. Rimnac, and A.A. Edidin. 2002b. Constitutive modeling of ultra-high molecular weight polyethylene under large-deformation and cyclic loading conditions. *Biomaterials* 23:2329–2343.

Bergström J.S., C.M. Rimnac, and S.M. Kurtz. 2003. Prediction of multiaxial behavior for conventional and highly crosslinked UHMWPE using a hybrid constitutive model. *Biomaterials* 24:1365–1380.

Christensen R.M. 2003. *Theory of viscoelasticity.* Mineola, NY: Dover Publications.

Gurtin M.E. 1981. *An introduction to continuum mechanics.* New York: Academic Press.

Hasan O.A., and M.C. Boyce. 1995. A constitutive model for the nonlinear viscoplastic behavior of glassy polymers. *Polym Eng Sci* 35:331–344.

Haward R.N. 1973. Physics of glassy polymers. Boston: Kluwer Academic Publishing.

Khan A.S., and S. Huang, 1995. *Continuum theory of plasticity.* New York: Wiley.

Khan A.S., and H. Zhang. 2001. Finite deformation of a polymer: Experiments and modeling. *Int J Plasticity* 17:1167–1188.

Kletschkowski T., U. Schomburg, and A. Bertram. 2002. Endochronic viscoplastic material models for filled PTFE. *Mech Mater* 34:795–808.

Kurtz S.M., M.L. Villarraga, M.P. Herr, et al. 2002. Thermomechanical behavior of virgin and highly crosslinked ultra-high molecular weight polyethylene used in total joint replacements. *Biomaterials* 23:3681–3697.

Lin L., and A.S. Argon. 1994. Review: Structure and plastic deformation of polyethylene. *J Mater Sci* 29:294–323.

Lubliner J. 1998. *Plasticity theory.* New York: Pearson Education POD.

Ogden R.W. 1997. *Non-linear elastic deformations.* Mineola, NY: Dover Publications.

Roark R.J., R.G. Budynas, and W.C. Young. 2001. *Formulas for stress and strain.* New York: McGraw-Hill.

Ward I.M., and D.W. Hadley. 1993. *An introduction to the mechanical properties of solid polymers.* New York: John Wiley & Sons.

Chapter 14. Reading Comprehension and Group Discussion Questions

14.1. Under what circumstances might an analyst consider using linear elasticity to simulate the material behavior of UHMWPE?

14.2. What are the main disadvantages of isotropic plasticity as a material model for UHMWPE?

14.3. For what types of mechanical behavior predictions is an advanced material model, such as the Hybrid Model, necessary?

14.4. How are advanced models, such as the Hybrid model, validated?

14.5. What are the current limitations of the Hybrid model as described in this chapter?

Chapter 15

Compendium of Highly Crosslinked and Thermally Treated UHMWPEs

Introduction

Radiation crosslinking and thermal treatment of ultra-high molecular weight polyethylene (UHMWPE) has aroused intense scientific and commercial interest within the orthopedic community since the late 1990s. The proliferation of crosslinking technology into hip replacements, and more recently knee replacements, has resulted in six new proprietary UHMWPEs, with trade names like Crossfire™, Longevity™, and Marathon™. Most of these new UHMWPEs are available for hips, one material (Prolong™) is available exclusively for knees, and one material (DURASUL®) can be found in both hip and knee products (Figure 15.1). The new product nomenclature, as well as the different techniques used in the irradiation, thermal processing, and sterilization of these new materials, may be confusing even to investigators well accustomed to UHMWPE technology in joint replacement.

This chapter is a compendium of the processing details, packaging information, as well as the most recent research findings related specifically to the six clinically available highly crosslinked and thermally treated UHMWPEs shown in Figure 15.1. The general characteristics of highly crosslinked and thermally treated UHMWPE materials have already been reviewed in the context of alternative bearings for hip replacement, as well as in the context of knee replacements, in Chapters 6 and 8, respectively. This chapter assumes prior knowledge about the basics of crosslinking and thermal treatment technology, so it is recommended that Chapters 6 and 8 be reviewed prior to reading this chapter if the reader has not done so already.

Scientific and commercial developments related to radiation crosslinking and thermal treatment of UHMWPE are evolving at a rapid pace. At the time of my first review on this topic, written between 1997 and early 1998, highly

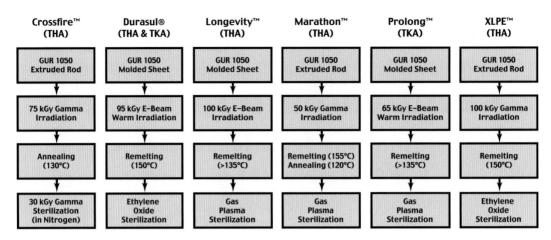

Figure 15.1

Comparison of starting stock materials and processing conditions used to produce contemporary thermally treated and highly crosslinked UHMWPE materials for TKA and THA.

crosslinked materials had not yet been clinically launched; that article was published in 1999 after the materials had been released for total hip arthroplasty (THA) (Kurtz et al. 1999). When I wrote my second review of this subject as a book chapter in collaboration with Orhun Muratoglu in the summer of 2001, highly crosslinked UHMWPEs had just been released for total knee arthroplasty (TKA) (2002). These previous works, along with a review by Gladius Lewis (2001), dealt primarily with the effects of crosslinking on the properties and *in vitro* wear performance of radiation crosslinked UHMWPE. A recent study by Collier and colleagues (2003) reviewed the physical and tensile properties of highly crosslinked UHMWPE materials for hip replacement before and after accelerated aging.

Because this chapter deals primarily with proprietary materials, the contents of this chapter were verified for accuracy with the individual manufacturers by the author. Some process details were not, however, provided or confirmed by the manufacturers. When manufacturers were reluctant to provide certain detailed information about their processes, we relied on previous published sources as the basis for the reported process conditions. A recent symposium on this topic, convened by the American Society and Testing Materials (ASTM) in November 2002, also provided a wealth of information about today's highly crosslinked UHMWPEs. The peer-reviewed proceedings of that symposium (Kurtz, Gsell, and Martell 2003), printed by ASTM as a special technical publication, were key references when compiling this chapter.

Honorable Mention

Two proprietary UHMWPE materials, Duration™ and Aeonian™, were not selected for inclusion among the six materials described in this compendium, but nonetheless deserve special mention for the sake of completeness on this topic. Duration was not included in the compendium because the material is thermally stabilized, but not highly crosslinked. Aeonian was not included in the compendium because, to my knowledge, this material is not distributed in the United States, and no data about its properties has been published in the peer-reviewed scientific literature.

Duration

In 1995, an oxidation-stabilized UHMWPE known as Duration was introduced to the orthopedics community by Howmedica (Rutherford, NJ). Duration is fabricated in a two-step process. In the first step, the UHMWPE is gamma irradiated in a package filled with inert gas; during the second step, the package containing the UHMWPE is heat treated at 37 to 50°C for up to 144 hours. The heat treatment step permits the free radicals to recombine in low oxygen environment. GUR 415 stabilized in this manner after a dose of 25 kGy has been reported with a gel content of 75%, as compared with 46% for air-irradiated UHMWPE, and 0% for virgin UHMWPE (Sun et al. 1996). Because Duration is thermally stabilized, but not highly crosslinked, this material is in many ways the precursor to many of the materials described in the compendium.

Wear testing by Essner and colleagues in a multidirectional hip simulator for up to 10 million cycles demonstrated that Duration exhibited 32% less volumetric wear than air-irradiated control UHMWPE (1997). However, in a recent radiostereometric study by Nivbrant and colleagues, the clinical wear rate of Duration was not significantly different than an air-irradiated control UHMWPE after 2 years of follow-up (2003).

Aeonian

Comparatively limited processing details of Aeonian (Kyocera, Japan) were reported by Muratoglu and Kurtz (2002). Aeonian starts as bar stock that has been crosslinked by 35 kGy of gamma radiation. Parts are machined from the crosslinked bar stock, packaged in nitrogen, and then terminally gamma sterilized (25–40 kGy). The total dosage for Aeonian thus ranges between 60 and 75 kGy. To the author's knowledge, the properties and clinical performance of this material have not yet been reported in the literature.

Crossfire

Development History and Overview

Crossfire, an annealed highly crosslinked UHMWPE, was developed by Scott Taylor and coworkers at Stryker Osteonics Corp. (Allendale, NJ). Crossfire was clinically introduced in the fall of 1998 for the Series II liner in the Omnifit acetabular cup design. After Osteonics acquired Howmedica in 1999, and

Table 15.1

Characteristics of Crossfire

Orthopedic manufacturer	Stryker Howmedica Osteonics, Inc. (Mahwah, NJ), www.howost.com
Joint applications	THA
Clinical introduction	1998
Starting stock material	GUR 1050 extruded rod
Irradiation crosslinking modality	Gamma
Irradiation temperature	Room temperature
Irradiation dosage (range)	75 kGy
Postirradiation thermal treatment process and duration	Annealed at 130°C for unknown duration
Sterilization modality	Gamma in nitrogen (N2-Vac), 30 kGy dose
Target total irradiation dosage	105 kGy
Detectable free radicals?	Yes
Minimum/maximum head size available	22/36 mm
Minimum thickness available	6 mm
Published in vitro properties/ performance data?	Yes
Published clinical performance data?	Yes, 2- to 3-year follow-up

formed Stryker Howmedica Osteonics, Crossfire was subsequently extended to the System 12 and Trident acetabular cup designs.

In the Crossfire process, extruded rod bar stock is irradiated with a nominal dose of 75 kGy and then annealed at 130°C (Kurtz et al. 2002–2003) (Figure 15.1). Acetabular components are then machined from the bar stock, barrier packaged in nitrogen, and then gamma sterilized with a nominal dose of 30 kGy. Consequently, components that have been through the Crossfire process have received a total dose of 105 kGy. An example of the N2-Vac packaging used for Crossfire is shown in Figure 15.2.

Properties and *in Vitro* Performance

Details about the *in vitro* test results with Crossfire have recently been published (Kurtz et al. 2002–2003). In comparison with conventional UHMWPE, Crossfire is associated with more than 90% improvement in median hip simulator wear rate across a broad range of implant designs. The crystallinity and

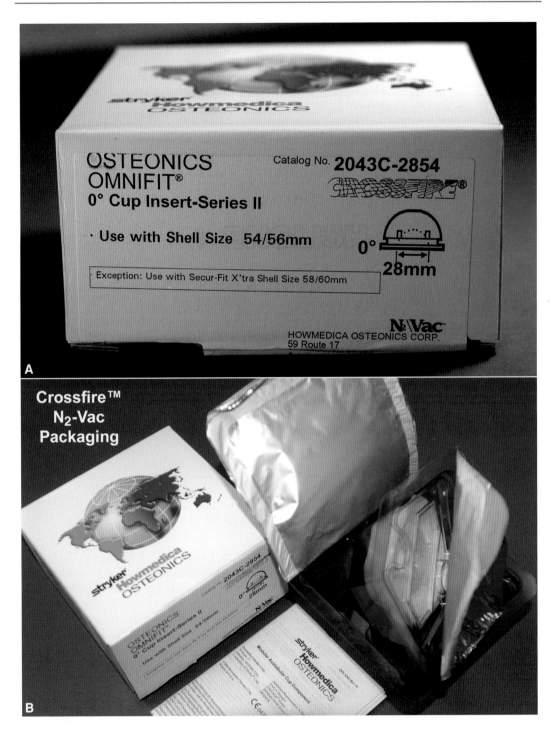

Figure 15.2
Packaging for Crossfire. (**A**) Outer packaging box. (**B**) N$_2$-Vac barrier packaging used for gamma sterilization of Crossfire.

microstructure, as well as the yield stress and ultimate strength of Crossfire, were reported to be comparable to conventional, barrier-packaged UHMWPE.

Because Crossfire is sterilized by gamma radiation, the material contains free radicals. However, the mechanical performance of the bearing was judged by the designers of Crossfire to be of greater importance than complete elimination of free radicals because of the use of barrier packaging, which minimizes oxidation during shelf storage before implantation.

Crossfire is not available for knee applications because of concerns raised during functional fatigue testing (Wang, Manley, and Serekian 2003). During patella testing, Wang and colleagues (2003) reported that components fabricated from Crossfire were observed to fracture through the support pegs.

Clinical Results

Follow-up data after 2 to 3 years for Crossfire in THA have been published at orthopedic meetings (Martell, Verner, and Incavo 2003, Nivbrant et al. 2003). After 2 to 3 years in a prospective, randomized study conducted by Martell and associates (2003), patients implanted with Crossfire showed significant (42–50%) reduction in wear when compared with the N_2-Vac control group. In an RSA study (Nivbrant et al. 2003), researchers reported 85% reduction in wear after 2 years of follow-up, as compared with patients in the control group in a consecutive series of patients. There was no significant difference in acetabular shell migration between patients implanted with Crossfire and conventional UHMWPE liners. Both clinical studies are ongoing, and future updates on their progress are expected.

DURASUL

Development History and Overview

DURASUL is based on warm irradiation with adiabatic melting (WIAM) technology developed by a team of researchers, including Orhun Muratoglu and William Harris at Massachusetts General Hospital. The WIAM process technology was licensed to Centerpulse Orthopedics, Inc. (formerly Sulzer, Austin, TX) and Zimmer, Inc. (see the entry for Longevity highly crosslinked UHMWPE, later). Centerpulse clinically introduced DURASUL in the Converge™ acetabular cup design in 1998 (Figure 15.3A). DURASUL was later introduced for TKA in the cruciate-retaining natural knee design (NK II) at the 2001 American Academy of Orthopedic Surgeons (AAOS). The DURASUL all-polyethylene patellar component was introduced in 2002 (Figure 15.3B).

In the WIAM process, the UHMWPE sheets or blocks are first machined into preforms (for acetabular liners) or slabs (for tibial inserts) and placed on a conveyor within an oven maintained at a temperature just below the melting temperature of the polymer (approximately 125°C). The warm UHMWPE pucks are exposed to a 10 MeV Rhodotron electron accelerator, which deposits

Table 15.2
Characteristics of DURASUL

Orthopedic manufacturer	Centerpulse, Inc. (Austin, TX), www.centerpulse.com
Joint applications	THA and TKA
Clinical introduction	1998 (THA), 2001 (TKA)
Starting stock material	GUR 1050 compression-molded sheet
Irradiation crosslinking modality	Electron beam
Irradiation temperature	~125°C, WIAM process
Irradiation dosage (range)	95 kGy
Postirradiation thermal treatment process and duration	Remelted at 150°C for 2 hours
Sterilization modality	Ethylene oxide
Target total irradiation dosage	95 kGy
Detectable free radicals?	No
Minimum/maximum head size available	22/44 mm (U.S.); 22/36 mm (Europe); 22/36 mm (Japan)
Minimum thickness available	5 mm
Published *in vitro* properties/performance data?	Yes
Published clinical performance data?	Yes, 2- to 3-year follow-up in THA; no published clinical data for TKA

the 95 kGy dose within seconds (Abt and Schneider 2003). The dosage rate is sufficiently high (~10 kGy/s) that the UHMWPE heats up, but not above the melt transition. Because the UHMWPE is about 50% amorphous at the start of the irradiation process, but partially melts to become nearly 100% amorphous by the end of the crosslinking process, DURASUL has a spatially varying crosslink density at a microscopic level. Following irradiation, the UHMWPE is maintained at 150°C for stabilization of free radicals (Figure 15.1). Components are then machined from the DURASUL material, enclosed in gas-permeable packaging, and sterilized by ethylene oxide gas (Abt and Schneider 2003, Muratoglu et al. 2001a).

Properties and *in Vitro* Performance

Because of its processing history, DURASUL has a biphasic crystalline microstructure that is manifested in two distinct thermal transitions during a

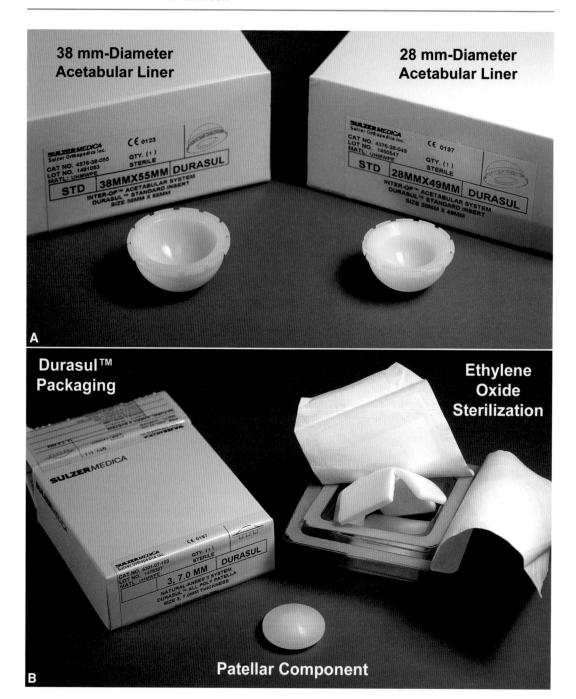

Figure 15.3
DURASUL components for hip and knee arthroplasty with packaging (Centerpulse Orthopedics, formerly Sulzer Orthopedics). (**A**) 28-mm and 38-mm diameter DURASUL acetabular liners. (**B**) DURASUL patellar component with packaging used for ethylene oxide sterilization.

differential scanning calorimetry experiment (Abt and Schneider 2003). Overall, DURASUL has a lower degree of crystallinity, elastic modulus, yield stress, and ultimate stress as compared with control UHMWPE. Nevertheless, the properties of DURASUL are maintained above the recommended guidelines of ASTM and ISO standards for conventional UHMWPE (Muratoglu et al. 2001a).

Muratoglu and colleagues (2001a) have verified the oxidative stability of WIAM UHMWPE. Electron spin resonance studies of DURASUL have reported that the material contains no detectable free radicals (Muratoglu et al. 2001a).

DURASUL has undergone extensive wear testing in hip and knee simulator studies, under both clean and abrasive lubrication conditions (Muratoglu et al. 2001b, 2002). Hip simulator studies have been conducted with DURASUL with femoral heads ranging from 22 to 46 mm in diameter for up to 27 million cycles (Muratoglu et al. 2001b). For all head sizes, DURASUL has exhibited significant reductions in wear in the hip simulator. Knee simulator studies also show a significant reduction of wear under normal and aggressive duty cycles (Muratoglu et al. 2002).

Clinical Results

Prospective studies are underway in the United States, Sweden, and Germany, and 2- to 3-year follow-up data has been reported for DURASUL in THA (Bragdon et al. 2002, Digas et al. 2003, Rabenseifner 2003, Manning et al. 2004). DURASUL has been reported to exhibit greater creep than conventional UHMWPE, which complicates the *in vivo* assessment of wear within 2 years, because wear and creep are indistinguishable radiographically. At 2 years, the magnitude of three-dimensional femoral head penetration is comparable between DURASUL and conventional UHMWPE, but the two-dimensional head penetration rate appears to be significantly lower for DURASUL (Bragdon et al. 2002, Digas et al. 2003). Longer follow-up from these studies, which are currently ongoing, will be helpful to confirm this finding.

Rabenseifner from Germany has also reported on histological findings around several DURASUL components that were revised for reasons other than wear (2003). In these case studies, the amount of wear debris particles was observed to be qualitatively lower around the DURASUL retrievals as compared with conventional control UHMWPE. In addition, based on histomorphometric analysis, the DURASUL did not appear to be stimulating a biological response. Clinical data is not yet available for DURASUL used in TKA.

Longevity

Development History and Overview

Longevity is based on WIAM technology patented by Massachusetts General Hospital. The WIAM technology was licensed to Zimmer, Inc. (Warsaw, IN).

Table 15.3
Characteristics of Longevity

Orthopedic manufacturer	Zimmer, Inc. (Warsaw, IN), www.zimmer.com
Joint applications	THA
Clinical introduction	1999
Starting stock material	GUR 1050 compression-molded sheet
Irradiation crosslinking modality	Electron beam
Irradiation temperature	~ 40°C
Irradiation dosage (range)	100 kGy (90–110 kGy target dose range)
Postirradiation thermal treatment process and duration	Remelted above the melt transition >135°C)
Sterilization modality	Gas plasma
Target total irradiation dosage	100 kGy
Detectable free radicals?	No
Minimum/maximum head size available	22/36 mm
Minimum thickness available	6 mm
Published *in vitro* properties/performance data?	Yes
Published clinical performance data?	Yes, 2- to 3-year follow-up

The Longevity process was developed by Zimmer and clinically introduced in the Trilogy™ acetabular cup design in 1999.

In the Longevity process, the UHMWPE bars are warmed, placed in a carrier on a conveyor, and are exposed to electron beam radiation, with a total dose of 100 kGy. The UHMWPE does not heat above the melt transition during the crosslinking. After irradiation, the UHMWPE is heated above the melt temperature (>135°C) for stabilization of free radicals. Components are then machined from the Longevity material, enclosed in gas-permeable packaging, and sterilized by gas plasma.

Properties and *in Vitro* Performance

The crystallinity, tensile properties, and fracture toughness of Longevity have been characterized before and after accelerated aging by Bhambri and coworkers (2003). In Bhambri's study, Longevity was also compared with conventional UHMWPE that was gamma sterilized in nitrogen. Bhambri and associates

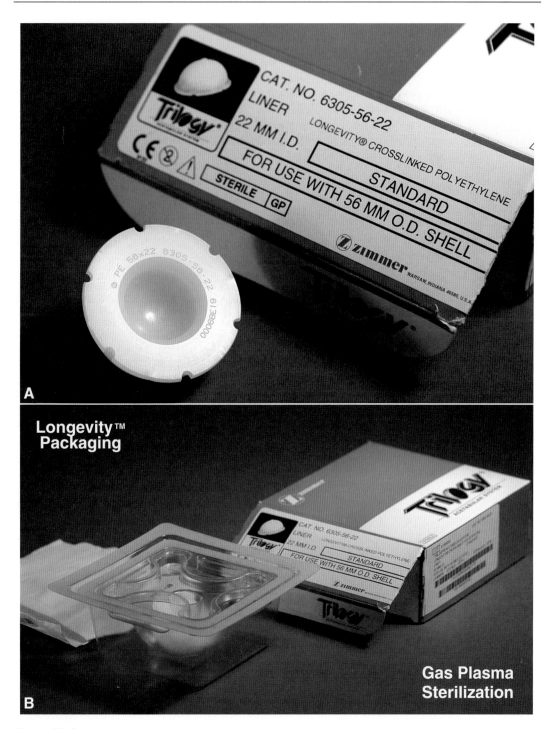

Figure 15.4
Longevity component for hip arthroplasty with packaging (Zimmer, Inc., Warsaw IN). (**A**) 22-mm diameter Longevity acetabular liner. (**B**) Packaging of Longevity.

observed that accelerated aging had no significant effect on the oxidation or mechanical properties of Longevity (2003).

The wear properties of Longevity have been studied under normal and abrasive conditions for 22 and 32 mm head sizes (Laurent et al. 2003). Laurent and colleagues (2003) found that Longevity maintained its superior wear resistance relative to conventional UHMWPE even under the presence of abrasive particles. Large-head Longevity liners have also been studied and demonstrate that Longevity wear characteristics are independent of head size.

Clinical Results

Favorable 2- to 3-year follow-up data for Longevity are documented in a study by Manning and colleagues (2004) presented at the 50th Annual Meeting of the Orthopedic Research Society.

Marathon

Development History and Overview

Marathon is based on the collaborative research efforts of Harry McKellop, Fu-wen Shen and coworkers, at the Los Angeles Orthopedic Hospital (McKellop, Shen, and Salovey 1998), and material scientists affiliated with DePuy-Dupont Orthopaedics. The process technology was licensed to DePuy Orthopedics, Inc. (Warsaw, IN) and clinically introduced in 1998. Marathon is currently available for the Pinnacle™ and Duraloc™ acetabular component systems.

In the Marathon process, extruded rod bar stock is irradiated with a dose of 50 kGy and then remelted at 150°C (McKellop, Shen, and Salovey 1998) (Figure 15.1). After remelting, the rods are then annealed at 120°C for 24 hours (Greer, King, and Chan 2003). Acetabular components are machined from the processed bar stock, enclosed in gas-permeable packaging, and then gas plasma sterilized (Figure 15.5).

Properties and *in Vitro* Performance

The *in vitro* test results for remelting of crosslinked UHMWPE were originally based on GUR 415 resin (McKellop, Shen, and Salovey 1998). Test results for Marathon, also based on GUR 415, were summarized in an abstract by DiMaio and colleagues (1998). In DiMaio's study, the fatigue behavior, mechanical properties, and hip simulator response of Marathon were evaluated in the aged and unaged condition. The mechanical properties of Marathon were judged to fall within the ASTM specifications for F648.

Table 15.4
Characteristics of Marathon

Orthopedic manufacturer	DePuy Orthopedics, Inc. (Warsaw, IN), www.depuy.com
Joint applications	THA
Clinical introduction	1998
Starting stock material	GUR 1050 Extruded Rod
Irradiation crosslinking modality	Gamma
Irradiation temperature	Room temperature
Irradiation dosage (range)	50 kGy
Postirradiation thermal treatment process and duration	Melted at 155°C for 24 hours and then annealed at 120°C for 24 hours, both in reduced oxygen atmospheres
Sterilization modality	Gas plasma
Target total irradiation dosage	50 kGy
Detectable free radicals?	No
Minimum/maximum head size available	22/36 mm
Minimum thickness available	6 mm
Published *in vitro* properties/performance data?	Yes
Published clinical performance data?	Yes, 2-year follow-up

More recently, Greer and colleagues (2003) published the *in vitro* test results for GUR 1020 and GUR 1050, processed with gamma and electron beam radiation, followed by Marathon thermal processing. Although the mechanical properties were found to be sensitive to molecular weight of the resin, the hip simulator wear rate, in contrast, did not appear to be affected by the resin type.

The oxidative stability of Marathon has been assessed by accelerated aging (DiMaio et al. 1998, McKellop, Shen, and Salovey 1998). In both McKellop's and DiMaio's studies, the properties of the crosslinked and remelted UHMWPE were not significantly affected by accelerated aging. Although Marathon is currently not available for knee applications, this material has been evaluated extensively in knee simulator experiments (McKnulty and Swope 2003).

Clinical Results

Hopper and colleagues have reported the clinical performance of Marathon (2003). After an average of 2.6 years of follow-up, Marathon liners were found

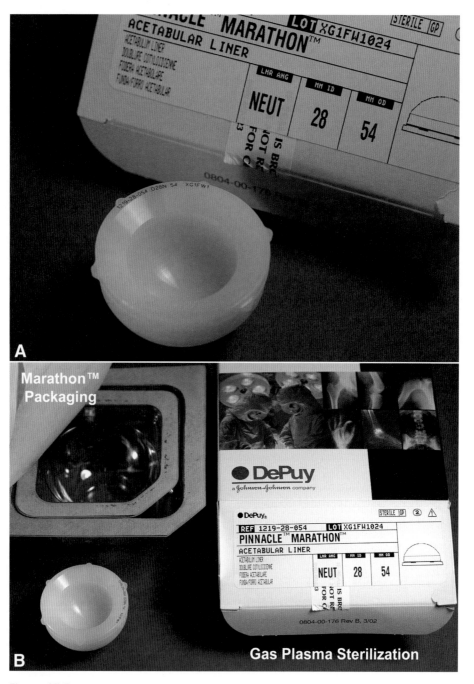

Figure 15.5
Marathon component for hip arthroplasty with packaging (DePuy Orthopedics, Inc., Warsaw, IN).
(**A**) 28-mm diameter Marathon acetabular liner. (**B**) Packaging of Marathon.

to wear 57% less than gas-plasma sterilized liners (0.09 ± 0.35 mm/year versus 0.21 ± 0.22 mm/year).

Prolong

Development History and Overview

Prolong is based on WIAM technology patented by Massachusetts General Hospital and licensed to Zimmer, Inc. (Warsaw, IN). Prolong was clinically introduced for cruciate-sparing, NexGen knee component systems in 2002.

The Prolong process is similar to the previously described DURASUL except that an electron beam irradiation dose of 65 kGy is used for Prolong and continuous compression-molded bars are processed as with Longevity. Components are machined from processed Prolong material, enclosed in gas-permeable packaging, and sterilized by gas plasma (Figure 15.6).

Table 15.5
Characteristics of Prolong

Orthopedic manufacturer	Zimmer, Inc. (Warsaw, IN), www.zimmer.com
Joint applications	TKA
Clinical introduction	2002
Starting stock material	GUR 1050 compression-molded sheet
Irradiation crosslinking modality	Electron beam
Irradiation temperature	~120°C
Irradiation dosage (range)	65 kGy (58 to 72 target dose range)
Postirradiation thermal treatment process and duration	Remelt above the melting temperature
Sterilization modality	Gas plasma
Target total irradiation dosage	65 kGy
Detectable free radicals?	No
Posterior-stabilized designs available?	No
Minimum polyethylene thickness available	6 mm
Published *in vitro* properties/performance data?	Yes
Published clinical performance data?	No

Properties and *in vitro* Performance

Prolong has been subjected to a battery of wear and functional fatigue testing (Laurent et al. 2003, Yao et al. 2003). The wear properties of Prolong have been studied in a knee simulator under normal and abrasive conditions (Laurent et al. 2003). Yao and colleagues have also tested Prolong in delamination and posterior condyle fatigue simulators (2003). The functional fatigue tests were conducted with aged Prolong and compared with conventional UHMWPE that was gamma sterilized in nitrogen. After accelerated aging, Prolong was found to have superior functional fatigue performance relative to conventional UHMWPE. A Food and Drug Administration–cleared claim for improved delamination resistance of Prolong over conventional polyethylene has been made based on the test data.

Clinical Results

Clinical results are not yet available for Prolong.

XLPE

Development History and Overview

Developed by Shilesh Jani, Victoria Good, and coworkers at Smith & Nephew, XLPE™ is the most recent highly crosslinked UHMWPE to be introduced into the U.S. orthopedic market for THA. XLPE is a remelted, highly crosslinked UHMWPE with a total dose of 100 kGy (Figure 15.1). Smith & Nephew did not provide a packaged example of their product to be included in this review.

Properties and *in vitro* Performance

The hip and knee simulator performance of XLPE has been reported against smooth and roughened counterfaces by Good and colleagues (2003). Good found that XLPE was more sensitive than conventional UHMWPE to abrasive wear. In the hip simulator, the XLPE still exhibited less wear than conventional UHMWPE with a roughened femoral head. This was not the case in the knee simulator, with XLPE exhibiting greater wear than conventional UHMWPE when articulating against a roughened femoral component. Based on these results, Smith & Nephew decided not to release XLPE for knee applications.

Clinical Results

Clinical results are not yet available for XLPE.

Table 15.6
Characteristics of XLPE

Orthopedic manufacturer	Smith & Nephew Orthopedics, Inc. (Memphis, TN), www.smith-nephew.com
Joint applications	THA
Clinical introduction	2001
Starting stock material	GUR 1050 extruded rod
Irradiation crosslinking modality	Gamma
Irradiation temperature	Ambient
Irradiation dosage (range)	100 kGy (range not available)
Postirradiation thermal treatment process and duration	Remelted at 150°C for unknown duration
Sterilization modality	Ethylene oxide
Target total irradiation dosage	100 kGy
Detectable free radicals?	No
Minimum/maximum head size available	22/36 mm
Minimum thickness available	6 mm
Published *in vitro* properties/performance data?	Yes
Published clinical performance data?	No

Current Trends and Prevalence in Total Hip and Total Knee Arthroplasty

The six highly crosslinked and thermally treated UHMWPE materials described in this chapter are gaining increased prevalence in THA, but they have thus far accomplished only limited penetration into TKA. The prevalence of these new materials is governed not only by physician acceptance, but also by the availability of patents and/or licenses to the technology, which are controlled by a handful of manufacturers.

At present, five companies have introduced hip and knee products manufactured from highly crosslinked UHMWPE in the United States (Tables 15.1 and 15.2). According to market literature and personal communications with individual manufacturers, these five companies produce approximately 82% of all the hip and knee replacements used in the United States today. Consideration of market data from Europe and Asia was beyond the scope of the present analysis, because of limited access to such information by the author.

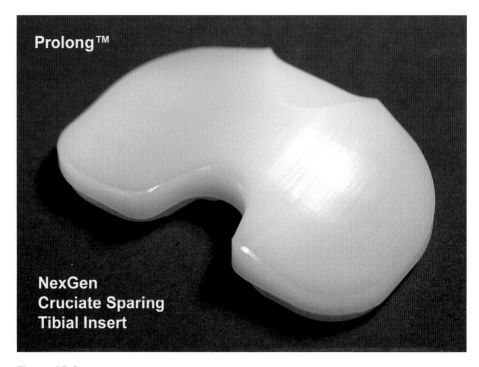

Figure 15.6
NexGen cruciate-sparing tibial component fabricated from Prolong (Zimmer, Inc., Warsaw, IN).

Prevalence of Highly Crosslinked UHMWPE in Total Hip Arthroplasty

Three noteworthy trends are apparent in the current use of highly crosslinked UHMWPE for THA, at least in the United States. These trends relate to the overall acceptance of the technology, the prevalence of gamma radiation as a crosslinking modality, and the prevalence of elevated radiation dosages for crosslinking.

First, despite their recent clinical introduction, highly crosslinked and thermally treated UHMWPE materials are now used in the majority of hip replacements in the United States. During 2003, highly crosslinked UHMWPE materials are projected to be used in an estimated 65% of U.S. hip arthroplasties (Table 15.7). Crossfire, Marathon, and Longevity have the greatest market share in the United States, and these three materials alone are estimated to account for 55% of hip replacements performed in 2003.

The second noteworthy trend is the prevalence of gamma radiation for crosslinking in these new materials for THA. Elevated crosslinking using gamma radiation accounts for an estimated 47% of THA in the United States for 2003. This proportion does not include the use of gamma radiation for sterilization, with historical doses of 25–40 kGy. As noted previously in Chapter 6, both gamma and electron beam radiation produce the same radiochemical reactions in UHMWPE, but the dosage rate is orders of magnitude faster in

Table 15.7

Highly Crosslinked Polyethylene Prevalence in Total Hip Arthroplasty for the United States in 2003 (Projected)

Manufacturer	Material	Radiation dosage (kGy)	THA market share[a]	Percentage conversion to crosslinking[b]	Total U.S. prevalence
CenterPulse	DURASUL	95	4.5%	80%	3.6%
DePuy	Marathon	50	25.0%	80%	20.0%
Smith & Nephew	XLPE	100	8.4%	70%	5.9%
Stryker Howmedica Osteonics	Crossfire	105	26.0%	80%	20.8%
Zimmer	Longevity	100	18.0%	80%	14.4%
Others	Conventional	0–40	18.1%		35.3%

[a]THA/TKA marketshare based on the 2001 market data from Orthopedic Network News. 2002. *2002 Hip Price Comparison/2002 Knee Price Comparison*.
[b]Based on personal communications with orthopedic manufacturers.

electron beam versus gamma radiation. Thus, the prevalence of gamma radiation for crosslinking in orthopedics is not the result of scientific factors that could be expected to influence clinical performance. Rather, gamma is preferred as a radiation source largely because of more practical production factors, such as cost, convenience, and ease of use.

The third noteworthy trend is the prevalence of doses in the range of 95–105 kGy for highly crosslinked UHMWPE. Four manufacturers, which produce materials using the dose range of 95–105 kGy, account for an estimated 45% of U.S. hip replacements. Marathon, produced with a dose of 50 kGy, has about 20% of the U.S. market share. It remains unclear whether 50 kGy is sufficient for significant wear and osteolysis reductions *in vivo*, or if the higher doses of 95–105 kGy used by the majority of orthopedic manufacturers will be necessary. During previous *in vitro* experiments, wear performance of crosslinked UHMWPE has been directly associated with radiation dosage (see Chapter 6). Consequently, future long-term clinical studies of highly crosslinked materials irradiated with 95–105 kGy may be expected to observe the maximum clinical benefit, if any, related to wear and osteolysis reduction.

Further inroads of highly crosslinked UHMWPE into U.S. population of THA patients are expected to occur at a much slower rate than was observed in the past 5 years. Restricted availability because of patent licenses, surgeon conservatism, and emerging alternative technologies are all expected to substantially curtail future expansion of highly crosslinked UHMWPE in orthopedics, at least within the short term for THA. Whereas conversion of previous surgeons to highly crosslinked UHMWPE has been based on *in vitro* experimental data and the generally positive short-term clinical studies, evidence of long-term clinical benefit is expected to be necessary to convince the remaining skeptical members of the orthopedic community to switch from conventional UHMWPE.

More importantly, patent licenses currently restrict highly crosslinked UHMWPE technology to a limited number of companies. Barring unforeseen developments on the licensing front, the number of companies with access to highly crosslinked UHMWPE is expected to remain relatively limited for at least the next decade, which coincides with the remaining lifetime of the first major patents on this technology. Even if existing manufacturers convinced 90% of their customers to switch (up from 60–80% estimated conversion in 2003), the total market share of highly crosslinked UHMWPE in THA would be expected to increase only from 65 to 74%, assuming the market share of existing companies remained the same. Thus, to increase the overall market share of highly crosslinked UHMWPE above 75%, either surgeons would need to switch manufacturers to gain access to crosslinking technology for their patients (a difficult choice, given the historic loyalty among some surgeons and manufacturers) or existing producers of highly crosslinked UHMWPE would need to gain market share by acquisition of competitors that currently do not have access to crosslinking technology.

The availability of metal-on-metal (MOM) and ceramic-on-ceramic (COC) alternative bearing designs will, to a certain extent, fragment the future THA market place. These alternative technologies will likely prevent highly crosslinked UHMWPE from reaching 90% penetration in THA, as suggested in the hypothetical scenario posed in the previous paragraph. However, the substantially increased cost of MOM and COC should prevent these technologies from gaining widespread use over UHMWPE, except perhaps in the young patient population. Not only clinical, but also epidemiolgical and cost-benefit research is needed to assist clinicians, insurance providers, and government agencies by providing objective data for decision making on this topic.

Prevalence of Highly Crosslinked UHMWPE in Total Knee Arthroplasty

The trends for highly crosslinked UHMWPE in TKA are very different than in THA. Unlike in THA, highly crosslinked UHMWPE has not yet become well accepted by orthopedic manufacturers and surgeons (Table 15.8). As of 2003, only two manufacturers produce highly crosslinked UHMWPE for TKA, and these manufacturers have limited their new materials to only cruciate-sparing designs, presumably because of concern about the performance of posterior-stabilizing posts in cruciate-sacrificing designs. Because of perceived risks associated with abrasion, fatigue, and fracture resistance of highly crosslinked UHMWPE, three other manufacturers with access to elevated crosslinking technology during the past 5 years have not yet introduced such products to the market place. Currently, in 2003, between 9 and 10% of TKAs performed in the United States incorporate highly crosslinked UHMWPE (Table 15.8).

Table 15.8

Highly Crosslinked Polyethylene Prevalence in Total Knee Arthroplasty for the United States in 2003 (Projected)

Manufacturer	Material	Radiation dosage (kGy)	TKA market share[a]	Percentage conversion to crosslinking[b]	Total U.S. prevalence
CenterPulse	DURASUL	95	6.7%	75%	5.0%
Zimmer	Prolong	65	22.1%	20%	4.4%
Others	Conventional	0–40	71.2%		90.6%

[a]THA/TKA marketshare based on the 2001 market data from Orthopedic Network News. 2002. *2002 Hip Price Comparison/2002 Knee Price Comparison.*
[b]Based on personal communications with orthopedic manufacturers.

Evidence-based clinical research will play a crucial role in influencing the future prevalence of highly crosslinked UHMWPE for TKA. Pending future clinical studies, surgeon acceptance remains the primary barrier to increasing the prevalence of highly crosslinked UHMWPE in TKA. Unlike in THA, there exist no alternative technologies equivalent to MOM or COC that would limit the diffusion of crosslinking technology in TKA.

Furthermore, the challenges facing clinical researchers in TKA are somewhat greater than in THA. In THA, accurate and repeatable methods of assessing the *in vivo* clinical performance of UHMWPE were established around the time highly crosslinked UHMWPE was clinically introduced. No such widely accepted methods exist for *in vivo* assessment of wear in TKA. Thus, even if short-term clinical reports with highly crosslinked UHMWPE turn out to be positive, the road to increased prevalence in TKA may take longer than in THA because the *in vivo* outcome measures for UHMWPE in knee replacement have not yet been standardized.

The Future for Highly Crosslinked UHMWPE

Innovations and developments in the field of highly crosslinked UHMWPE are still far from over. As this book goes to press in the winter of 2003, Centerpulse Orthopedics, Inc., the producer of DURASUL material used in THA and TKA, has been acquired by Zimmer, Inc. In addition, several orthopedic manufacturers are continuing to develop new highly crosslinked materials for hip and knee replacement, with the goal of introducing them within the next 12 to 24 months. The precise details of potential future product launches is not yet publicly available, and hence could not be included in this review. Future advances in highly crosslinked UHMWPE for joint arthroplasty should be considered a near certainty.

Although the short-term clinical results for highly crosslinked UHMWPEs in THA have thus far been encouraging, further follow-up is still necessary to confirm the long-term patient outcomes with these promising new UHMWPE bearing materials. The clinical introduction of highly crosslinked UHMWPE in the knee is still too recent for there to be any published long-term clinical data. It will be 10 years until sufficient evidence has been collected to test the hypothesis that these materials significantly reduce the incidence of revision in hip replacement, and the outcomes in the knee could take another decade to reach consensus among orthopedic surgeons and researchers. For all of these reasons, highly crosslinked UHMWPE is expected to be a major clinical research topic in orthopedics throughout the first decades of the 21st century.

Acknowledgments

This chapter was made possible by the assistance of colleagues at the five orthopedic manufacturers of highly crosslinked UHMWPE, who provided access to representative products and processing informaton: Jorge Ochoa and Don McKnulty (DePuy Orthopedics); Janet Krevolin (Centerpulse Orthopedics, Inc.); Cheryl Blanchard, Ray Gsell, and Dale Swarts (Zimmer, Inc.); Paul Serekian (Howmedica Osteonics, Corp.); and Shilesh Jani and Victoria Good (Smith & Nephew). Many thanks are also due to Orhun Muratoglu (Massachusetts General Hospital, Boston, MA) and Clare Rimnac (Case Western Reserve University) for many helpful discussions and assistance with this chapter.

References

Abt N.A., and W. Schneider. 2003. Influence of irradiation on the properties of UHMWPE. In *Highly crosslinked and thermally treated ultra-high molecular weight polyethylene for joint replacements*. S.M. Kurtz, R. Gsell, and J. Martell, Eds. West Conshohoken, PA: American Society for Testing and Materials.

Bhambri S.K., R. Gsell, L. Kirkpatrick, et al. 2003. The effect of aging on mechanical properties of melt-annealed highly crosslinked UHMWPE. In *Highly crosslinked and thermally treated ultra-high molecular weight polyethylene for joint replacements*. S.M. Kurtz, R. Gsell, and J. Martell, Eds. West Conshohoken, PA: American Society for Testing and Materials.

Bragdon C.R., G. Digas, J. Karrholm, et al. 2002. RSA evaluation of wear of conventional vs. highly crosslinked polyethylene acetabular component in vivo. *Transactions of the American Association of Hip and Knee Surgeons* 12:23.

Collier J.P., B.H. Currier, F.E. Kennedy, J.H. Currier, G.S. Timmins, S.K. Jackson, and R.L. Brewer. 2003. Comparison of cross-linked polyethylene materials for orthopaedic applications. *Clin Orthop* 414:289–304.

Digas G., J. Kärrholm, J. Thanner, et al. 2003. Highly crosslinked polyethylene in hip arthroplasty. Randomized study using radiosterometry; preliminary report. *Transactions of the 7th EFORT Congress* 7:1136/359.

DiMaio W.G., W.B. Lilly, W.C. Moore, and K.A. Saum. 1998. Low wear, low oxidation radiation crosslinked UHMWPE. *Transactions of the 44th Orthopedic Research Society* 23:363.

Essner A., V.K. Polineni, G. Schmidig, et al. 1997. Long-term wear simulation of stabilized UHMWPE acetabular cups. *Transactions of the 43rd Orthopedic Research Society* 22:784.

Good V., K. Widding, M. Scott, and S. Jani. 2003. The sensitivity of crosslinked UHMWPE to abrasive wear: Hips vs. knees. In *Highly crosslinked and thermally treated ultra-high molecular weight polyethylene for joint replacements*. S.M. Kurtz, R. Gsell, and J. Martell, Eds. West Conshohoken, PA: American Society for Testing and Materials.

Greer K., R. King, and F.W. Chan. 2003. Effects of raw material, irradiation dose, and irradiation source on crosslinking of UHMWPE. In *Highly crosslinked and thermally treated ultra-high molecular weight polyethylene for joint replacements*. S.M. Kurtz, R. Gsell, and J. Martell, Eds. West Conshohoken, PA: American Society for Testing and Materials.

Hopper R.H., Jr., A.M. Young, K.F. Orishimo, and J.P. McAuley. 2003. Correlation between early and late wear rates in total hip arthroplasty with application to the performance of highly crosslinked polyethylene liners. *J Arthroplasty* 18(7 Suppl 1): S60–67.

Jani S., and M. Scott. 2003. Wear debris generation in joint simulator testing of crosslinked UHMWPE. In *Highly crosslinked and thermally treated ultra-high molecular weight polyethylene for joint replacements*. S.M. Kurtz, R. Gsell, and J. Martell, Eds. West Conshohoken, PA: American Society for Testing and Materials.

Kurtz S.M., O.K. Muratoglu, M. Evans, and A.A. Edidin. 1999. Advances in the processing, sterilization, and crosslinking of ultra- high molecular weight polyethylene for total joint arthroplasty. *Biomaterials* 20:1659–1688.

Kurtz S.M., M. Manley, A. Wang, et al. 2002–2003. Comparison of the properties of annealed crosslinked (Crossfire) and conventional polyethylene as hip bearing materials. *Bull Hosp Jt Dis* 6:17–27.

Kurtz S.M., R. Gsell, and J. Martell. 2003. *Highly crosslinked and thermally treated ultra-high molecular weight polyethylene for joint replacements*. West Conshohoken, PA: American Society for Testing and Materials, ASTM STP.

Laurent M.P., C.R. Blanchard, J.Q. Yao, et al. 2003. The wear of highly crosslinked UHMWPE in the presence of abrasive particles: Hip and knee simulator studies. In *Highly crosslinked and thermally treated ultra-high molecular weight polyethylene for joint replacements*. S.M. Kurtz, R. Gsell, and J. Martell, Eds. West Conshohoken, PA: American Society for Testing and Materials.

Lewis G. 2001. Properties of crosslinked ultra-high-molecular-weight polyethylene. *Biomaterials* 22:371–401.

Manning D.W., P.P. Chiang , J.M. Martell, J.O. Galante, and W.H. Harris. 2004. *In vivo* wear of traditional vs. highly crosslinked polyethylene. *Transactions of the 50th Orthopedic Research Society* 29:1478.

Martell J., J.J. Verner, and S.J. Incavo. 2003. Clinical performance of a highly crosslinked polyethylene at two years in total hip arthroplasty: A randomized prospective trial. *J Arthroplasty* 18(7 Suppl 1):S55–59.

McKellop H., F.W. Shen, and R. Salovey. 1998. Extremely low wear of gamma-crosslinked/remelted UHMW polyethylene acetabular cups. *Transactions of the 44th Orthopedic Research Society* 23:98.

McKnulty D, and S. Swope. 2003. The effect of crosslinking UHMWPE on *in-vitro* wear rates of fixed and mobile bearing knees. In *Highly crosslinked and thermally treated ultra-high molecular weight polyethylene for joint replacements*. S.M. Kurtz, R. Gsell, and J. Martell, Eds. West Conshohoken, PA: American Society for Testing and Materials.

Muratoglu O.K., and S.M. Kurtz. 2002. Alternative bearing surfaces in hip replacement. In *Hip replacement: Current trends and controversies*. R. Sinha, Ed. New York: Marcel Dekker.

Muratoglu O.K., C.R. Bragdon, D.O. O'Connor, et al. 2001a. Markedly improved adhesive wear and delamination resistance with a highly crosslinked UHMWPE for use in total knee arthoplasty. *Transactions of the 27th Annual Meeting of the Society for Biomaterials* 24:29.

Muratoglu O.K., C.R. Bragdon, D.O. O'Connor, et al. 2001b. Larger diameter femoral heads used in conjunction with a highly cross-linked ultra-high molecular weight polyethylene: A new concept. *J Arthroplasty* 16:24–30.

Muratoglu O.K., C.R. Bragdon, D.O. O'Connor, et al. 2002. Aggressive wear testing of a cross-linked polyethylene in total knee arthroplasty. *Clin Orthop* 404:89–95.

Nivbrant B., S. Roerhl, B.J. Hewitt, M.G. Li. 2003. *In vivo* wear and migration of high crosslinked poly cups: An RSA study. *Transactions of the 49th Orthopedic Research Society* 28:358.

Rabenseifner L. 2003. *In vivo* results with highly cross-linked polyethylene in the hip. *Transactions of the 7th EFORT Congress* 7:1137/380.

Sun D.C., A. Wang, C. Stark, and J.H. Dumbleton. 1996. The concept of stabilization in UHMWPE. *Transactions of the Fifth World Biomaterials Congress* 1:195.

Wang A., M. Manley, P. Serekian. 2003. Wear and structural fatigue simulation of crosslinked ultra-high molecular weight polyethylene for hip and knee bearing applications. In *Highly crosslinked and thermally treated ultra-high molecular weight polyethylene for joint replacements*. S.M. Kurtz, R. Gsell, and J. Martell, Eds. West Conshohoken, PA: American Society for Testing and Materials.

Yao J.Q., M.P. Lu, T.S. Johnson, et al. 2003. Improved resistance to wear, delamination and posterior loading fatigue damage of electron beam irradiated, melt-annealed, highly crosslinked UHMWPE knee inserts. In *Highly crosslinked and thermally treated ultra-high molecular weight polyethylene for joint replacements*. S.M. Kurtz, R. Gsell, and J. Martell, Eds. West Conshohoken, PA: American Society for Testing and Materials.

Chapter 15. Reading Comprehension and Group Discussion Questions

15.1. Why might an implant manufacturer choose to thermally treat a highly crosslinked UHMWPE material either above or below the melt temperature? What are the benefits and tradeoffs associated with each choice?

15.2. Why might an implant manufacturer choose to crosslink UHMWPE with 50 as opposed to 100 kGy?

15.3. Why are highly crosslinked UHMWPE materials currently more widely accepted for hip replacement as opposed to knee replacement?

15.4. Let us assume that you or a close relative needs an artificial hip replacement, and that waiting for improved future technologies is not an option. Let us further assume that you have chosen the best surgeon in the particular country in which you reside, and your chosen surgeon champion will implant any type of joint material combination you desire. Having reviewed the biomaterials options that are currently available, as summarized in this chapter as well as in Chapter 6, would you recommend highly crosslinked over conventional UHMWPE, or would you prefer metal-on-metal, or ceramic-on-ceramic technologies? If you *would* recommend highly crosslinked UHMWPE, which formulation would you select, and why? If you would choose MOM or COC for yourself or relative instead of UHMWPE, what factors led to that decision?

15.5. How would your response to question (4) change if you or your relative needed a total knee replacement?

UHMWPE
Lexicon

www.UHMWPE.org

Online Teaching Tools!

Visit

www.UHMWPE.org/classmaterials.html
Input Passcode (Upper Case Letters)**: KPMSA5K**
Download Class Materials!

Available UHMWPE Lexicon Class Materials:

- Graphics for Graduate or Undergraduate Instruction that can be Downloaded as Powerpoint Presentations for Classroom Instruction

- Links to Related Biomaterials and Orthopedic Sites

- Overview of State-of-the-Art Research in Several Key Polyethylene Related Problems of Clinical Significance

DATA REQUIREMENTS FOR
ULTRAHIGH MOLECULAR WEIGHT POLYETHYLENE
(UHMWPE)
USED IN ORTHOPEDIC DEVICES

DRAFT

March 28, 1995

Call 301-443-9435 (flash fax)
for the latest version of this document

Orthopedic Devices Branch
Division of General and Restorative Devices
Center for Devices and Radiological Health
U.S. Food and Drug Administration

9200 Corporate Blvd.
Rockville, MD 20850

(301) 594-2036

I. INTRODUCTION

The purpose of this document is to list the data needed for orthopedic devices containing UHMWPE. These data should be included in premarket notifications (510k), Investigational Device Exemptions (IDE) applications, Premarket Approval (PMA) applications, reclassification petitions, and master files to aid FDA in determining the substantial equivalence and/or safety and effectiveness of UHMWPE in implantable orthopedic devices. In this document UHMWPE is referred to as polyethylene (PE).

For specific applications, FDA may require information in addition to that outlined in this document. In other instances, data requirements in this document may be omitted with sufficient justification. Suggestions and recommendations presented in this document are not mandatory requirements and the words "should", "must" and "shall" are not used in a regulatory sense and should not be construed as such.

FDA may periodically update this document to reflect modifications in the data requirements for evaluating PE.

II. TEST REQUIREMENTS

All submissions with PE components must provide the data in Stage 1 below. If the PE is similar to PE on the market, no further material data are required. However, additional design specific testing, such as surface contact stresses, may be needed. If the data from Stage 1 demonstrates that the PE differs from PE on the market, additional Stage 2 data may be required. Which Stage 2 data are needed will be determined for each individual submission. For example, any claims for the PE must be substantiated. Stage 3 data may be required if the Stage 1 and Stage 2 data show the new material to be significantly different than other PE components. Stage 3 is also required if the chemical composition of the product differs from ASTM F 648: UHMWPE Powder and Fabricated Form for Surgical Implants. Note: all testing must be performed on the final, sterilized material.

Stage 1   Mechanical Properties:
- Ultimate Tensile Strength
- Yield Strength
- Young's Modulus (Modulus of Elasticity)
- Poisson's Ratio
- % Elongation

Other Properties
- Molecular Weight
- Density and Porosity
- Crystallinity
- Glass Transition Temperature, T_g
- Crystallization Temperature Range, T_c
- Melting Temperature, T_m
- Oxidation Temperature, T_o

Stage 2   Mechanical Tests:
                      - Creep
                      - Wear
                      - Fatigue
                      - Crack Propagation
                      - J Integral,
                    Other Tests:
                      - Thin Sectioned Photomicrograph
                      - IR Spectra and Chemical Structure
Stage 3   Clinical Data
                    Biocompatibility

It is up to the investigator to determine which initial tests are necessary in stages II and III.

The tests should report the storage time and test environment.

III. CONTROLS

Li and Howard ("Characterization and Description of an Enhanced Ultra High Molecular Weight Polyethylene for Orthopaedic Bearing Surfaces." *Transactions of the 16th Annual Meeting of the Society of Biomaterials*, p. 190, May, 1990) have demonstrated that significant property variations occur between the four sources of PE in the world, the different grades of PE and different PE lots. Wright, T.M. ("Polyethylene: Mechanisms of Wear and Enhanced Forms." *The Hip Society 20th Open Scientific Meeting at the AAOS*, p. 26, February, 1992) reported that testing of various PE implant components produced differences in creep properties of 400% and differences in ductility of 111%. Therefore, the PE must be compared to a legally marketed PE that has been clinically successful under similar physiologic loading conditions.

IV. REPORTING

To help FDA evaluate the substantial equivalence and/or safety and effectiveness of new PE components, submitted test reports should contain the information listed below.

1. Report title
2. Investigators' names
3. Facility Performing the test
      - Name
      - Address
      - Phone Number
4. Dates
      - Test initiation
      - Test completion
      - Final report completion

5. Objectives/Hypothesis
6. Test and control samples
   - Sample selection criterion
   - Design
   - Materials
   - Processing methods
   - Differences between test and control samples and marketed device
7. Methods and Materials
   - Test setup schematic or photograph
   - Description of grips or potting medium interfacing with samples
   - List of dependent, independent and uncontrolled variables, i.e.:
     - Test and control sample parameters
     - Environment composition, pH, volume, flow, temperature
     - Electromagnetic fields, applied charges, irradiation
     - Load directions, points of application and magnitudes
     - Times (e.g. rates, frequencies, number of cycles)
   - Methods of specimen examination (e.g., failure analysis)
   - Chronological description of the test procedures
   - Deviations from referenced protocols and standards
8. Results
   - Time from manufacturing until commencement of testing
   - Discussion of the data
   - Conclusions, including statistical analysis
   - Discussion of the objective/hypothesis
   - Clinical implications of results (including a discussion of assumptions)
9. Appendices
   - Experimental data
   - Calculations
   - Bibliography of all references pertinent to the report

The following voluntary standards may be useful in preparing data outlined in this document:
   - ASTM D 621: Standard Test Methods for Deformation of Plastics Under Load.
   - ASTM D 638: Standard Test Methods for Tensile Properties of Plastics
   - ASTM D 671: Standard Test Methods for Flexural Fatigue of Plastics by Constant-Amplitude-of-Force.

Index

A

Accelerated aging tests, 284, 291–293
Acid formation, 252
Aeonian™, 338–339
Aequalis™/Aequalis™ Fracture
    shoulder prosthesis system
    components, 202
Aging tests, 284, 291–293
Air permeable packaging. *See* Gas
    permeable packaging
Alkyl macroradicals (R•), 254–256
Alumina ceramic(s)
    femoral heads, 105–106
        in vivo fracture risk, 108–109
    hip bearings, 102, 103t, 104
    introduction of, 53
Alumina composite material, 103t, 105
American Society for Testing and
    Materials (ASTM)
    standard D4020-01A, 265
    standard F648, 15, 265
Analytical closed-form solution
    methods, 310, 311t
Anatomical Shoulder™ system
    components, 198f, 199
Anterior-posterior (A-P) radiographs,
    176–177, 178f
A-P radiographs, 176–177, 178f
ArCom™
    barrier packaging, 42f
    processing, 27, 28f, 29
Arthritis
    osteoarthritis
        hip complications, 132
        shoulder complications, 190–191
    shoulder complications, 193

Arthritis (*Continued*)
    osteoarthritis, 190–191
    rheumatoid arthritis, 190
Artificial disc replacement. *See* Total disc
    arthroplasty/replacement
    (TDA/TDR)
Aseptic loosening, 73–74
ASTM standard D4020-01A, 265
ASTM standard F648, 15, 265
Average radiographic wear, 78
Averill, Robert, 194

B

Balloon lesions, 176
Barrier packaging
    air permeable packaging, replacement
        of, 39
    gamma sterilization in, 41–44
Basell Polyolefins, 16–17
Bi-Angular® shoulder prosthesis system
    components, 197f, 198
Bicondylar knee arthroplasty, 125
    cruciate-sacrificing designs, 134,
        136–141
    cruciate-sparing designs,
        134–136
Bigliani/Flatow® humeral prostheses,
    202
Bio-Modular® shoulder prosthesis
    system components, 197
BiPolar shoulder prosthesis system
    components, 198
Bolland's cycle, 250
"Bow-tie" wear scar, 182, 183f
Branched polymers, 3
Bryan, Richard, 129

C

Calcium stearate, 21–22
Ceramic-on-ceramic (COC) alternative
  hip bearings, 93–94, 101
  alumina ceramics, 102, 103t, 104
    femoral heads, 105–106
    in vivo fracture risk, 108–109
  alumina composite material,
    103t, 105
  contemporary designs, 106–108
  historical overview, 101–102
  in vivo fracture risk, 108–109
  zirconia, 102, 103t, 104–105
    failure rate, 109
Chain folding, 6
Change in enthalpy, 8
Characteristic material response, 311–317
Charnley, Sir John, 53. *See also*
    Wrightington Hospital
  artificial joint design, 55
  filled PTFE experimentation, 64–65
  hip arthroplasty
    pink dental acrylic cement, use of,
      56–57
    wear performance study, 79–82
  hip arthroplasty designs
    first design with PTFE, 56
    second, third and fourth designs
      with PTFE, 58
    fifth and final design with PTFE,
      58–60
  knee replacement design, 129, 135f
  Thompson prostheses, implantation
    of, 65
  UHMWPE, first reaction to, 66
Chas. F. Thackray Ltd., 67
Chemical characterization, 274
Chemical testing
  electron spin resonance spectroscopy,
    276–277, 278f
  Fourier transform infrared
    spectroscopy, 274–276
  gel permeation chromatography,
    270–271
  swell ratio testing, 278–280
  trace element analysis, 274
CHIRULEN®, 16–17, 24
  SB Charité™ III implants, 227
COC alternative hip bearings.
    See Ceramic-on-ceramic (COC)
    alternative hip bearings

Cofield™/Cofield²™, Monoblock
    shoulder prosthesis system
    components, 201
Compression molding, 24–25
  direct compression molding, 27, 29
Compressive response, 313–314
Computer-assisted radiographic wear
    measurement
  Martell technique, 84
  three-dimensional techniques,
    83–84
Computer modeling and simulation,
    310–311
  analytical closed-form solution
    methods, 310, 311t
  characteristic material response,
    311–317
  FE analysis, 310–311
  handbook solution, 311t
  hybrid model, 326–332, 333f
  hyperelasticity, 320–321
  isotropic J_2-plasticity, 324–326
  linear elasticity, 318–320
  linear viscoelasticity, 321–324
  material modeling, 317–334
Consolidation. *See* Conversion/
    consolidation
Consolidation defects, 24
Conversion/consolidation, 22–24
  ArCom™ UHMWPE processing, 27,
    28f, 29
  compression molding, 24–25
  defects, 24
  direct compression molding, 27, 29
  extruded *versus* molded UHMWPE,
    29–31
  grain boundaries, 23
  intergranular diffusion, 22–23
  ram extrusion, 25–27
  self-diffusion, 22
Copolymers, 3
Craven, H.
  UHMWPE cup machine(s), 61–62
  UHMWPE testing, 66
  wear testing rig, 62–64
Creep, 283
Crossfire™, 339–342
Crosslinked HDPE components, 53–54
Crosslinking, 245–246
  H-crosslinking mechanism,
    249–250

Crosslinking (*Continued*)
   highly crosslinked UHMWPE.
      See Highly crosslinked/thermally
      stabilized UHMWPE
   isolated radicals, reaction of, 247–248
   mechanical behavior and wear, effect
      on, 295–298
   radicals
      formation during irradiation,
         246–247
      isolated radicals, reaction of,
         247–248
   Y-crosslinking mechanism, 248–249
Cruciate and collateral knee ligaments,
   153, 154f
Crystalline lamellae, 6–7, 8f

D
DCM (direct compression molding),
   27, 29
Delta® shoulder prosthesis, 212
Density measurements, 272–273
Density properties, 30
Differential scanning calorimetry
   (DSC), 8
Dilute solution viscometry, 266t
Direct compression molding (DCM),
   27, 29
Disc replacement. *See* Total disc arthro-
   plasty/replacement (TDA/TDR)
Disk bend test. *See* Small punch test(ing)
Dislocated shoulder, 191
DSC (differential scanning calorimetry),
   8, 266–267, 268f
Duracon total knee prostheses, 152f
DURASUL™, 342–345
Duration™, 338–339

E
E-beam irradiation-induced oxidation,
   253
Eius unicondylar prostheses, 152f
Electron spin resonance (ESR)
   spectroscopy, 276–277, 278f
Equibiaxial small punch data, 314,
   317f
ESR (electron spin resonance)
   spectroscopy, 276–277, 278f
Ester formation, 252
Ethylene gas, 4
   polymerization to UHMWPE powder.
      See Polymerization

Ethylene oxide sterilization (EtO), 38t,
   44–45
Extruded UHMWPE
   versus molded UHMWPE, 29–31
   ram extrusion, 25–27

F
Fatigue testing, 282–283
   small punch, 301–304, 305f
FDA testing guidelines, 264–265
FE analysis, 310–311
Fick's law, 255
Fixed-bearing knee designs, 144, 151,
   152f
FLEXICORE™ TDR, 239
Flow temperature (T_f), 7–9
Fluoroscopy-guided A-P radiographs,
   177–178
Food and Drug Administration (FDA)
   testing guidelines, 264–265
Foundation®/Foundation® fracture
   humeral prostheses, 199–200
Fourier transform infrared (FTIR)
   spectroscopy, 274–276
Freeman-Swanson knee prosthesis, 134f,
   135f, 140–141
FTIR (Fourier transform infrared)
   spectroscopy, 274–276
Fusion assessment, 271
Fusion defects, 24

G
Gamma irradiation-induced oxidation,
   253
Gamma sterilization
   in air permeable packaging, 38–41
   in barrier packaging, 41–44
Gamma Vacuum Foil (GVF) barrier
   packaging, 43f
Gas permeable packaging
   barrier packaging, replacement
      with, 39
   ethylene oxide sterilization, 38t
   gamma sterilization in, 38–41
   gas plasma sterilization, 38t
Gas plasma sterilization, 38t, 44–47
Gel permeation chromatography (GPC),
   270–271
Geomedic knee prosthesis, 132, 133f,
   134f, 135
Geometric knee, 135–136
Geometric strain hardening, 289

Geometric strain softening, 289
Glass transition temperature (T_g), 7–8
Glenohumeral forces, 195
Global™ Advantage® humeral prostheses, 199
Global™ FX humeral prostheses, 199
Global™ humeral prostheses, 199, 206
Gluck, 123
GPC (gel permeation chromatography), 270–271
Grain boundaries, 23
Griffith wear performance study, 79–82
Guépar hinged knee replacement, 127f
Gunston, Frank, 123
GUR resins, 16–17
    versus 1900 resin, 19–20
GVF (Gamma Vacuum Foil) barrier packaging, 43f

H
H-crosslinking mechanism, 249–250
HDPE (high-density polyethylene), 4
    crosslinked HDPE components, 53–54
Hemiarthroplasties. *See also* Shoulder arthroplasty/replacement
    bipolar prosthesis, 209
    procedures, 191–192
    results and rates, 209
    UHMWPE's role in, 209
Hercules Powder Company, 17
High-density polyethylene (HDPE), 4
    crosslinked HDPE components, 53–54
Highly crosslinked/thermally stabilized UHMWPE, 93, 337–338
    Aeonian™, 338–339
    Crossfire™, 339–342
    current trends, 353
    DURASUL™, 342–345
    Duration™, 338–339
    future for, 357–358
    hip arthroplasty/replacement. *See* Hip arthroplasty/replacement
    knee arthroplasty/replacement, 182, 184
    Longevity™, 345–348
    Marathon™, 348–351
    prevalence
        in total hip arthroplasty, 354–356
        in total knee arthroplasty, 356–357
    Prolong™, 351–352
    XLPE™, 352, 353t

Hip arthroplasty/replacement
    age of persons receiving, 71, 72f
    alumina ceramics, 102, 103t, 104
        femoral heads, 105–106
        in vivo fracture risk, 108–109
    alumina composite material, 103t, 105
    aseptic loosening, 73–74
    average radiographic wear, 78
    ceramic-on-ceramic alternative bearings, 93, 94, 101
        alumina ceramics, 102, 103t, 104
        alumina composite material, 103t, 105
        contemporary designs, 106–108
        historical overview, 101–102
        in vivo fracture risk, 108–109
        zirconia, 102, 103t, 104–105, 109
    ceramic on UHMWPE, 105–106
    highly crosslinked/thermally stabilized UHMWPE, 53, 109–110
        contemporary designs, 110, 111f
        current clinical outlook, 114
        historical clinical experience, 110
        prevalence in THA, 354
        thermal treatment, effect of, 111–114
    historical developments, 53–55. *See also* Charnley, Sir John; Wrightington Hospital
        alumina ceramic, 53
        crosslinked HDPE components, 53–54
        highly crosslinked UHMWPE, 53
        Hylamer, 54
        McKee-Farrar prosthesis, 97, 99f
        McKee prostheses, 96–97, 98f
        Wiles, 96
    linear wear rate, 77, 85t
    metal-on-metal alternative bearings, 93–96
        biological risks, 100–101
        contemporary designs, 98–99, 100f
        historical overview, 96–97, 98f
    osteolysis, 74, 93
    projected increase in, 72, 73f, 74
    radiographic lysis, 74
    stresses in UHMWPE components, 156–157
    timeline of developments, 54t
    volumetric wear rate, 78, 85t

Hip arthroplasty/replacement
(*Continued*)
   wear measurement
      computer-assisted radiographic
         wear measurement, 83–84
      Livermore circular templates, 83
      radiostereometric analysis, 83–84
   wear performance/rates
      average radiographic wear, 78
      in cemented acetabular components,
         75–77
      Charnley/Griffith studies, 79–82
      Isaac study, 82–83
      linear wear rate, 77, 85t
      in modular acetabular components,
         85–86
      volumetric wear rate, 78, 85t
   zirconia, 102, 103t, 104–105
      failure rate, 109
HIPing (hot isostatic pressing), 27,
   28f, 29
Hip simulators, 284
HM (hybrid model), 326–332, 333f
Hoechst, 16
Homopolymers, 3
Hot isostatic pressing (HIPing), 27,
   28f, 29
H radicals, 246–247
H transfer reactions, 246–248
Hybrid model (HM), 326–332, 333f
Hydroperoxide decomposition (ROOH),
   254–255
Hydroperoxides, 251–252
   decomposition, 254–255
Hylamer, 54
   glenoid component wear, 206
Hyperelasticity, 320–321

I
Insall-Burstein (IB) knee prosthesis,
   160, 161f
Inspection of knee UHMWPE
   components, 180
Integral work to failure (WTF), 288–289
Integrated® shoulder prosthesis system
   components, 198
Intergranular diffusion, 22–23
Intrinsic viscosity (IV) measurements,
   17–18, 269
Irradiation. *See* Sterilization
Isaac study (wear performance), 82–83

Isolated radicals, reaction of, 247–248
ISO standard 5834-1, 15
Isotropic J_2-plasticity, 324–326
IV (intrinsic viscosity) measurements,
   17–18, 269

J
J-integral testing, 280–281

K
Kenmore, 194
Ketone formation, 251
Knee anatomy, 153–154
Knee arthroplasty/replacement
   abrasion, 171–172
   age of persons receiving, 71, 72f
   anatomical considerations, 153–154
   articulating surface damage modes,
      167–172
   backside wear, 180–181
   bicondylar knee arthroplasty, 125
      cruciate-sacrificing designs, 134,
         136–141
      cruciate-sparing designs, 134–136
   biomechanics of, 153–160
   burnishing, 171
   deformation at surface, 170–172
   delamination, 170, 172
   embedded debris, 168, 169f, 170, 172
   fixed-bearing knee designs, 144, 151,
      152f
   Gunston's cemented implant design,
      123, 127–129
   highly crosslinked and thermally
      stabilized UHMWPE, 182, 184
   historical developments, 126–129
   infections, 165
   loosening, 165
   metal backing, incorporation of, 142,
      143f
      fixed bearing designs, 144
      mobile bearing designs, 139f,
         144–146
   mobile bearing knee designs, 139f,
      144–146, 151
   osteolysis, 172–176
   patellar complications, 165
   patellar component implants, 125
   patellar resurfacing, 125
   patello-femoral arthroplasty, 141–142
   pitting, 167–168, 169f, 172
   polycentric knee arthroplasty, 129–132

Knee arthroplasty/replacement (*Continued*)
  post damage, in posterior-stabilized tibial components, 181–182, 183f
  projected increase in, 72, 73f, 74
  revision surgery, reasons for, 165–166
  scratching, 169f, 170, 172
  semiconstrained hinged knee design, 125
  survivorship of, 162–163, 163f–165f
  total condylar knee, 135f, 136–141
  "tufting," 171–172
  UHMWPE component stresses, 156–160
  unicondylar knee arthroplasty, 125
  unicondylar polycentric knee arthroplasty, 132–134
  wear or surface damage, 165–167
    articulating surface damage modes, 167–172
    backside wear, 180–181
    post damage, in posterior-stabilized tibial components, 181–182, 183f
    in vivo wear assessment methods, 176–180
    "wear polishing," 171
Knee joint loading, 154–156

L
LCS mobile bearing knees, 145–146
LDPE (low-density polyethylene), 4
Linear elasticity, 318–320
Linear low-density polyethylene (LLDPE), 4
"Linear lytic defect," 176
Linear polymers, 3
Linear viscoelasticity, 321–324
Linear wear rate (LWR), 77, 85t
Lipid absorption, 257, 258f
Livermore circular templates, 83
LLDPE (linear low-density polyethylene), 4
Longevity™, 345–348
Low-density polyethylene (LDPE), 4
LWR (linear wear rate), 77, 85t

M
Machining, 31–32
Machining marks, 31
MacIntosh tibial plateau, 126, 127f
Macroradicals, 246–247, 250
  alkyl, 254–256
  peroxy, 254

Marathon™, 348–351
Mark-Houwink equation, 18
Marmor knee prosthesis, 132, 135f
Martell technique, 84
Material behavior
  computer modeling, 311–317
  testing of. *See* Chemical testing; Mechanical testing; Physical testing
Material modeling, 317–318, 334
  hybrid model, 326–332, 333f
  hyperelasticity, 320–321
  isotropic J_2-plasticity, 324–326
  linear elasticity, 318–320
  linear viscoelasticity, 321–324
MAVERICK TDR, 239
McKee-Farrar prosthesis, 97, 99f
McKee prostheses, 96–97, 98f
McKeever tibial plateau, 126, 127f
Mechanical characterization, 280
Mechanical testing
  creep, 283
  fatigue testing, 282–283
  J-integral testing, 280–281
  Poisson's ratio, 280
  small punch. *See* Small punch test(ing)
  tensile testing, 281, 282f
Medical grade powder requirements, 15
Melt temperature (T_m), 7–8
Meniscal knee bearings, 144–145
Metal-on-metal (MOM) alternative hip bearings, 93–96
  biological risks, 100–101
  contemporary designs, 98–99, 100f
  historical overview, 96–97, 98f
METASUL, 98, 100f
Miller-Gallante (MG) knee prosthesis, 160, 161f
Mobile bearing knee designs, 139f, 144–146, 151
Modeling. *See* Computer modeling and simulation
Modular Shoulder System, 201
Molded UHMWPE
  compression molding, 24–25
  direct compression molding, 27, 29
  versus extruded UHMWPE, 29–31
Molecular weight, 17–19
MOM alternative hip bearings. *See* Metal-on-metal (MOM) alternative hip bearings

Monomers, 3
Montell Polyolefins, 17
M_V (viscosity average molecular
  weight), 18

N
Neer, Charles, II, 193–194
Neer II/Neer III shoulder prosthesis
  system components, 194,
  200–201
Nu-Life dental cement, 56–57
N_2-Vac barrier packaging, 43f

O
OA (osteoarthritis)
  hip complications, 132
  shoulder complications, 190–191
OIT (oxidation induction time)
  measurements, 267
Osteoarthritis (OA)
  hip complications, 132
  shoulder complications, 190–191
Osteolysis, 74, 93
Oxidation, 250–251
  after implant manufacture, 256–257,
    258f
  aging tests, 284, 291–293
  critical products of, 251–252
  E-beam irradiation-induced, 253
  gamma irradiation-induced, 253
  rate, 255–256
  sterilization, effects of, 253–257
  in vivo oxidation, 294, 295f
Oxidation induction time (OIT)
  measurements, 267

P
Packaging, 37–38
  barrier
    gamma sterilization in, 41–44
    replacement of air permeable
      packaging, 39
  gas permeable
    barrier packaging, replacement
      with, 39
    ethylene oxide sterilization, 38t
    gamma sterilization in, 38–41
    gas plasma sterilization, 38t
Patellar component implants, 125
Patellar resurfacing, 125
Patello-femoral arthroplasty,
  141–142

Patello-femoral joint loading, 155t
PCL (posterior cruciate ligament),
  153–154
Péan, 193
Peroxy macroradicals (ROO•), 254
Perplas Medical, 24
Peterson, Lowell, 129
Photo-oxidation, 250
Physical properties
  HDPE, 5t
  UHMWPE, 5t, 265
Physical testing
  density measurements, 272–273
  differential scanning calorimetry,
    266–267, 268f
  dilute solution viscometry, 266t
  fusion assessment, 271
  intrinsic viscosity, 269
  oxidation induction time
    measurements, 267
  scanning electron microscopy, 267,
    268f, 269
  transmission electron microscopy,
    271–272
Poisson's ratio, 280
Polycentric knee arthroplasty, 129–132
Polyethylene, 4–5
Poly Hi Solidur Meditech, 24
Polymerization, 14–16
  calcium stearate, 21–22
  GUR resins, 16–17
  GUR resins *versus* 1900 resin, 19–20
  and molecular weight, 17–19
  1900 resins, 16–17
Polymers, 2–4
Polytetrafluoroethylene (PTFE)
  Charnley's hip arthroplasty designs.
    See Charnley, Sir John
  debacle, 71
Posterior cruciate ligament (PCL),
  153–154
Posterior-stabilized total condylar
  prosthesis II (TCP II), 140
PRODISC implants, 226–227
  biomaterials, 234–235
  biomechanics of performance, 236–237
  clinical performance, 238–239
  design concept, 234, 235f
  historical development, 234
  shock absorption capacity, 235–236
Prolong™, 351–352

PTFE (polytetrafluoroethylene)
  Charnley's hip arthroplasty designs.
      See Charnley, Sir John
  debacle, 71

R
Radicals
  formation during irradiation, 246–247
  H radicals, 246–247
  isolated radicals, reaction of, 247–248
  macroradicals, 246–247, 250
    alkyl, 254–256
    peroxy, 254
Radiographic lysis, 74
Radiostereometric analysis (RSA), 83–84
R• (alkyl macroradicals), 254–256
Ram extrusion, 25–27
RA (rheumatoid arthritis), 190
RCH-1000, 5, 24
Resins
  conversion to consolidated form.
      See Conversion/consolidation
  GUR resins, 16–17
  GUR resins *versus* 1900 resins, 19–20
  1900 resins, 16–17
Reverse™ Shoulder Prosthesis system,
      211f, 212
Reverse total shoulder prosthesis design
      concept, 211–212
Revision
  knee arthroplasty/replacement,
      165–166
  rate(s), 73–74
  shoulder arthroplasty/replacement,
      193
Rheumatoid arthritis (RA), 190
ROOH (hydroperoxide decomposition),
      254–255
ROO• (peroxy macroradicals), 254
Rotating platform knees, 144
RSA (radiostereometric analysis), 83–84
Ruhrchemie AG, 14–15

S
Savastano knee prosthesis, 132, 133f
SB Charité™ III implants, 226–227
  abrasive wear on contact zones, 230,
      231f, 232f
  biomaterials, 227–229
  biomechanics of performance, 230,
      233, 234t
  clinical performance, 237–238

SB Charité™ III implants (*Continued*)
  core deformation, 229–230
  design concept, 227
  historical development, 227
Scanning electron microscopy (SEM),
      267, 268f, 269
Scorpio PS total knee prostheses, 152f
Select® shoulder prosthesis system
      components, 199
Self-diffusion, 22
Semiconstrained hinged knee
      design, 125
Semiconstrained reverse shoulder
      prosthesis, 212
SEM (scanning electron microscopy),
      267, 268f, 269
Shear punch testing, 298–301
Shelf life of components, 47–48
Shelf storage, oxidation during, 256, 258f
  small punch tests, 291–293
Shiers knee, 126, 127f
Shoulder arthroplasty/replacement, 189
  annual number of, 192
  biomechanics of, 195–196
  controversies in, 207, 209–210
  glenoid component materials, 209–210
  hemiarthroplasties
    bipolar prosthesis, 209
    procedures, 191–192
    results and rates, 209
    UHMWPE's role in, 209
  history of, 193–195
  load magnitudes and directions,
      195–196
  patient age, 193, 203
  procedures, 191–192
  revision of, 193
  stresses in UHMWPE components,
      195–196
  success rates, 203–204, 209
  total. *See* Total shoulder
      arthroplasty/replacement
      (TSA/TSR)
Shoulder complications
  arthritis, 193
    osteoarthritis, 190–191
    rheumatoid arthritis, 190
    TSA success rates, 203
  dislocations, 191
  fractures/trauma, 191, 193
    TSA success rates, 203

Shoulder complications (*Continued*)
 ligament abrasions and ruptures, 190
 osteoarthritis, 190–191
 rheumatoid arthritis, 190
 tendon abrasions and ruptures, 190
Shoulder joint, 190
Simulation
 generally. *See* Computer modeling and
  simulation
 hip simulators, 284
Small punch test(ing), 283, 288–291
 aging of UHMWPE, 291–293
 crosslinking's effect on mechanical
  behavior and wear, 295–298
 fatigue punch testing, 301–304, 305f
 geometric strain hardening, 289
 geometric strain softening, 289
 metrics of, 288–289
 shear punch testing, 298–301
 in vivo changes of UHMWPE, 294, 295f
Solar® humeral prostheses, 200
Song's model, 32
Spinal discectomy, 219
Spinal disc replacement. *See* Total disc
  arthroplasty/replacement
  (TDA/TDR)
Spinal fusion, 219–221
Sterilization, 37–38
 ethylene oxide sterilization, 38t, 44–45
 gamma sterilization
  in air permeable packaging, 38–41
  in barrier packaging, 41–44
 gas plasma sterilization, 38t, 44–47
 and oxidation, 253–257
 radical formation during, 246–247
 temperature effects during, 253–255
 at Wrightington Hospital, 66–67
Stillbrink, 194
Sulzer Orthopedics' MOM hip designs,
 98, 100f
Swedish Knee Arthroplasty Register,
 163
Swell ratio testing, 278–280

T
TCP (total condylar prosthesis), 139
TCP II (total condylar prosthesis II),
 140
TDA/TDR. *See* Total disc
  arthroplasty/replacement
  (TDA/TDR)

TEM (transmission electron microscopy),
 271–272
 crystalline lamellae, 6–7, 8f
Tensile properties, 30
Tensile testing, 281, 282f
T_f (flow temperature), 7–9
T_g (glass transition temperature), 7–8
Thackray, 67
THA/THR. *See* Total hip
  arthroplasty/replacement
  (THA/THR)
Thermally stabilized UHMWPE. *See*
  Highly crosslinked/thermally
  stabilized UHMWPE
Thermal transitions, 7–8
"3-D/2-D matching," 178
Tibiofemoral joint
 anterior shear, 155t
 compression, 155t
Ticona, 15–17, 24
TKA/TKR. *See* Total knee
  arthroplasty/replacement
  (TKA/TKR)
Total condylar knee, 135f, 136–141
Total condylar prosthesis (TCP), 139
Total condylar prosthesis II (TCP II),
 139–140
Total disc arthroplasty/replacement
  (TDA/TDR), 219–221
 biomechanical considerations, 222–226
 design goals, 221
 FLEXICORE™ TDR, 239
 indications for, 221
 interfaces for devices, 222
 kinematic considerations, 222–223,
  224f
 kinetic considerations, 223, 225
 load-sharing considerations, 225–226
 MAVERICK TDR, 239
 PRODISC implants, 226–227
  biomaterials, 234–235
  biomechanics of performance,
   236–237
  clinical performance, 238–239
  design concept, 234, 235f
  historical development, 234
  shock absorption capacity, 235–236
 SB Charité™ III implants, 226–227
  abrasive wear on contact zones, 230,
   231f, 232f
  biomaterials, 227–229

Total disc arthroplasty/replacement
  (*Continued*)
    biomechanics of performance, 230, 233, 234t
    clinical performance, 237–238
    core deformation, 229–230
    design concept, 227
    historical development, 227
    versus spinal discectomy, 219
    versus spinal fusion, 219–221
    UHMWPE alternatives, 239
    UHMWPE designs, 226–239
Total hip arthroplasty/replacement (THA/THR), 71, 73
    generally. *See* Hip arthroplasty/replacement
    highly crosslinked UHMWPE, prevalence of, 354–356
Total knee arthroplasty/replacement (TKA/TKR), 123, 124f. *See also* Knee arthroplasty/replacement
    evolutionary stages for UHMWPE in, 125
    highly crosslinked UHMWPE, prevalence of, 356–357
    osteolysis, 172–176
    tricompartmental, 124f, 125
    in vivo wear assessment in, 176–180
Total shoulder arthroplasty/replacement (TSA/TSR)
    abrasion, 207
    Aequalis™/Aequalis™ Fracture system components, 202
    Anatomical Shoulder™ system components, 198f, 199
    Bi-Angular® system components, 197f, 198
    Bigliani/Flatow® humeral prostheses, 202
    biomechanics of, 195–196
    Bio-Modular® system components, 197
    BiPolar system components, 198
    burnishing, 207
    cobalt chromium alloy, use of, 212
    Cofield™/Cofield²™, Monoblock system components, 201
    complete wear-through, 207
    complications with, 204t
        glenoid loosening, 204–205
        instability, 204t, 205
        wear or damage, 205–207, 208f

Total shoulder arthroplasty/replacement
  (*Continued*)
    contemporary designs, 197–203
    deformation, 207
    delamination, 207
    Delta® prosthesis, 212
    embedded debris, 207
    Foundation®/Foundation® fracture humeral prostheses, 199–200
    fractures, 207
    future directions
        in design, 211–212
        in materials, 212
    glenoid loosening, 204–205
    Global™ humeral prostheses, 199, 206
    history of, 193–195
    instability, 204t, 205
    Integrated® system components, 198
    Modular Shoulder System, 201
    Neer II/Neer III system components, 200–201
    pitting, 207
    procedures, 192
    Reverse™ Shoulder Prosthesis system, 211f, 212
    reverse total shoulder prosthesis design concept, 211–212
    scratching, 207
    Select® system components, 199
    semiconstrained reverse prosthesis, 212
    Solar® humeral prostheses, 200
    success rates, 203–204, 209
    wear or damage, 205–207, 208f
Townley knee prosthesis, 134f, 135f, 136
Trace element analysis, 274
Transmission electron microscopy (TEM), 271–272
TSA/TSR. *See* Total shoulder arthroplasty/replacement (TSA/TSR)

U
Ubbelohde viscometer, 269, 270f
UKA (unicondylar knee arthroplasty), 125
Ultrasound for knee wear assessment, 178–179
Uniaxial compressive response, 313–314
Uniaxial tension response, 313
Unicondylar disease, 132

Unicondylar knee arthroplasty (UKA), 125
Unicondylar polycentric knee
  arthroplasty, 132–134

V

Viscosity average molecular weight
  (M_v), 18
Volumetric wear rate (VWR), 78, 85t
von Mises stresses
  hip replacements, 156
  knee replacements, 158–159
VWR (volumetric wear rate), 78, 85t

W

Walldius knee, 126, 127f
Wear performance/rates
  crosslinking, effect of, 295–298
  HDPE, 5, 6f
  hip arthroplasty. *See* Hip
    arthroplasty/replacement
  knee arthroplasty/replacement,
    165–167
    articulating surface damage modes,
      167–172
    in vivo wear assessment methods,
      176–180
  from machining, 32

total shoulder arthroplasty/
  replacement, 205–207, 208f
UHMWPE, 5, 6f
Wrightington Hospital
  hip arthroplasty/replacement. *See also*
    Craven, H.
    implant fabrication at, 61–62
    UHMWPE cup sterilization at,
      66–67
  knee arthroplasty/replacement
    Charnley's design, development
      of, 129
    Gunston's design, development of,
      123, 127–129
  UHMWPE's arrival at, 66
WTF (integral work to failure), 288–289

X

XLPE™, 352, 353t

Y

Y-crosslinking mechanism, 248–249

Z

Zipple, 194
Zirconia, 102, 103t, 104–105
  failure rate, 109